Physics

by Paul V. Pancella, PhD, and Marc Humphrey, PhD

A member of Penguin Random House LLC

ALPHA BOOKS

Published by Penguin Random House LLC

Penguin Random House LLC, 375 Hudson Street, New York, New York 10014, USA • Penguin Random House LLC (Canada), 90 Eglinton Avenue East, Suite 700, Toronto, Ontario M4P 2Y3, Canada (a division of Pearson Penguin Canada Inc.) • Penguin Books Ltd., 80 Strand, London WC2R 0RL, England • Penguin Ireland, 25 St. Stephen's Green, Dublin 2, Ireland (a division of Penguin Books Ltd.) • Penguin Random House LLC (Australia), 250 Camberwell Road, Camberwell, Victoria 3124, Australia (a division of Pearson Australia Group Pty. Ltd.) • Penguin Books India Pvt. Ltd., 11 Community Centre, Panchsheel Park, New Delhi—110 017, India • Penguin Random House LLC (NZ), 67 Apollo Drive, Rosedale, North Shore, Auckland 1311, New Zealand (a division of Pearson New Zealand Ltd.) • Penguin Books (South Africa) (Pty.) Ltd., 24 Sturdee Avenue, Rosebank, Johannesburg 2196, South Africa • Penguin Books Ltd., Registered Offices: 80 Strand, London WC2R 0RL, England

001-278656-July2015

International Standard Book Number: 978-1-61564-789-7
Library of Congress Catalog Card Number: 2014959609

17 16 15 8 7 6 5 4 3 2 1

Interpretation of the printing code: The rightmost number of the first series of numbers is the year of the book's printing; the rightmost number of the second series of numbers is the number of the book's printing. For example, a printing code of 15-1 shows that the first printing occurred in 2015.

Printed in the United States of America

Publisher: *Mike Sanders*
Associate Publisher: *Billy Fields*
Senior Acquisitions Editor: *Brook Farling*
Development Editor: *Ann Barton*
Production Editor: *Jana M. Stefanciosa*
Cover Designer: *Laura Merriman*
Book Designer: *William Thomas*
Indexer: *Tonya Heard*
Layout: *Brian Massey*
Proofreader: *Amy Borrelli*

Contents

Introduction

Let's face it, physics has a certain reputation. You probably know what we're talking about. Many people think physics is the most difficult, most esoteric science there is. Only the smartest kids in the class have any chance of understanding it. You need supercomplex, high-level math to solve physics problems. It has nothing to do with the real world, but for some reason they make you take physics in school if you want to be an engineer or a doctor or even an architect.

Well, we disagree with almost all of that. Physics is cool because it is the basis for all of the rest of science and technology. It is actually fun and interesting because it lets you figure out how various parts of the world—like tides and sunsets and bumper cars—actually work. Physics explains a lot of things we observe happening in the real world, not just in careful lab experiments but also in ordinary everyday life. Sometimes we can even use that knowledge to predict the future! Physics is the most fundamental science, but it doesn't have to be intimidating. Since you're reading this book, we're guessing that you are at least willing to give it a try, and we appreciate that. In this brief introduction, we're going to describe our goals in writing this guidebook, and how we put it together to achieve them.

The Goals of This Book

The main purpose of this book is to help you grasp physics at an introductory level. That means we won't be able to cover all of the different fields of study that fall under the big banner of physics; there is just too much. It also means that we won't be talking much about the latest discoveries, or speculating about future possibilities like time travel or teleportation.

The physics we will be talking about may be less exotic, but that doesn't mean it's any less interesting. The stuff we'll be covering is the absolutely necessary foundation for all the rest of it. Physics is truly vast, because its subject is the entire material universe. But you have to start somewhere. Our main goal with this book is to make it as easy as possible for you to get that good start and to build a solid foundation.

In order to do that, we'll spend most of the book explaining the basic concepts using ordinary language. We'll try to find interesting examples to which you can relate, but which also illustrate the key points we're trying to get across. We will have to use some math, but nothing beyond basic algebra and trigonometry. We'll try hard to put these concepts into context, so they are easier to remember and use correctly. We'll show you how the different parts of physics connect with and reinforce each other. We will also provide you with a few practice exercises to try yourself, so you know when you're ready to move on to the next topic.

All of the concepts in this book have been known for well over a hundred years by now, so they have been taught over and over again to many generations of students. Through all of that time, teachers have developed a pretty logical sequence in which to present introductory physics topics.

This same sequence is found in almost every physics textbook used in colleges and advanced high school courses. We have adopted that same sequence for this book, partly because all those previous educators probably knew a thing or two, but also because this book will then be easier for you to use if you find yourself taking one of those classes. The section and chapter headings should be a pretty good match for most of the topics you will encounter in a typical introductory physics course.

If you happen to be taking a class in physics now, feel free to skip around this book, referring to specific sections when you need a different explanation or some additional help with this or that particular topic. If, on the other hand, you are using this book as your primary source for learning physics, you will probably be better off reading it in sequence, starting with the first chapter where we lay down most of the ground rules.

Unlike some other areas of study, physics is not just a collection of facts. In physics there is a definite structure, with anchor points and branches and interlocking girders. Certain concepts provide the support for other ideas, and more advanced topics are incomprehensible without a solid understanding of the basic foundations. These foundational concepts are the ones we'll be dealing with in this Idiot's Guide.

How This Book Is Organized

Part 1, Mechanics, is the traditional way to begin a serious study of physics. In this case, "mechanics" doesn't mean fixing cars. Rather, it refers to all kinds of motion that happens in the world, how to describe that motion precisely, and what causes changes in motion. It's a good place to start because you can usually visualize the kinds of motion we will be talking about, and get a feel for the forces that influence that motion. Later parts of this book deal with phenomena that are partly or completely invisible. Even though Part 1 is the longest section of this book, we won't be going into great detail. We'll concentrate on pointlike particles with definite mass, and then so-called "rigid bodies" that have size but don't change shape. Ideas we introduce and develop in this part, like energy and momentum, have far-reaching consequences in the rest of physics and other sciences.

Part 2, Waves and Fluids, extends the basic ideas of mechanics beyond individual particles and rigid objects. Lots of interesting phenomena occur when oscillations propagate in some sort of continuous medium. Sound is one important and interesting example. This part will also be our opportunity to talk about other cool properties of fluids, which are used to explain why boats float, cannonballs sink, and airplanes fly.

Part 3, Thermodynamics, is an introduction to the important concept of heat, a very special and ubiquitous form of energy. An understanding of heat as energy has allowed humans to tap into various fuels that provide power for a great many tasks, saving us from pushing plows, digging ditches, carrying cargo, and all kinds of manual labor. Thermodynamics is also relevant in a lot of other ways, helping us stay warm in the winter and keeping our fish sticks frozen. Beyond that, it is also important for understanding the eventual fate of the entire universe!

Part 4, Electricity and Magnetism, marks the point where a second introductory course in physics usually starts. Here we dive into the realm of electric charges and the forces between them. These forces may seem less familiar, but as we'll see they turn out to be absolutely everywhere. Here we'll develop and use the somewhat abstract concept of a force field. It turns out that electric charges and forces are intimately connected with magnetic phenomena, so we'll take a look at magnetism, too. In this part, we'll also get a start on understanding the wonderful world of electric circuits and the basic components used to build them.

Part 5, Light and Optics, introduces the physics that allows us to see the world around us. We'll learn how the phenomenon of light makes a connection between the waves we studied in Part 2 and the electromagnetism of Part 4. Visible light turns out to be a wave phenomenon that is both similar to and differs from the mechanical waves studied earlier. It is part of a wide spectrum of electromagnetic waves that also includes radio waves, x-rays, microwaves, and lots of other radiation that we have learned to use for our own purposes.

Extras

At no extra charge, we are including many additional features that you may find helpful. We call them sidebars because they are optional in a way. If you are really on a roll, you can skip any of these without missing anything essential or losing continuity as you read. On the other hand, if you need a pause to digest, or want to go a little deeper, these offer you some additional insights that may help your understanding.

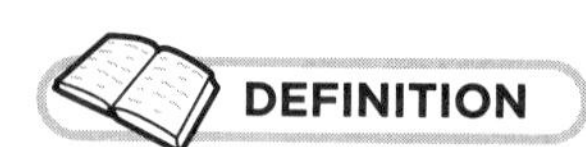

DEFINITION

The most essential terms defined. Some of these words may be unfamiliar to you at first. Most of them, though, are everyday words that physics uses in a certain specific way. These terms also appear in Appendix A, but these sidebars occur in the text right where you need them.

RED ALERT!

Common errors and misconceptions to avoid. These are pitfalls that we have seen many students (including ourselves) topple into over the years, so we give them some special attention.

CONNECTIONS

Some anecdotes and tidbits to spur your interest through connections to daily life, references to historical events, or links to some of the most cutting-edge topics in physics.

TRICKS AND HACKS

Shortcuts you can use to bypass complicated calculations, or at least check that those calculations are giving you reasonable results.

TRY IT YOURSELF

Along the way we'll also provide some quantitative exercises that you can try for yourself. If you really want to learn basic physics, you should find a calculator and try to do these. The answers can be found in Appendix E. If you can get to the right answer without peeking, then you'll know you are making progress.

Acknowledgments

PVP wants to acknowledge his parents, who always supported his interests in science and technology, allowing him to take mind-blowing extra classes at the local planetarium at a very young age, for example. He also appreciates his current employer, Western Michigan University, both for helping him hone his teaching skills, as well as allowing him time to write this book. And, of course, his loving wife, who has been a steadfast source of comfort and support.

MAH would like to acknowledge the many great physics teachers he had while learning this subject himself for the first time—Paul Nevins, Dean Kaul, and Gerald Hardie, to name but a select few. He also owes a debt of gratitude to his thesis advisor, Ron Walsworth, who taught him not only how to apply the physics he'd learned in the classroom, but also how to communicate physics in written form. Most of all, he'd like to thank his wife and two small children for reminding him on a daily basis about the wonders of our natural world.

Trademarks

CHAPTER

1

The Preliminaries

In This Chapter

- How physicists think about the world
- The math skills you need to understand physics
- Dimensions, units, and the metric system
- A few other important tips before getting started

Physics can be a daunting topic for many, but it doesn't need to be. In this chapter we'll cover some general topics that will help smooth your path through the chapters to come. Before we dive into the physics itself, we want to talk about some of the rules and conventions that will guide our efforts.

We'll start by discussing how physicists view the world. If you understand a little bit about how a physicist thinks, you may be able to avoid some misunderstandings when we try to explain things later on. We'll give a very brief overview of the main types of mathematics we will rely on throughout the book. We'll also introduce you to the metric system, a system of units that is used by scientists all over the world, with the exception of a few English-speaking countries (like the United States). We'll briefly compare the metric system to the English system, though we won't put a lot of emphasis on how to convert between the two systems of units.

In the last part of this chapter, we'll give you some general hints on how to approach physics problems and not only solve them correctly, but use them to cement your understanding of the concepts as they are applied.

How Physicists Think

Physicists have a certain way of looking at the world. First of all, physicists view the world as an objective reality, independent of culture, location, or personal opinions. The reality physics seeks to explain is the same for everyone at all times. If it seems like science changes as time goes by, it is not because the natural world or the physical laws have changed, it is because physicists are getting better (hopefully) at understanding and explaining that world.

But what do we mean by "explaining" something in the world, anyway? In simple terms, we mean that a *theoretical model* has been developed that behaves the same way that some part of the world does. The model may be as simple as a single formula, or it might require a complex system of interrelated mathematical elements. Either way, mathematics is the language that physicists use to both describe and to model physical systems. Why do we use so much math? That's easy: because it works.

A **theoretical model** is a well-defined mathematical system that can be used to predict how a physical system will change over time. It may be as simple as a single equation that relates the physical effects of some system to their causes and prevailing conditions.

For all this to work, we must talk about aspects of nature that are both observable and quantifiable. We will have to define various physical quantities that can take on exact numerical values. The formulas and equations that make up our models express definite relationships between these quantities. These relationships exist in nature, and it is the continuing job of physics to discover and refine these relationships, through careful observations and measurements. Whatever mathematical models we come up with, the only way to judge whether or not they are correct is to compare them carefully with nature itself. No matter how elegant a model is, or how much we might like it, if nature doesn't behave the way that the model does, then the model has to be changed or discarded.

Physicists are also known for their ability to distill the essentials when considering a physical situation. This skill is necessary even for introductory physics, but it's one that is difficult to teach. What do we mean? Real physical situations can appear to be very complex, even when the physics involved is relatively simple. Sometimes the presence of extra, unimportant information distracts us from what is crucial for our understanding. Some features are incidental and have no bearing on the underlying physics.

For example, when calculating how far a cannonball will fly, the color of the ball has no effect, while its size and launch speed are critical. This is a pretty obvious example, but you get the point. When the difference between essential and incidental gets more subtle, we'll try to point it out.

If you try to explain every little detail about some situation that you observe, it is easy to get overwhelmed, confused, and frustrated. Lots of different kinds of physics might be happening at the same time. For the beginner, it is often hard to see the connections between phenomena that look on the surface to be very different. The ability to see below the surface to the real fundamental concepts comes only with experience. Assuming you are just beginning your study of physics, this is one of the things we're going to help you with. We will guide you by discussing the essential features, and simply ignoring a lot of smaller effects. The downside is that it will make some of our predictions slightly inaccurate when compared to the real world. This does not mean that our simple models are wrong; it just means we've temporarily ignored some other physical effects that are smaller and may require more detailed analysis than we want to get into at the introductory level.

Math Skills for Physics

The universe is a complex place, even when we dig deeply into the most fundamental interactions. Mathematics is the language of science, and physics as a whole ends up using some pretty high-level math.

If you're just beginning to learn physics, however, you should be able to get by with some more basic math. You'll be able to handle all of the formulas and exercises in this book if you have mastered algebra at the level taught in high schools and are familiar with basic *trigonometry*. We're going to assume you have sufficient algebra skills, and we'll review the necessary trigonometry when necessary.

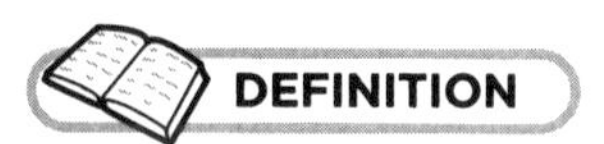

Trigonometry is the branch of mathematics that studies the relationships between lengths and angles in triangles. In physics, this math is used mainly to understand physical quantities for which direction is important.

That means we assume that you can handle equations that contain multiple unknown quantities. You should be able to manipulate these equations to "solve" them symbolically for one of the variables. That usually just involves a series of steps in which you carry out the same mathematical operation on both sides of an equation. You should understand ratios, and be able to perform all the basic operations with fractions. You should understand what it means to raise a quantity to a power, even if that power is not a whole number (or integer). If you need a little review on exponents and logarithms, you'll find a short primer in Appendix C.

A Few Basic Conventions

Throughout this book, we will be using symbols to represent physical quantities. In general, these symbols are treated as variables in algebraic expressions. We will mostly stick with single letters for our variables, which may be lowercase or uppercase. Sometimes we will use Greek letters. Most of the time, we'll use the same symbols that other teachers and texts tend to use (x, y, and z for positions; t for time; v for velocity; etc.). When we have more than one quantity of the same type, we'll generally use subscripts to distinguish them (like x_1 and x_2), so be sure to pay attention to those. In any case, we will always try to define our symbols when we use them, and you should, too.

Another conventional use of a symbol is the Greek *delta*, which looks like this: Δ. This symbol will always be used in conjunction with another variable, such as a location x. When we write something like Δx, it stands for the change in the location x over some defined interval. Literally, it means $(x_2 - x_1)$, where x_1 stands for the location of something at some particular time, and x_2 represents the location of that same thing some definite time later. The use of Δ almost always means you should take the later value minus the earlier value. This is done so that the delta of the quantity has a positive value when something is increasing, and negative when it is decreasing.

Dimensions and Units of Measurement

Another important ground rule you must take to heart is this: every physical quantity has *dimensions*. Any numerical value assigned to a physical quantity must therefore have *units* attached to it, and those units must be appropriate for that dimension. If a number is supposed to indicate the value of some physical quantity, then the number without units is meaningless. It is easy to confuse the concepts of units and dimensions, so let's be clear from the start.

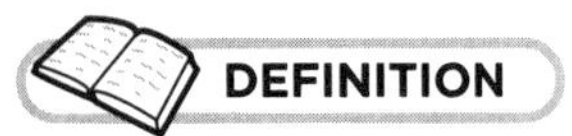

The **dimension** of a quantity refers to the kind of physical quantity that's being described (e.g., length, mass, or energy).

Units refers to the reference standards in which quantities are measured. The same physical quantity can be measured in different units, which means that the same quantity can take on different numerical values, depending on the units being used.

When we refer to dimensions, we are really talking about the actual nature of the physical quantity; what kind of quantity it is. The dimension can be length, time, acceleration, force, mass, etc. These are all fundamentally different kinds of things, and we have to distinguish them from each other at all times.

In particular, it never makes any physical sense to add two quantities together that have different dimensions. It is perfectly fine to add two masses, for example, to arrive at a total mass, but it makes no sense to add a mass to a time. The same thing goes for subtraction, of course. What would you get if you subtracted a length from a temperature? Nonsense, so we recommend you don't do it! For the same reason, if you have an equation, no matter how complicated it looks, the quantities on both sides of the equals sign must have the same dimensions.

On the other hand, it *is* okay to multiply or divide quantities with different dimensions. If you take a product of two things that have different dimensions, the result is a quantity with yet another dimension. This will become clearer when we actually start working with some real examples.

RED ALERT!

In everyday language, the term "dimension" often has a different connotation—that being a fundamental property of space. For example, a circle has two dimensions while a sphere has three. Be careful not to confuse this use of "dimension" with the way it is used in this section.

For every dimension, there is a standard unit of measurement. The act of measurement is essentially comparing the thing to be measured with the standard unit, also known as a reference. For example, when we speak about the distance between two trees, the dimension under consideration is length while the units could be feet or meters. As you can see, the units aren't the same as the dimensions (although the units do tell you what the dimension ought to be).

It is easy to see why having everyone agree on a single common system of units would be helpful for doing science. Physics is an international endeavor, and physicists in one part of the world are frequently comparing their results with work done in another. It would be a huge waste of time if they constantly had to convert between different units just to compare results. So in 1960, scientists worldwide agreed to a comprehensive set of standards and units known as the *SI* system.

DEFINITION

SI is the abbreviation for Système International, French for the international system of units and standards. This system relies heavily on powers of ten, so it is also commonly referred to as the metric system.

Since the SI system is the standard for physicists worldwide, we will stick with it here, too. The only downside to working in the metric system is that some of you may not be as familiar with the size of a centimeter or meter as you are with an inch or a foot. But don't worry, you'll soon get a feel for how long a meter is and what a kilogram's worth of mass feels like, and then you'll be fine.

The following table lists just a few examples of dimensions we will be using and the standard units that go with each. This is not the complete list of all the units we will need, but it provides some examples of what we have been talking about. It also illustrates another point: there are only a few fundamental dimensions, which can be combined by multiplication and division to create other dimensions. For example, volume is a dimension that is derived from the more fundamental dimension of length. What's more, the first four dimensions in the table below are the only fundamental dimensions that we'll need in the first half of this book. Note the nifty ways that the others are constructed from these.

SI Units for Common Dimensions

Dimension	Basic SI Unit	Abbreviation
time	second	s
length	meter	m
mass	kilogram	kg
temperature	kelvin or Celsius degree	K or °C
	Derived Units	
speed	meters / second	m/s
acceleration	meters / second / second	m/s^2
area	meters × meters	m^2
volume	meters × meters × meters	m^3
force	kilograms × meters / second / second	N (newtons)
energy	kilograms × meters × meters / second / second	J (joules)

In the case of time, the scientific system of units is no different from the commonly used system. The fundamental unit is the second that you are all familiar with. Sixty of those make a minute, sixty minutes make an hour, etc. We will also use days and years where appropriate.

For the other units, when we want to use larger or smaller units than the standard ones, we use powers of ten or a thousand, which are naturally handled by something called scientific notation. If you are rusty or unfamiliar with scientific notation or the abbreviations used for common powers of ten, have a look at Appendix C before starting the next chapter.

What You Can Do to Help Yourself

The idea that a symbol represents a physical quantity is crucial to doing physics. Many beginning students get bogged down by all of the different symbols that appear in the various formulas, so much that they forget that physics is connected to the world we experience every day. This connection is very strong and direct. You would do well to keep it at the front of your mind. Whenever you see a variable used in a mathematical expression in this book, you should immediately call to mind the physical quantity that is represented by that symbol. We'll do our part by clearly defining our variables, and using the most standard terminology that we can.

Physics is a precise science, and we tend to use a lot of abbreviations. That means every little symbol is important. Be careful not to overlook minus signs, subscripts, powers (that look like superscripts), etc., especially when you are working out a problem. Be patient with yourself and make sure you can justify every step that you take. By all means, feel free to make an algebraic substitution when solving a problem, but make darn sure the thing you are putting in is really always equal to the thing you are replacing.

When working through the mathematics of a problem, it pays to keep track of the units of the various quantities in the intermediate steps—especially when the problems get a little more complicated. Remember that the units reflect the dimensionality of every physical quantity, so the units on both sides of every equation have to match exactly. Also, keep in mind that units themselves obey algebraic rules. For example, if you divide two numbers that have the same units, the units cancel out. Keeping track of the units can help you spot (and correct) many common algebra errors, like putting a ratio in upside-down.

CONNECTIONS

On September 23, 1999, the Mars Climate Orbiter passed a little too close to the red planet and burned to a crisp. The reason for this was a mismatch between the SI units used by one team of engineers and the English units used by another. Failing to check your units when solving physics problems can cost you, but hopefully not as much as this infamous, $125 million mistake.

Actually doing problems is extremely helpful if you really want to cement in your mind the concepts you've learned. It takes time and effort, and many students recoil at the idea of doing extra problems that are not assigned for credit. But practice really helps, especially when you are just starting. That includes thinking carefully about each step in any examples that are worked out for you.

Another key practice is to try to anticipate what a reasonable answer to a problem would look like. You should at least know what units to expect, based on the type of quantity that you are asked to calculate. But often the general size of the answer, or even the direction (in the case of a vector quantity) can be guessed at, since the problems and examples will be taken from the real world. If you make a guess and then your worked-out answer is very different, it may be a hint that something went wrong in your solution. There could be a problem with either your guess or your math; either way it indicates some more thought is called for.

Finally, a little bit of imagination can be a big help in studying physics, so by all means try to visualize situations wherein the concepts we'll be talking about are at work. Thinking about "limiting cases" can be very instructive. That means picking one of the variables and thinking about what would happen if it took on an extreme value. For example, when considering a formula containing speed, what does it imply for speeds that are extremely fast or extremely slow? Is that what you'd expect? If you imagine any realistic inputs to your formulas, the outputs should also make sense. You may be surprised at how helpful a little common sense as well as a reality check or two can be in learning basic physics.

With those preliminaries out of the way, we are ready to move into the vast world of physics. We will begin by studying all aspects of matter in motion. Fasten your safety belt.

The Least You Need to Know

- In order to learn physics, you need to know and be able to use all the tools of high school algebra.
- Like most scientific works, this book will use the international system (SI) of units, known informally as the metric system.
- Be patient with yourself as you try to learn physics.
- Be prepared to use your imagination to visualize new concepts.
- Never forget that physics must always be connected with the real world.

PART

1

Mechanics

This first part is the meat and potatoes of any introduction to the vast world of physics. The topic is matter in motion, and there is no shortage of familiar examples to illustrate the many aspects of this topic. We start by using basic math and some of the tools from Chapter 1 to precisely describe motion. We need to agree on our terms and how we describe motion before we can make sense out of why things move.

We will quickly move toward understanding what causes motion, introducing the concept of force and the laws that connect forces and motion. Among many examples of forces we'll spend some extra time on gravity, the natural tendency for objects with mass to attract each other. For most of Part 1, we'll be looking at situations where the size of an object with mass doesn't matter. This makes our analysis a lot simpler, since we can treat the mass of an object as if it were all located at a single point. Toward the end of this part, we will need to go beyond this to consider size and shape when we look at bodies that rotate.

Two extremely important general concepts come out of this basic study of motion: energy and momentum. These quantities have important ramifications in all fields of physics and beyond, so you'll want to pay special attention to them when they come up.

CHAPTER

2

Linear Motion

In this chapter, we'll take a quantitative approach to some very basic physics: the description of motion. We start simple, by only considering motion along a single straight line. This restriction still allows for motion in two directions, which are opposite to each other (as in back and forth, or to and fro).

We'll take our time and define our quantities very carefully, even at the risk of boring you. Many of these definitions will seem simple and obvious, but it is crucial at this beginning stage that our foundation be firm, with no cracks or misunderstandings, so that it will be able to support all of our later work.

In This Chapter

- Specifying the precise position of an object in one dimension
- Speed and velocity defined
- The way to quantify changes in velocity: acceleration
- Formulas to help calculate positions, velocities, and accelerations in one dimension

Position in One Dimension

For our purposes, it is not good enough to say that some object is located "over there," or "just behind the house," or even "on the far left end of the third shelf from the bottom." If we want to build up a scientific description of motion, we have to be a lot more precise when we talk about location.

To help you get a handle on how all this works, we'll first consider only the location of an object in one spatial dimension. Soon enough, we'll add other dimensions in order to get a more realistic description of the space we live in, and the same terminology and tools we develop here will extend nicely to the other dimensions as well.

For example, imagine a little red caboose sitting on a long, perfectly straight railroad track. Forget, for now, that Earth is round; we'll say this track is on a perfectly flat surface. We'll also assume that it extends as far as you can see in both directions, with no curves or branches whatsoever. The caboose's wheels fit the track perfectly so that it can glide along the track without ever derailing. The caboose may be at any location along the track, but, being a caboose, must always stay on the track.

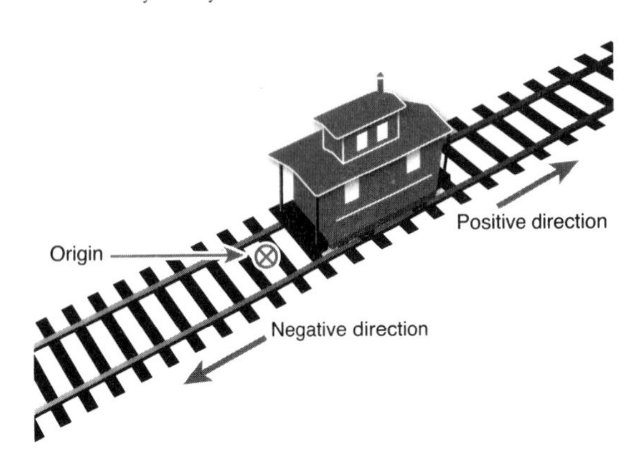

Figure 2.1

Imagining a caboose on a track is one way to visualize one-dimensional motion.

Once we have our caboose and our track, we need three more ingredients before we can clearly specify where the caboose is located. First, we must designate some fixed location on the track as special, a point to serve as a reference. We call this point the origin of our *coordinate system.* Whenever the caboose happens to be located at that point, we will say its *position coordinate* is exactly zero.

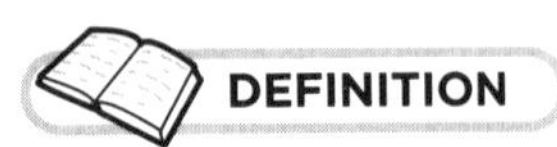

A **coordinate system** is a concept in geometry that allows us to assign a unique number or set of numbers to any position in space. It consists of an origin point and a set of coordinate axes.

A **position coordinate** is the number (with units) that indicates how far from the origin something is located along one of the axes.

What we are actually doing is constructing a coordinate system on top of the railroad track, but only in our minds. It is the simplest kind of coordinate system, because we only need one axis, and that axis line goes straight along the track. The coordinate system is an abstract idea from geometry; it doesn't exist until we decide to imagine it there. It's also worth noting that we can put the origin wherever we want to, so we usually pick the most convenient spot.

The second ingredient we need is a decision about directions. Since there are only two possible directions along the track, we can simply assign positive numbers to one direction and negative to the other. Standing at the origin, we look along the track one way and call that direction positive and the opposite direction negative. Here again we're free to select whichever we like, but once we've made our choice we have to stick with it—at least as long as we are talking about the same physical system. To remind ourselves (and keep it straight for others), it also helps to label which way is which.

The third and final ingredient we need is a unit of measurement. The units set the scale of our coordinate system. Distances and lengths are really the same dimension, and by selecting the metric system we've already decided to use units of meters to measure both. Location is relative; it just amounts to the distance between the object and the origin (with the algebraic sign, positive or negative, telling us which direction).

Putting this all together, we have a system in which a single number, which may be positive or negative, uniquely describes any single position that our little red caboose can have. If we say that it is located at +125.3 meters, we know exactly what that means. It means that the caboose is located 125.3 meters from the origin point in the positive direction along the track.

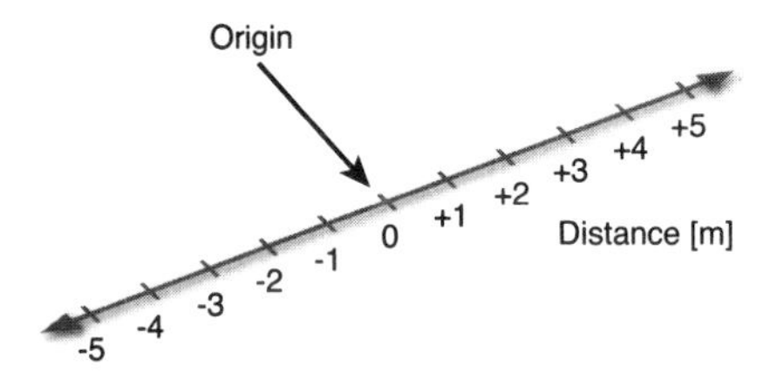

Figure 2.2

A one-dimensional coordinate system, complete with origin, direction, and units of measurement.

If you remember your basic geometry, and you've been following along carefully, you may have spotted a little problem. An exact location corresponds to a single point. A geometric point is very, very small. In fact, it doesn't have any size at all. A caboose, on the other hand, has plenty of size. What does it even mean to say that a little red caboose (or any other real object) is located at a point? In reality, it is located at a lot of points at once, and different parts of the caboose are located at different places.

We'll get around that problem by designating a single, fixed point *on the caboose* as our reference point for measuring its location. We could take that point to be the front of the car, or the very center, or whatever, as long as we communicate our choice to anyone else with whom we want to talk about position. Even if we don't explicitly talk about such a reference point in all of our future examples, we will be assuming that one has been chosen. (Incidentally, some books will refer to this as the "particle approximation" since it treats a large object as if it were just a small particle, for location purposes.)

One more note about positive and negative directions: when a coordinate axis is pictured horizontally, it is customary to make the positive direction go to the right, and negative to the left. Similarly, for a vertical axis, up is usually called positive and down negative. But you should not think of these as the only "correct" choices.

Think of the train track again. You look at the track from where you are standing, and you say to yourself, "Obviously, I should follow convention and choose the direction to my right as the positive direction." But now think about your friend on the other side of the track. When she looks at the same track, the direction *she* would call "to the right" would be the one *you* called "to the left." The choice for which direction to make positive is an arbitrary one. But if people are going to agree on the exact numbers that correspond to real positions, then they also have to agree on which direction is positive.

Velocity

We are now equipped to talk about changes in position in a precise, scientific way. By change we simply mean that an object may have a different position at a later time. In order to make quantitative sense of this, we need to not only measure positions carefully, we also need to be able to measure time intervals precisely.

The rate at which an object's position is changing is called its *velocity.* Let's use the letter x to designate a position for now, and the letter v will stand for the velocity. Let's also imagine a certain period of time, which has a definite beginning and end. Let x_1 be an object's location at the beginning of the time interval, and x_2 be its location at the end. Then we will use the abbreviation Δx to represent the change in position during the specified time interval: $\Delta x = x_2 - x_1$.

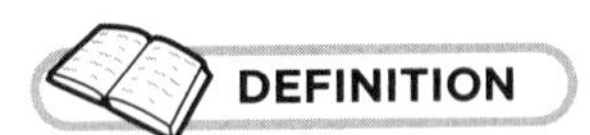

DEFINITION

Velocity is the time rate of change of an object's position. Over a definite interval of time Δt, the average velocity in the x direction is defined as $v_{avg} = \Delta x/\Delta t$.

Notice that the earlier position is always subtracted from the later position when using the Δ notation, as we said in Chapter 1. Recall that each of the x coordinates themselves may be positive or negative, and that the sign plays an important role in determining the exact positions. Depending on the signs and which number is larger, Δx may also be positive or negative, once you have done your arithmetic and accounted properly for all the signs.

It turns out that the sign of Δx tells us the direction of the velocity in a very straightforward way. If Δx is positive, then the object is moving in the positive direction along the coordinate axis, and the velocity v is also positive. If Δx is negative, it means that x_2 was farther in the negative direction than x_1. (Remember that "more negative" means the same thing as "less positive.") In this case v is negative and the motion is toward the negative end of the axis.

The dimensions of velocity are length divided by time. Since our standard length unit is the meter, and our basic time unit is the second, the preferred units for velocity are meters per second, or m/s.

TRICKS AND HACKS

Here is one case where it might be useful to memorize a conversion factor. Using the fact that there are 1,609 meters in a mile and 3,600 seconds in an hour, you can calculate that 1 m/s is equal to 2.237 miles per hour (mph). When you are working problems later, and you want to know if a velocity in m/s is reasonable for the situation, just multiply the answer in m/s by two to get a rough feeling for the velocity in mph.

In general, the velocity v can also change as time goes on. It is important to be aware of whether or not v is changing at any particular time or in any physical situation. Let's think a little harder about the definition for velocity. We'll go back to our caboose, and take a particular interval of time (let's say 10 minutes, or 600 seconds). Let's also say that the position of the caboose at the start was +1,000 m, but after 10 minutes of motion, it is located at -5,000 m.

If we only know the positions at those two times, then the only thing we can calculate is the *average* velocity over that time period. The average velocity would be $v_{avg} = \Delta x/\Delta t =$ (-6,000 m)/(600 s) = -10 m/s. The average velocity provides interesting information, but it is just an average. The caboose could have taken lots of different velocities during that time.

It is possible, however, to define a velocity for a single instant in time. Not surprisingly, this is called the *instantaneous velocity*. All we have to do is reduce the time interval Δt in our definition of average velocity. If a particle or body is moving, and you want to know its velocity at a specific time, just consider smaller and smaller intervals around the time of interest. As you reduce the size of the time interval Δt, you also reduce the distance traveled in that time. But that doesn't mean the ratio of Δx to Δt has to shrink to zero. In fact, that ratio approaches the instantaneous velocity just as the time interval approaches zero.

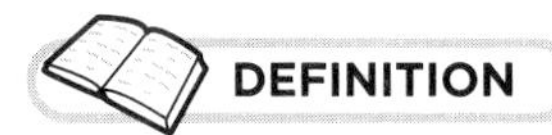

DEFINITION

Instantaneous velocity is the limiting value of the average velocity as the averaging time interval is made vanishingly small. It is defined for a single moment in time.

There is one more thing we should mention about velocity in one dimension. Sometimes we are less interested in the direction of motion than we are in how fast something is moving. The positive number that corresponds to the absolute value (or magnitude) of the velocity is what we refer to as speed.

In physics, the terms speed and velocity are not interchangeable. Velocity includes information about direction, while speed does not. Speed is simply the absolute value of an object's velocity. An object with a velocity of –42 m/s has a speed of 42 m/s.

If you drive a car, you already have a feel for what this means. Every car has a gauge that the driver can easily see which is called a speedometer. It is not called a "velocity-o-meter" because it only tells you how fast you are moving, nothing about your direction (a compass or even your GPS is needed for that). The speedometer never reads a negative number, even though your car is perfectly capable of driving in reverse.

The same distinction between average and instantaneous can be made for speed as was done for velocity. Unless indicated otherwise, you should assume a velocity given by v indicates an *instantaneous* velocity. If we want to talk about an average velocity or speed, we will make that explicit. In fact, we will make the same assumption for any other physical quantity that is changing, and this same recipe (the limiting value of a ratio) is used for other quantities besides velocities and speeds.

Acceleration

We have just mentioned that velocity can change over time, just like position. To finish our description of linear motion, we need to be able to talk quantitatively about this type of change as well.

The physical quantity that reflects a change in velocity is called *acceleration.* The way we calculate acceleration from velocity is completely analogous to the way we calculate velocity from position.

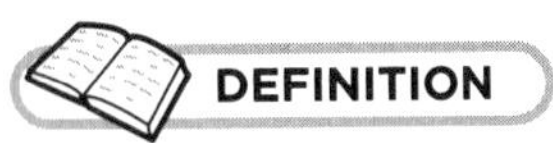

Acceleration is the time rate of change of an object's velocity. Over a definite interval of time Δt, the average acceleration in the x direction is defined as $a_{avg} = \Delta v/\Delta t$.

The letter a is often used to designate acceleration. As you can see from the definition, the dimension for acceleration is distance divided by time, divided by time again (Δv has the same dimensions as v, of course). The standard units that go with this dimension are m/s^2. Whatever symbols are used to represent acceleration and velocity, you will never mix up these two things if you pay close attention to the units used to measure them. Different units mean different dimensions, which means that acceleration and velocity are different physical quantities.

Just as with velocity, we can't talk about average acceleration without talking about a definite time interval. If the velocity of the object is the same at the end of the interval as it was at the beginning (including the direction), then the average acceleration would be zero during that time. By definition, if the velocity is constant during some time, then the acceleration is exactly zero. If acceleration changes over time, we can define the instantaneous acceleration just as we did for velocity by considering the limiting value of the defining ratio as the time interval shrinks to nothing.

An acceleration equal to zero does not mean there is no motion, or that the velocity is equal to zero. It only means the velocity is constant (unchanging) so the velocity could be zero or any constant, non-zero value.

The more interesting cases occur when the acceleration is not zero. Non-zero acceleration means that the velocity *is* changing. Just like velocity, acceleration can be positive or negative. However, the sign of the acceleration does not have to be the same as the sign of the velocity. The algebraic sign of the acceleration is determined by the sign of Δv, or in other words, the direction in which the velocity is *changing*.

For example, if our little red caboose is moving in the positive x direction at a certain time, then later is found to be moving even faster in the positive direction, then the acceleration was positive. This is the classic example of positive acceleration, when something is "speeding up" and moving in the direction we call positive.

Alternatively, if the caboose is moving in the positive direction initially, then slows down as time goes by, that will be a case of negative acceleration. (Some books may use the word "deceleration" to describe this case, but we think that can be confusing, so we will avoid that term.) If you follow the mathematical definition, you can easily see where this negative sign comes from. The later v is a smaller positive number than the earlier v, so $\Delta v = v_2 - v_1$ will end up negative. (We will never consider any negative time intervals Δt.)

But what if the velocity is negative through the whole time interval? Everything would then get turned around. If the caboose speeds up as time goes on (but is moving in the negative direction) then the acceleration is also negative. Again, this automatically comes out correctly if you use the mathematical definition. The later velocity is even more negative than the earlier one. Finally, to be complete, what is the sign of the acceleration if an object moving in the negative direction is slowing down? If you said positive, you are correct.

CONNECTIONS

Suppose our little red caboose is moving in the positive direction with positive acceleration. Its velocity will just continue to increase. Based on what we've said so far, you might conclude that it could increase to arbitrarily high speeds. In fact, for extremely high speeds (approaching about 300,000 km/s) a new type of physics kicks in that slows objects so they can't exceed a sort of universal speed limit. This special physics is called special relativity, which Albert Einstein first conceived while thinking about—you guessed it—trains!

Something particularly interesting happens in cases where the initial velocity and acceleration are in opposite directions: the final velocity could have a different sign from the initial velocity! Imagine our caboose initially moving at a good speed in the positive direction. If there is a constant negative acceleration, it will start slowing down. Since it is slowing down, the velocity will eventually reach zero. If the acceleration stays constant and negative even past that time, the velocity will keep changing and actually change signs, crossing through zero to become negative.

Incidentally, this also occurs whenever you throw a ball straight up in the air. At the beginning it is moving upward, but after a while it will turn around and move down. We'll get into the source of the acceleration in this case in Chapter 6.

Examples of Linear Motion

Before moving to the next chapter, let's graph a few examples to try and make all of this clear. We'll plot the various quantities as functions of time to illustrate what is going on.

Let's begin by revisiting our definition of average velocity:

$$v_{avg} = \frac{\Delta x}{\Delta t} \qquad \textbf{Equation 2.1}$$

Using this equation, you can calculate an average velocity if you have a definite time interval, and you know the location of the object at the start and end of that time. But because of the way algebra works, we know that the same equation can be rearranged to allow the calculation of a different unknown. For example, if you already know the average speed of an object and where it starts, you can easily predict where it will be at a later time.

You can see this just by substituting for $x_2 - x_1 = \Delta x$ into Equation 2.1. Solving the result algebraically for x_2 results in this equation:

$$x_2 = x_1 + v(\Delta t) \qquad \textbf{Equation 2.2}$$

This equation doesn't contain any new information, but it's nice for predicting how far something will go in a certain time at a certain speed. It is valid only if v really represents the average velocity (including the sign) over the entire time interval Δt. Of course, if the velocity is constant during that time and equal to v, that works, too (and notice we have dropped the "avg" subscript). In words, the (average or constant) velocity multiplied by the time interval, then added to the starting position, gives the position at the end of the time interval.

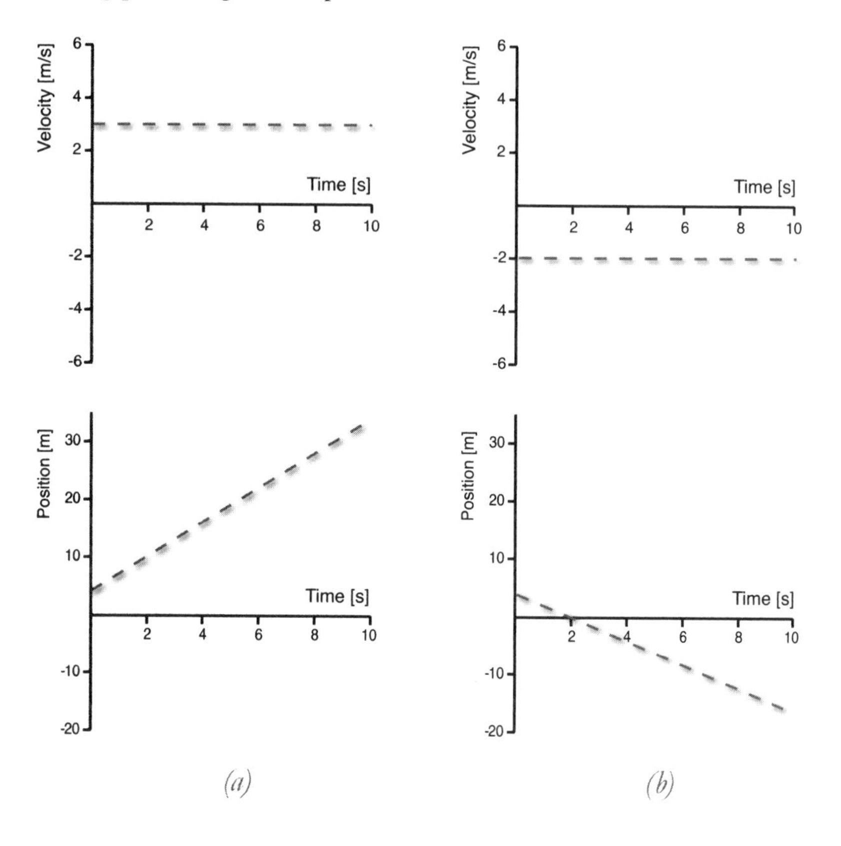

Figure 2.3

If the position vs. time graph is a straight line, it means the velocity has some constant value, either positive (a) or negative (b).

If you know how far you need to go, and how fast you will be going, you can rearrange this again to figure out how long it will take to get there:

$$\Delta t = \frac{\Delta x}{v} \qquad \textbf{Equation 2.3}$$

Here again, v is assumed to be the constant or average velocity. You may have actually used this equation before, even if you didn't know it, if you ever tried to figure out when you would arrive at the next town during a road trip.

Let's see how all this works. Consider a bike trail that is exactly 20 kilometers long. If you can ride that trail from start to finish in 50 minutes, what would your average speed be, in m/s? All you need to do is divide the distance by the time it takes. You know that a kilometer (km) is equivalent to 1,000 meters, and that every minute has 60 seconds, so the velocity will be:

$$v = 20{,}000 \text{ m} \div (50 \text{ min} \times 60 \text{ s/min}) = 6.67 \text{ m/s} \qquad \textbf{Equation 2.4}$$

Now suppose the best speed you can really pedal is 5 m/s. How far would you get in 50 minutes? Just multiply the 5 m/s by the 3,000 seconds (50 minutes) and you will see that the answer is 15,000 meters, or 15 kilometers.

TRY IT YOURSELF

1. How long would it take to bicycle the whole 20 km trail at your 5 m/s pace?

In later chapters, we will see a lot of cases where the acceleration of an object is constant. In other words, the velocity of the object is changing at a constant rate. How would cases like this look?

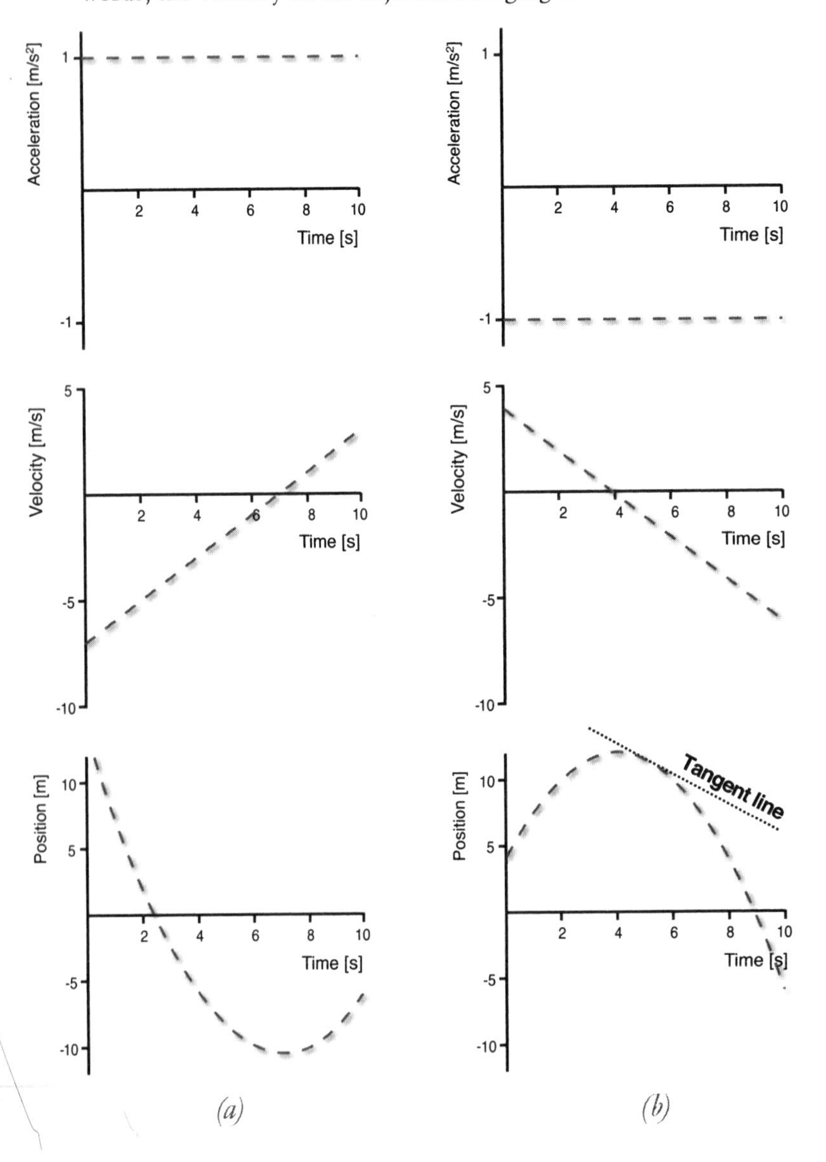

Figure 2.4

Plots of position, velocity, and acceleration vs. time for positive acceleration (a) and negative acceleration (b).

Figure 2.4(a) shows position, velocity, and acceleration for an object over a given time interval, in a case where the acceleration has a constant positive value. Just to make it interesting, we start the interval with the object moving in the negative direction. You can tell the acceleration is constant because the line showing acceleration is horizontal, having the same value at all times. The graph of velocity is also a straight line, but it is sloping upward, showing that the velocity is changing in the positive direction as time goes on.

The slope of that velocity vs. time graph is determined by how big the acceleration is. Remember, acceleration is the rate at which the velocity is changing. A larger acceleration means a larger change in velocity for each time interval, so the slope of the line would be steeper. If acceleration were smaller, then the velocity wouldn't change as fast and the slope of the line would be less steep. (For fun, think about what this graph would look like if the acceleration were zero.)

Notice that the plot showing position is curved. It doesn't really matter where on the position axis the object starts, but since we start with a negative velocity, it has to be moving toward the negative direction at the beginning of the time period. The curve is an indication that the velocity is not constant, but is changing. That initial negative velocity gets less negative as time goes by, because of the positive acceleration. Eventually (around $t = 7$ s in Figure 2.4[a]) that velocity will be zero (just for a moment). After that moment passes, the velocity becomes positive, and we see that the object turns around and starts moving in the positive direction.

This graph also helps us to visualize what the instantaneous velocity really means. When the velocity was constant (Figure 2.3), we saw that the x vs. t graph was a straight line, and the slope of that line was the velocity. In that case, focusing on any part of the line would show the same slope. Now with a changing velocity, the slope of the x vs. t graph changes too, resulting in a curved line. If you choose a single point on the curve, corresponding to any instant in time, you can draw a straight line that contacts the curve only at that single point. The slope of that tangent line is the instantaneous velocity at the corresponding time. Such a tangent line is drawn in Figure 2.4(b), and shows that the instantaneous velocity at about $t = 5$ s is negative.

In Figure 2.4(b), we show the plots with a constant *negative* acceleration. We arbitrarily chose to start the object off this time with a positive velocity. Notice the differences between this and Figure 2.4(a). The negative acceleration means the velocity graph now slopes downward. Again, we give it enough time to pass through zero, so that the velocity changes from positive to negative, all with the same constant acceleration. The graph for the position is still curved, but in the opposite direction.

If you know the velocities at two different times, then you can easily use the definition to calculate the average acceleration between those two times:

$$a_{\text{avg}} = \frac{\Delta v}{\Delta t} \qquad \textbf{Equation 2.5}$$

But let's say you know the acceleration, you know it is constant, and you know the velocity at some particular time. Could you figure out the velocity at a later time?

Let's imagine a car this time, going in the negative direction on a long straight country road. Now, assume the car has a constant positive acceleration of +3.0 m/s^2 and an initial velocity of -20 m/s. What is the velocity 12 seconds later? Well, because the acceleration is +3 m/s^2, we know that the velocity is increasing at a rate of 3 m/s every second. So the total change in velocity Δv is 3 m/s^2 × 12 s = 36 m/s (notice that we just solved Equation 2.5 for Δv). Add this change to the original velocity, -20 m/s, and you will get the final velocity, 16 m/s.

Often we are more interested in how far an object travels when it has some constant acceleration. This is a little more difficult, but a fairly simple trick will help us. We really just have to remember that distance and time are related by the average velocity. The handy trick is how to figure out average velocity when the velocity is changing.

Recall that in the constant acceleration case, a velocity vs. time graph is a straight line (Figure 2.4). When that is true, for any time period, the average velocity is exactly halfway between the starting velocity and the velocity at the end of the period. In our notation, we used v_1 and v_2 for these quantities, respectively. The average velocity is then given by:

$$v_{avg} = \frac{v_1 + v_2}{2} \qquad \textbf{Equation 2.6}$$

Since $\Delta x = v_{avg} \times \Delta t$, and $v_2 = v_1 + a \times \Delta t$, then we can do a little algebra to find:

$$\Delta x = v_1(\Delta t) + \frac{1}{2}a(\Delta t)^2 \qquad \textbf{Equation 2.7}$$

Therefore, if you know the acceleration and the starting velocity, you can calculate how far an object will go in any interval of time, as long as the acceleration is constant during that time.

TRY IT YOURSELF

2. A drag racer wants to finish his race in 8.5 seconds, and the length of the course from start to finish is 400 meters. Beginning with zero velocity, what constant acceleration will the car have to have during that time? What will his final velocity be as he crosses the finish line?

Finally, in some situations you are interested in relating velocity and position, but you don't care so much about time. In that case we can derive another useful expression:

$$v_2^2 = v_1^2 + 2a(\Delta x) \qquad \textbf{Equation 2.8}$$

We don't need to go through a lengthy derivation to be able to use this equation. We just need to make sure we understand what all the symbols mean, and that it is only valid if the acceleration has the constant value a the whole time that the object is moving through the interval Δx. As always, v_1 and v_2 are velocities at the beginning and end, respectively. This equation will tell you the ending velocity any distance after a point where the starting velocity was v_1, if you know the acceleration.

The Least You Need to Know

- In order to describe motion quantitatively, you need to set up a coordinate system with an origin, a specific orientation, and a scale.
- Velocity is the quantity that describes how fast something is moving and in what direction.
- When all the motion is along a single line, the two possible directions can simply be called positive and negative.
- Acceleration is a directional quantity that corresponds to the rate of change of velocity.
- Using the definitions and a little algebra, you can calculate all sorts of things about the motion of an object, for cases where either the velocity or the acceleration is constant.

CHAPTER

3

Motion in Two and Three Dimensions

In order to deal with the real world, we must go beyond one dimension. In this chapter we will add an important new tool that will allow us to do so: the vector. We will explain what this means and what's new as a result, and even give you the basics of vector mathematics. The beauty is that, once you have mastered this, the same concepts you worked so hard on in the last chapter will remain directly applicable: the coordinate system, velocity, speed, and acceleration.

By the end of this chapter, you will be fully equipped to consider the causes of motion, especially things that cause motion to change, all of which we will tackle in the next chapter.

In This Chapter

- Scalars vs. vectors
- Vector mathematics
- Vectors for velocity and acceleration
- Uniform circular motion

Vectors and Their Properties

In physics we must distinguish between two classes of physical quantities. One class includes all of the quantities that do not require any directional information. These quantities are called *scalars*. A good example of a scalar quantity is mass. Mass quantifies the amount of stuff that is present. When you have 15 kilograms of something, that is its mass, whether it is moving or not. A single number is sufficient to specify the mass, and this is true for all scalar quantities. Other important scalars we will be using are time, energy, and temperature. There is no direction associated with the value of any of these.

For a lot of other physical quantities, direction matters just as much as the amount. The name used for these kinds of quantities is *vectors*. For a vector quantity like velocity, it is not enough to specify how much velocity you have; the direction of the motion also matters. Say you leave home and you plan to travel 34 kilometers in the next hour, we won't know where to find you unless you also tell us the direction in which you plan to travel. Direction is an intrinsic property of all vectors.

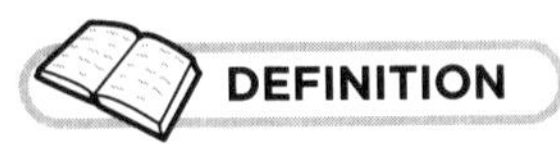

DEFINITION

Scalar is the name for the class of physical quantities that do not have any direction.

Vector is the name for the class of physical quantities for which direction is essential information.

For the purposes of this book, there are no other classes besides these two. Every physical quantity we discuss will be either a scalar or a vector quantity.

It is really important to understand the difference between scalars and vectors, and to know, for every physical quantity, which class it falls into. In most textbooks, the type of quantity being discussed is indicated by the typeface used to represent it. If we are using a variable to stand for a scalar, we just use regular type, as we did with x, v, and a in the previous chapter. Conversely, if a variable represents a vector, the symbol will appear in **boldface** type. Thus from now on in this book, a velocity vector should appear as $\mathbf{v}$. When writing by hand, however, it's not easy to make bold characters. Instead, the custom when using pencil or pen or chalk is to put an arrow over any symbol that represents a vector. A general velocity vector written by hand would look like this: $\vec{v}$.

Position as a Vector

The world we live in has three spatial dimensions. That means, unless there is some physical restriction, there are three independent directions in which we can move. The clearest way to see this is to construct another coordinate system, this time with three coordinate axes instead of just the one we used in the previous chapter. These three axes are lines that meet at a single point in space, a point we again call the origin of our coordinate system. Each of these three lines must be perpendicular to (at right angles to) the other two, or "mutually orthogonal," in the language of mathematics.

If you are sitting indoors, it is easy to picture such a coordinate system. Pick any corner of the room, and look down at the point where the two walls join the floor. Call that point the origin. The line where the floor joins the adjacent wall to the left can be the x-axis, and the line that joins the floor to the other adjacent wall will be called the y-axis. These two axes are both horizontal, and the "xy-plane" would correspond to the floor of the room. Then the vertical line that joins the two walls at that same corner would be called the z-axis. Notice that (in most rooms) there is a 90-degree angle between the x- and y-axes, and that the z-axis is perpendicular to both of the other two axes. Such a system is often called a *Cartesian coordinate system.*

A **Cartesian coordinate system** is any coordinate system based on simple rectangular coordinates, i.e., whose axes are mutually perpendicular straight lines. This term is applied to both two-dimensional and three-dimensional systems.

The purpose of the coordinate system is to enable us to specify the precise position of any particle or object. As before, the origin is where the position is zero in all three directions. Pick a random location somewhere in the room that is not at the origin. The position vector is like an arrow that starts at the origin and ends at the point you picked. The arrow is a good way to visualize vectors because it not only has a certain length but it points in a definite direction in space.

At this point, we should get rid of the walls and floor of the room, (in our minds) and just keep the x-, y-, and z-axes. Imagine that each of these axes is infinitely long in both directions from the origin. As before, in addition to locating the origin of a coordinate system in the real world, we must also specify which direction is positive for each of the axes. Once you have done this, three simple numbers are all you need to precisely describe any location in the universe!

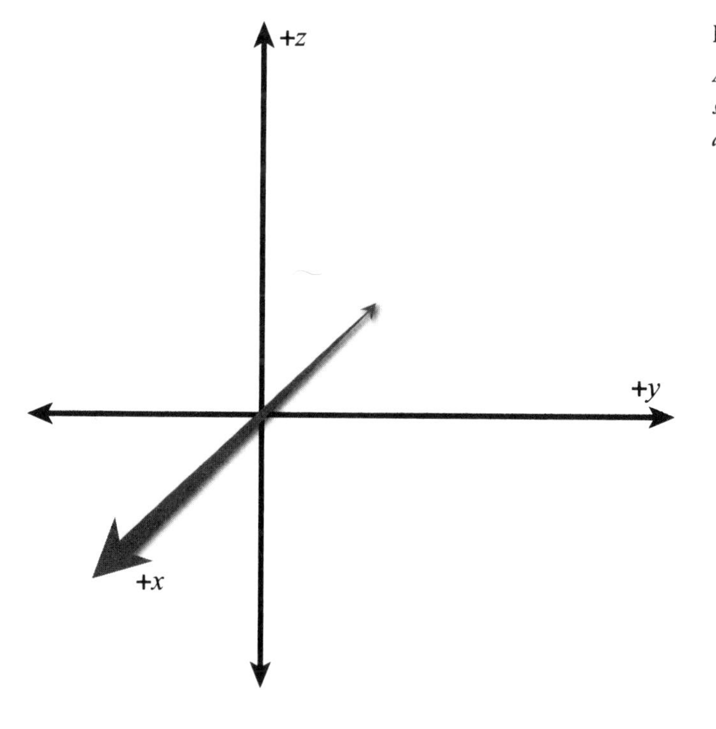

Figure 3.1

A three-dimensional Cartesian coordinate system will have three mutually orthogonal axes that we call x, y, *and* z.

The three numbers (in correct distance units, of course) are called the coordinates x, y, and z of a point. The z coordinate, for example, is how far above or below the xy-plane you are (the floor from our previous example). Above is customarily called positive, while below is considered to be negative. The z coordinate is your position projected onto the z-axis, which is where you would end up if you moved toward the z-axis parallel to the floor (xy-plane). Likewise, the x coordinate is the distance of the point from the yz-plane. It is located either in front of or behind it, depending on whether x is positive or negative.

In order to look at the position as a vector, we must remember that this is a relative position, in the sense that it is referenced to the origin. We will often use the symbol $\boldsymbol{r}$ for a general position vector. A certain position vector $\boldsymbol{r}$ has three components, which correspond to the three dimensions and the three position coordinates (x, y, and z) described above. In most of the rest of this book, we will restrict ourselves to cases of two-dimensional motion, for which we will usually use the xy-plane (keeping $z = 0$, for example). This is not too bad, it turns out, because there is a lot of interesting motion in the world that happens in only two dimensions. It's also nice because 2D motion is a lot easier to illustrate on the flat pages of a book than 3D motion would be.

The position vector gives us an easy way to visualize the general behavior and properties of vectors. But right from the start, we want to help you generalize these ideas and be able to apply them to all kinds of vectors with different dimensions. It is important to know that vectors need not be anchored to any certain position (like a coordinate system's origin). A vector is completely defined by its magnitude and direction. The magnitude is like the length of the arrow (with the appropriate units) and the direction tells us which way it is pointing. You could imagine two arrows that are located in different places, but which have the same length and each point in the same direction. These two arrows would represent exactly the same vector.

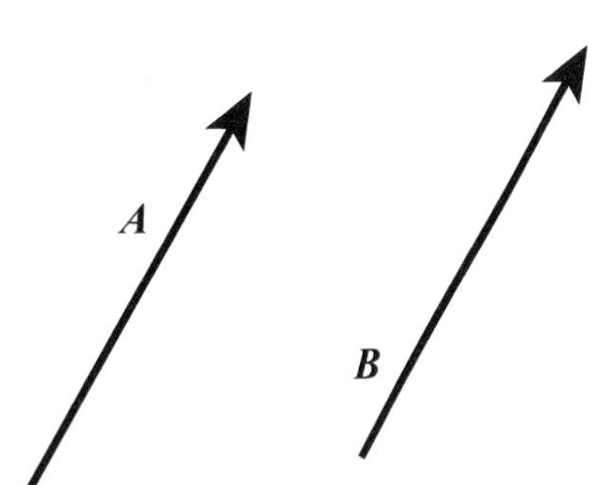

Figure 3.2

These two arrows represent the exact same vector, because they have exactly the same length and point in exactly the same direction. In mathematical terms, we can simply write $\mathbf{A} = \mathbf{B}$.

Vector Components, Magnitude, and Direction

Recall the 3D position vector that we talked about before. This describes the position of a point in space using an arrow that starts at the origin of our coordinate system and ends at the point. The pointy end of the arrow is located exactly at the point we are interested in. In three-dimensional space, this vector has three components, one for each axis. The x component is numerically equal to the x coordinate of the point, the y component corresponds to the y coordinate, and the z component corresponds to the z coordinate. Like the coordinates, any of the vector components can be positive or negative.

The components are related to both the magnitude and the direction of the vector. You can think about the components literally as directions telling you how to get to the destination point, the head of the arrow. Starting at the origin, move along the x-axis by a distance equal to the x component (in the positive or negative direction depending on the sign). Then turn by 90 degrees and move in the plus or minus y direction for a distance specified by the y component. Finally, go straight up or down by a distance given by the z component. Following these three moves, you will find yourself at the destination point. If you think about it, you could have made these three moves in any order and still have arrived at the same place.

We use the term "magnitude" to describe how big (or how long) a vector is, without regard to the direction. In this case, it is the length of the arrow, the straight-line distance between the origin and the ending point. It is not the sum of the lengths of the components, even though we used them as a way to get there from the origin. The magnitude of a vector turns out to be the square root of the sum of the squares of the components. For our position vector $\boldsymbol{r}$, this is:

$$|\boldsymbol{r}| = r = \sqrt{x^2 + y^2 + z^2} \qquad \textbf{Equation 3.1}$$

If you confine yourself to the xy-plane (keeping $z = 0$ at all times), you see that Equation 3.1 reduces to the familiar Pythagorean theorem, with r as the length of the hypotenuse of a right triangle.

Remember that we are using boldface type to indicate vectors. The absolute value sign when applied to a vector is a fancy way of saying the magnitude of the vector. In Equation 3.1 you see another conventional use of notation. When the symbol for a vector is *not* in boldface type, that also means we are talking about the magnitude. Since the magnitude of a vector is a scalar, this notation is consistent. Also, note that the magnitude of a vector cannot be negative; we use only the positive square root. For any vector, the components must have the same dimensions and units as the vector itself, and so the magnitude of the vector must also have those same units.

As for vector direction, it is typical to specify this using angles. In two dimensions, one angle is all that's needed, and we'll call this angle θ. Just as for other coordinates, this requires us to agree on a reference line from which to measure the angle as well as a rule for determining positive or negative angles. The most common convention is to use the positive x-axis as the reference direction. Any vector parallel to the x-axis and pointing in the positive x direction would have an angle $\theta = 0°$. And, we will take the counter-clockwise direction to correspond to positive angles relative to this reference.

Any position vector $\boldsymbol{r}$ in two dimensions can be uniquely described by giving its magnitude r and its direction θ. The angle θ can be any angle from zero to 360° (or larger, or negative, with the understanding that there is more than one way to label the same angle).

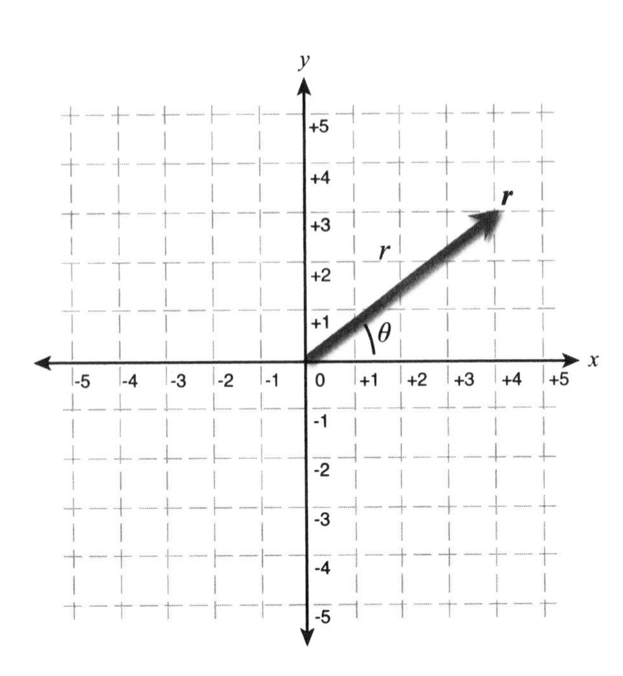

Figure 3.3

A general **r** *position vector. This vector has an* x *component equal to +4 units, and a* y *component equal to +3 units. The magnitude* r *works out to be 5 units, and* $\theta \approx 37°$.

Trigonometry helps us relate the two different ways of representing specific vectors. If we follow the convention for defining the angle of a vector θ, then the x component of a vector will always be its magnitude (r) multiplied by the trigonometric function $\cos(\theta)$. Similarly, the y component of that vector will be the magnitude (r) multiplied by $\sin(\theta)$.

Any vector in two dimensions can be represented in two ways: either by giving its two Cartesian components (x and y), or by giving its magnitude and direction (r and θ). These two methods of specifying a vector contain exactly the same information. In each case, two numbers are required, because we are working in two dimensions. (Note that in three dimensions, we would need three numbers, such as the three components. There is still only one magnitude, but the direction would require two numbers to specify it.)

We have seen how to get the components if we know the magnitude and the angle. What about going the other way around? We already showed how to use the Pythagorean theorem to figure out the magnitude if we know the components. Can we also find the angle θ if we only know the components? It's easy if you have a calculator that can do inverse trigonometric functions.

Let's now switch to a more general notation, which will apply to all vectors, even if they are not position vectors. Take, for example, the two-dimensional vector $\boldsymbol{B}$, which has x and y components B_x and B_y respectively and whose magnitude is B. Suppose this vector makes an angle θ with the x-axis in the conventional way. If you are given this angle and the magnitude of the vector, just as above, you can calculate the components by:

$$B_x = B\cos(\theta) \quad \text{and} \quad B_y = B\sin(\theta) \qquad \textbf{Equation 3.2}$$

On the other hand, if you are starting with the components, then you find the magnitude and direction with these two equations:

$$B = \sqrt{B_x^{\,2} + B_y^{\,2}} \quad \text{and} \quad \theta = \tan^{-1}\left({B_y}/{B_x} \right)$$ **Equation 3.3**

This latter expression comes from the definition of the tangent function in trigonometry, which states that in a right triangle, the tangent of an angle is equal to the sine of the angle divided by the cosine. Note that the y component has to be in the numerator of the fraction for this to work.

In the following exercise, notice that the vector in question is a velocity. Direction and magnitude (speed) are given, so you should be able to figure out which of the expressions above are used to calculate one of the components.

TRY IT YOURSELF

3. A motorcycle has a speed of 24 m/s, and is moving at an angle of 130° with respect to the x-axis. What is the x component of its velocity vector?

Adding and Subtracting Vectors

Vectors can't be added as simply as scalars can. If you know the magnitudes of two vectors, you may be tempted to say that the sum of the two vectors is equal to the sum of their magnitudes, but that will almost always be wrong.

To show you why this is the case, we will illustrate the addition of vectors in Figures 3.4 and 3.5. When we add two vectors, we first picture one vector as an arrow with the correct magnitude and direction. Then we draw the second vector such that its tail starts at the head end of the first vector. The result of adding the two vectors is a new vector, one that starts at the tail of first one and ends at the head of the second one. This process can be repeated to illustrate the sum of any number of vectors. Just as for scalar addition, the order in which the terms are added makes no difference.

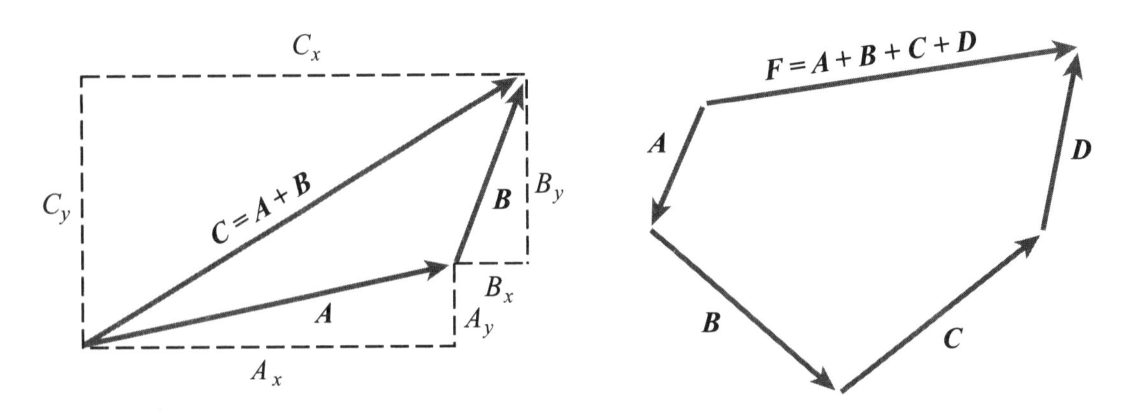

Figure 3.4

Vector **C** *is the sum of vectors* **A** *and* **B** *(left). Vector* **F** *is the sum of vectors* **A**, **B**, **C**, *and* **D** *(right).*

The components of a vector are key to getting the right numerical results for vector addition. After all, it would be rather annoying if you had to whip out your ruler and protractor every time you needed to add vectors to do physics. Fortunately, components make this very easy.

The only thing you need to remember is that every component of the sum of two vectors is equal to the sum of the components of the individual vectors you are adding. Take the example on the left of Figure 3.4, where $\boldsymbol{C} = \boldsymbol{A} + \boldsymbol{B}$. Here, we've added in a few dashed lines to indicate that $C_x = A_x + B_x$ and $C_y = A_y + B_y$. The same rule would hold for the *z* coordinates, if there were any. Although the vectors in the figure all have positive components, this works just as well if any of the components are negative.

The definition of subtraction is just the addition of the opposite. In the world of vectors, the opposite of one vector means a vector with the same magnitude, but pointing in the exact opposite direction. Thus we can write $\boldsymbol{A} - \boldsymbol{B} = \boldsymbol{A} + -\boldsymbol{B}$. Figure 3.5 shows what this looks like in two dimensions.

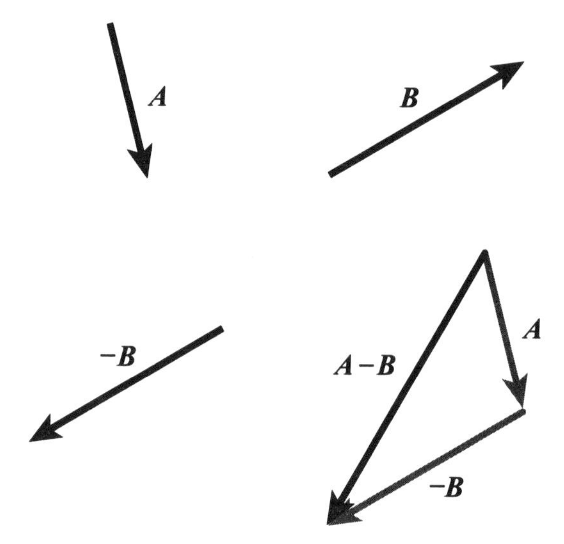

Figure 3.5

Subtracting vector **B** *from* **A** *is the same as adding the two vectors* **A** *and negative* **B**.

Before moving on, we should probably mention one more operation with vectors. Just as you can't add or subtract quantities of different dimensions (e.g., length, mass, or time), it makes no sense to add or subtract scalars and vectors. Since a scalar has no direction, there is no sensible way to add a scalar to a vector. Moreover for any equation, both sides of the equals sign must be either scalar or vector. If you ever find yourself equating a scalar quantity with a vector quantity, then you had better go back and look for your mistake!

It is permissible, however, to multiply a vector and a scalar together. For the position vector $\boldsymbol{r}$ for example, when we write $3\boldsymbol{r}$ we mean that $3\boldsymbol{r}$ is another vector, three times as long as $\boldsymbol{r}$ and in the same direction as $\boldsymbol{r}$. You can also think of the scalar (3 in this example) as multiplying all of the components of the vector. If the scalar is negative, then the resulting vector has a direction exactly opposite the one that got multiplied. We were already doing this, implicitly, when we talked about the negative of a vector. The vector $-\boldsymbol{B}$ is the same as what you'd get by multiplying the vector $\boldsymbol{B}$ by the scalar quantity -1. No matter what, the result of multiplying a scalar times a vector is another vector.

The multiplication we've introduced here is sometimes referred to as "scalar multiplication," even though the result is a vector. This should not be confused with the "scalar product," which is one of two possible ways to multiply vectors together.

When multiplying a vector by a scalar, the scalar may or may not have dimensions of its own. If it doesn't have its own dimensions (or units), the multiplication does not change the dimension (or units) of the vector. However, the scalar could have its own units. In that case, the result of the multiplication would be to change the dimension of the vector itself. This should become clearer when we do some real examples shortly.

Velocity and Acceleration in a Single Plane

As with position, the general ideas of velocity and acceleration are the same in 2D or 3D space as they were in the 1D (single-axis) discussion of the previous chapter. The difference is that there is a lot more freedom in the directions of these two vector quantities. And directions matter, because they are essential to the nature of vectors.

We start again with velocity. The same definition applies, now in vector form:

$$\boldsymbol{v}_{\text{avg}} = \frac{\Delta \boldsymbol{r}}{\Delta t}$$ **Equation 3.4**

As before, for instantaneous velocity we just let the time interval shrink to a very small time interval around the time we are interested in.

Now imagine a hummingbird that is buzzing along some curvy path in space. In any brief time interval, the hummingbird changes its position in a certain direction, moving from its original position ($\boldsymbol{r}_1$) to a new position ($\boldsymbol{r}_2$) a little further down the path. It is that change in position that we are calling $\Delta \boldsymbol{r}$. Just as before, $\Delta \boldsymbol{r} = \boldsymbol{r}_2 - \boldsymbol{r}_1$, but now we are subtracting two position vectors. Make a sketch to convince yourself that for a short time period, the resulting short vector $\Delta \boldsymbol{r}$ has to point right along the path in the direction that the particle (or hummingbird) is moving.

Recall that time is a scalar quantity. In our definition of $\boldsymbol{v}_{\text{avg}}$, we effectively multiplied the vector $\Delta \boldsymbol{r}$ by the scalar $1/\Delta t$. Therefore, the average velocity is another vector $\boldsymbol{v}_{\text{avg}}$ which must point in the same direction as $\Delta \boldsymbol{r}$. The dimensions of $\boldsymbol{v}_{\text{avg}}$ and its components have gone from meters (m) to meters per second (m/s), as expected. You can thus say generally that the instantaneous velocity vector for a particle points in a direction that is tangent to the path of the particle in space.

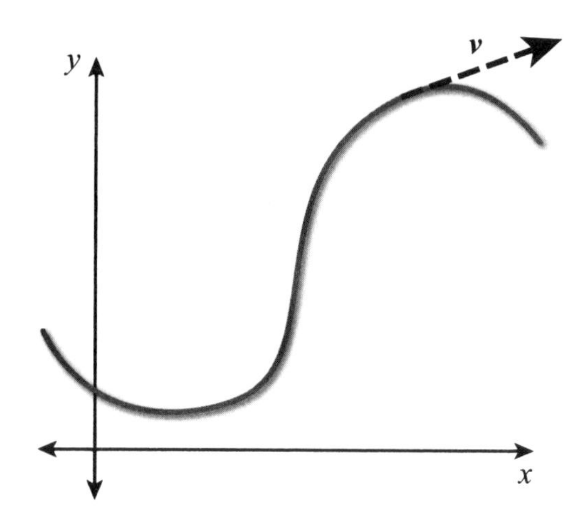

Figure 3.6

The instantaneous velocity vector is always tangent to the path a particle takes through space, as shown here for 2D motion in the xy*-plane.*

Now if velocity can also change over time, we can easily generalize the definitions of average and instantaneous acceleration that we saw in Chapter 2. Acceleration is still the rate at which velocity is changing, exactly equal to the change in velocity divided by time. It will still have standard units of m/s^2. Since velocity is a vector quantity, the change in velocity is also a vector $\Delta\boldsymbol{v}$. Dividing the vector $\Delta\boldsymbol{v}$ by the scalar Δt results in an acceleration that must be a vector:

$$\boldsymbol{a}_{\text{avg}} = \frac{\Delta \boldsymbol{v}}{\Delta t}$$ **Equation 3.5**

Notice now that there are two ways that the vector velocity can change. It can change in magnitude or direction (or both). Either one of these alone still counts as a change. If a velocity only changes its magnitude and maintains its direction, we have the same sort of cases we discussed in the previous chapter. But it's easy to forget that an object moving with constant speed but with varying direction is also accelerating. Don't make this mistake!

The direction of the acceleration vector is the same as the direction in which the velocity vector is changing. If our hummingbird is increasing its speed while continuing to move in the same direction, then the acceleration is in the same direction as the velocity. But if it is tracing out a curved path (say, flying forward while veering off to the left), it means that the velocity is also changing direction, and the acceleration vector must also have some component corresponding to the directional change (to the left).

Uniform Circular Motion

Let's look a little deeper at the case of motion that occurs in one plane but along a curved path. Consider a particle that moves along a path that is a perfect circle. Let's assume for now that the speed of the particle on that path is constant and equal to v. We will use R to stand for the radius of this circle, to avoid confusion with the general position vector $\boldsymbol{r}$. R is thus a constant with the dimension of length.

If the speed is constant, it means that there is no component of the acceleration vector in the direction of motion. However, as we've just seen, it does not mean that the acceleration is zero! Since velocity is a vector, we know for sure that there is some change in velocity as the particle moves around the circle. The change in velocity is due only to the velocity changing direction. Since we know that the velocity vector is always tangent to the path, the direction of the change must be toward the center of the circle (that is, always making a 90-degree angle with the direction of motion).

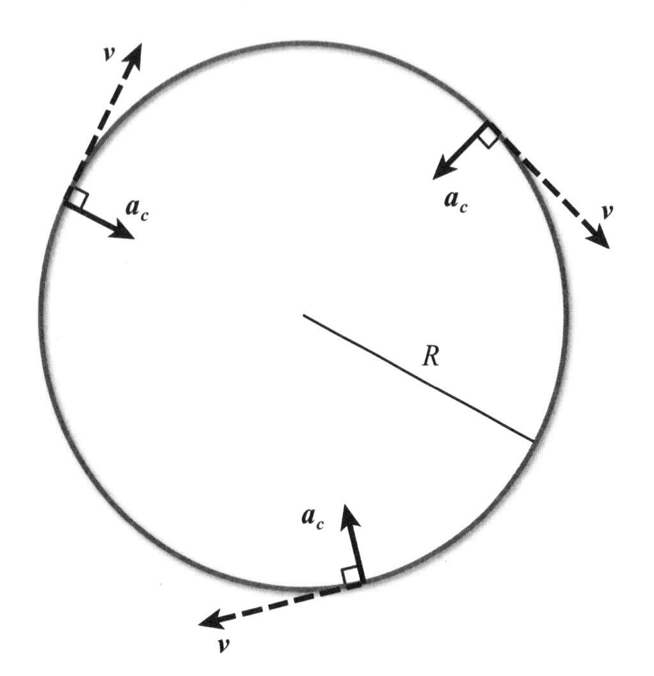

Figure 3.7

Motion along a circle of radius R *at a constant speed* v.

Physics teachers love to call this type of motion "uniform circular motion." The word "uniform" refers to the constant speed. You must remember that constant speed does not mean constant velocity, because the velocity can change direction without changing its magnitude. We will not prove it here, but the magnitude of the acceleration in this case can always be calculated, if you know the speed and the radius of the circle being traveled, using the formula $a_c = v^2/R$. We will call this special kind of acceleration *centripetal*, and use the label $\boldsymbol{a}_c$.

Centripetal acceleration is the acceleration that is due to a velocity that changes direction. The simplest case is uniform motion along part of a circle with radius R. In that case, the magnitude of this acceleration is $a_c = v^2/R$, where v is the constant speed of the moving object, and its direction is toward the center of the circle.

You can convince yourself that the formula for a_c makes sense if you simply remember that the acceleration is the rate of change of the velocity. The faster you are moving in a circle, the faster your velocity will be changing direction, right? That's why v^2 is in the numerator. But if the radius of the circle is larger, then for the same speed, the velocity is not changing as fast. So the R in the denominator is also logical. If you think about driving a car or a bike, then you can also appreciate that a smaller R corresponds to a sharper turn.

TRY IT YOURSELF

4. You're pedaling your bike at a speed of 13 m/s when you go into a sharp right turn. Your speed remains constant, but since you are turning, you experience an acceleration to the right equal to 8.4 m/s^2. What is the radius of your path?

At this point, your toolbox for describing motion of any object, in any number of dimensions, is complete. For an object that has been set in motion, you can now calculate where it will go in the future, and how quickly it will get there. But how do you get an object to go in the first place? It is to this important question that we'll turn in the next chapter.

The Least You Need to Know

- Vectors are distinguished from scalar quantities by including direction as well as magnitude.
- Addition of vectors differs from that of scalars, but is easily handled if you decompose the vectors into their components relative to a fixed coordinate system.
- Position, velocity, and acceleration are all vector quantities that are defined in three-dimensional space. Restricting ourselves to cases in two-dimensions makes it easier for the beginning physicist to visualize and understand what is going on.
- Uniform circular motion is one simple example of 2D motion in which the acceleration is related in a definite way to the constant speed and the radius of the circle.

CHAPTER

4

Force and the Laws of Motion

In this chapter, we really get down to business. Thanks to an amazing genius named Isaac Newton, who did most of his work way back in the 1600s, we can not only describe motion, but we have a beautiful framework for explaining why and how motion actually happens in the world.

As we shall see, any time the velocity of an object changes, there must be something called a force involved. There are many different kinds of forces, but they can all be described by vectors, and they can all lead to accelerations of massive objects in the same fundamental way.

In This Chapter

- Newton's famous laws of motion
- The relationship between force, mass, and acceleration
- Tools and tips for applying Newton's laws

Newton's Laws of Motion

Isaac Newton was one of the greatest physicists of all time, and he made it look easy. He understood and appreciated the work done by many scientists who went before him. He could look at a myriad of observations and measurements of the complex world and distill them down to their essence. He could see the simple relationships between the most important quantities in nature, and express those relationships in the elegant language of mathematics.

This is precisely what he did in formulating three simple "laws" that govern motion. He gave us the tools to figure out what causes things to move, and thus to predict future motions—and not just here on Earth, but across the entire universe. Newton's laws apply to objects with mass, which pretty much describes any physical object. We've mentioned the quantity mass in passing, when we talked about units and also as an example of a scalar quantity. We hope you already have some feel for what mass is, as a measure of how much matter is present in an object. We can only define mass more precisely when we have a basic understanding of Newton's three laws of motion.

The First and Least Interesting Law

Newton's first law of motion is not exactly revolutionary. It states that an object at rest will remain at rest, and an object in motion will remain in motion at a constant velocity, unless a force acts on that object.

This gives us a qualitative definition of a new quantity in physics: *force*. Newton recognized that something had to influence an object in order to change its velocity. He also knew that velocity is a vector quantity. Motion in a straight line at constant speed is a case where velocity is not changing. No motion at all is similar, the only difference being that the constant velocity just happens to be zero. To Newton, these two states—no motion or constant velocity motion—are basically the same. The first law says that constant velocity (including possibly zero) is what you get when there is no force acting on an object. Force is thus defined as that mysterious thing which is required to change an object's velocity.

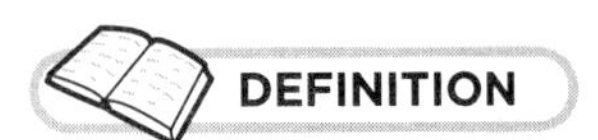

Force is the general term for any physical cause which has the potential to change the motion of an object with mass.

You probably already have some feel for what a force is. If some object is not moving, like a book on a table, you can get it to move by pushing it with your hand. Your hand provides a force and the book, which was at rest, is no longer at rest. But if you stop pushing it, you don't see it continue to move in a straight line with a constant speed. The reason for that is that there are other forces acting besides the push from your hand. Disentangling the effects of all those different forces is a lot of what we'll be doing in the next few chapters.

The Second and Most Important Law

Newton's second law is where most of the action is. It allows us to do quantitative calculations for cases where forces are present. We'll state the second law mathematically, then explore its meaning in some detail. Here goes:

$$\mathbf{F}_{net} = m\mathbf{a} \qquad \textbf{Equation 4.1}$$

First of all, remember that we are talking about one object at a time, an object that has some fixed mass m. This simple expression says that at any instant in time, if an object has an acceleration equal to $\boldsymbol{a}$, then there is a net force acting on that object equal to $\boldsymbol{F}_{net}$. Conversely, if the object is acted upon by some force $\boldsymbol{F}_{net}$, then it will experience an acceleration $\boldsymbol{a}$. We know that acceleration is a vector quantity, as discussed in the previous chapters, and this $\boldsymbol{a}$ is understood to be the instantaneous acceleration. Mass is a scalar quantity, so the product on the right side of Equation 4.1 must be a vector. Moreover, $\boldsymbol{F}_{net}$ and $\boldsymbol{a}$ must be pointing in the same direction.

Part of what this law does is provide a more precise definition of force. It tells us that force in general must be a vector quantity. It also gives us the dimensions for force, which must be mass times length divided by time squared. Thus our units for force will be kg m/s^2. In honor of the great man himself, these force units are called newtons, abbreviated N. So, it follows that 1 N = 1 kg m/s^2.

It is common for more than one force to act on one mass at the same time. When we write $\boldsymbol{F}_{net}$, we mean the total of all the forces that are acting on the object at the same instant in time. One of the many observations summed up in this simple law is that when multiple forces act, they add to each other just the way vectors add. Sometimes $\boldsymbol{F}_{net}$ is written as $\Sigma \boldsymbol{F}$, which is a mathematical shorthand way of saying "add up all of the forces that are acting on the object." The strength of a force corresponds to the magnitude of the force vector. Every force also has a direction associated with it, and that direction is just as important as its magnitude (as we learned in the previous chapter).

We can use this improved understanding of force to gain a better idea of what mass really is. Newton's second law says that acceleration and net force are directly proportional to each other. When viewed this way, we see that the (scalar) mass is simply the constant of proportionality. In other words, the more mass an object has, the more force is needed to change its motion.

Recall that acceleration is just a measure of how fast (and in what direction) the velocity is changing. If you have a bowling ball and a ping pong ball sitting motionlessly in front of you, you already know which one you'll have to push harder if you want to get them both moving at the same speed. To produce the same acceleration on an object with 2,500 times the mass requires 2,500 times the force! *Mass* is really just a measure of how difficult it is to change an object's velocity. This is the sense in which mass is related to inertia. In everyday language, inertia is a general resistance to change. In physics, inertia measures the resistance to a change in motion.

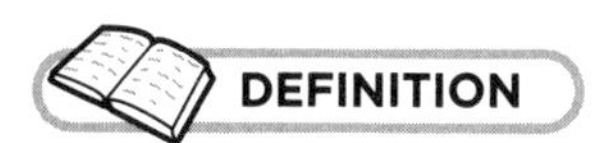

Mass is a measure of how much matter is contained in the object and also how difficult it is to change the motion of the object. It is an intrinsic property of almost all physical objects.

We can now see that Newton's first law of motion is really contained in his second law. The first law is a special case of $\boldsymbol{F}_{\text{net}} = m\boldsymbol{a}$, the particular case where $\boldsymbol{F}_{\text{net}}$ happens to be zero. In this case $\boldsymbol{a}$ is also zero, which just means that $\boldsymbol{v}$ is not changing, regardless of what $\boldsymbol{v}$ is. It is also worth pointing out that $\boldsymbol{F}_{\text{net}}$ can be equal to zero even if there are non-zero forces acting. All that is required is that all the forces add up to zero, when you add them as vectors. For example, as we'll see in Chapter 5, there are a number of forces that act on, say, a potted plant sitting on a windowsill. But these typically cancel each other out so that it sits nicely at rest.

The Third and Most Confusing Law

You may have heard about Newton's third law before, something about every action having an equal but opposite reaction (or words to that effect). While that sounds simple enough, this important law is often misunderstood, so let's consider it carefully.

A complete and accurate statement of the third law requires a lot more words, and maybe a picture or two. The third law can only be invoked when *two* bodies or objects are interacting with each other by means of forces. When two bodies interact, each one exerts some force on the other. Let's consider two objects, which we'll label Thing 1 and Thing 2. Let's use the vector $\boldsymbol{F}_{21}$ to describe the force that Thing 1 exerts on Thing 2. Then the force that Thing 2 exerts on Thing 1 would be labeled $\boldsymbol{F}_{12}$. There may be other forces present in the system, in which case these vectors $\boldsymbol{F}_{21}$ and $\boldsymbol{F}_{12}$ each represent only the total force acting on one of the bodies due to the other.

Newton's third law states that $\boldsymbol{F}_{12} = -\boldsymbol{F}_{21}$. Once you have the setup, it is a pretty simple law. It doesn't matter which thing is Thing 1; neither of the two forces is negative in any absolute sense. This is a vector equation; the minus sign just means that the two forces are in opposite directions. One of the forces can be in any direction, but then the direction of the other force has to be exactly opposite. Whatever force Thing 1 exerts on Thing 2, Thing 1 feels a force *from* Thing 2 that is equal in magnitude and opposite in direction.

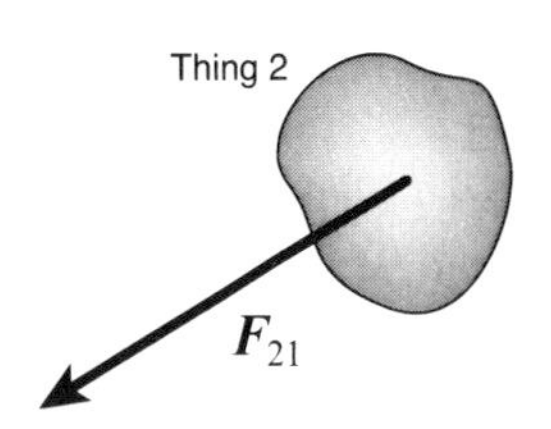

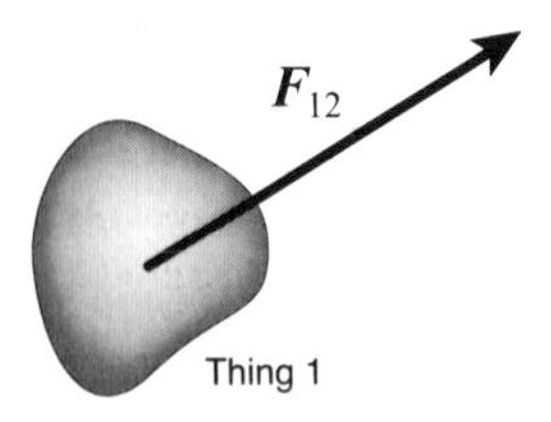

Figure 4.1

Thing 1 exerts a force $\mathbf{F}_{21}$ *on Thing 2.* $\mathbf{F}_{12}$ *is the force exerted by Thing 2 on Thing 1. The third law states that* $\mathbf{F}_{12} = -\mathbf{F}_{21}$.

For example, if you are on roller skates and you push against a wall, the wall automatically pushes back on you by the same amount, causing you to accelerate away from the wall. The wall doesn't have to actively do anything. Newton's third law takes care of everything, making sure there is a reaction to your act of pushing on the wall.

Remember that whenever you invoke the third law, two bodies must be participating. We often call these two bodies an "action-reaction pair." It's important to note that the two forces in a third law example are being applied separately, one force to each of the two different objects! This may seem clear now, but we'll see in Chapter 5 that it is often a source of confusion.

Different Kinds of Forces

There are a lot of different forces out there. They act on all kinds of masses in all kinds of situations. A lot of the forces we'll be dealing with require there to be some contact with the mass being forced. These are the pushes and pulls we are most familiar with. Sometimes the push lasts for a while, sometimes it is brief. A hockey stick changes the motion of the puck when a player takes a slap shot toward the goal. A child pulls her wagon by grabbing the handle. A skydiver deploys his chute by pulling on the ripcord. In every case, there is physical contact between whatever is applying the force and the object which is receiving the force. Also, we mustn't forget, every one of these forces has a definite direction associated with it.

When looking for examples of forces, physics teachers often like to use ropes or chains or cables to apply a pulling force. One reason for doing this is that just by looking at the situation, you can tell the direction of the force being applied. Ropes and strings can only pull, never push. And the direction of the force is exactly along the length. Forces exerted by strings and ropes and the like are often called *tension forces*. In this book, whenever we use something like a rope to apply a force, it will be idealized a little bit. The rope will be perfectly straight, completely inelastic (that is, unstretchable), and we will ignore any mass that it might have.

We will often encounter forces exerted on a body due to contact with a flat surface, like a road or table or wall or ramp—a so-called *contact force*. In such cases, we usually want to distinguish between two components: one perpendicular to the surface and one parallel to the surface. A surface can always exert a force perpendicular to itself, pushing on an object that is in contact with it. This force is so common that we have a specific term for it—the *normal force*—and it is always a push rather than a pull (assuming the object is not glued to a surface).

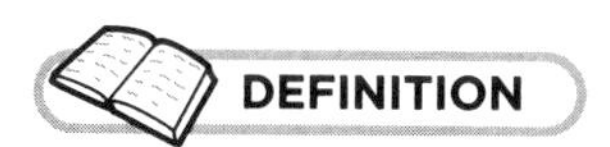

A **tension force** is the pull exerted by a rope or cable or similar thing attached to an object. A **contact force** is a force exerted on one object by another object at their point of contact. A **normal force** is a contact force exerted on an object by a flat surface, in a direction perpendicular to the surface.

If a surface also exerts a portion of its force parallel to itself, that must involve some kind of friction. The net force due to an object's contact with a surface would then be at some angle with the surface that is not equal to 90 degrees.

The physics of friction is interesting enough that we generally want to treat it separately from the normal force. We'll look at friction forces in more detail in the next chapter. In some simple examples, we will want to ignore friction completely. When a surface is described as "very smooth" or having "negligible friction," it means that you only have to consider the normal force due to contact with the surface; the surface exerts no force parallel to itself.

When referring to the normal force, the word "normal" does not mean ordinary. It is used instead in the geometric sense, meaning perpendicular to some surface.

There are some forces, however, that are capable of acting without any physical contact. You are probably familiar with at least one of them already—the force of gravity. In playing with magnets, you may have encountered another force that can act over a distance without touching. Like gravity, electric and magnetic forces are embodied in what we call fields. We're going to postpone any real discussion of force fields until much later in this book, but gravity is so ubiquitous that we might as well talk about it now.

Chapter 6 of this book will be devoted entirely to gravity, but it won't hurt to give you a little preview here. Living on Earth, we can't escape the pull of gravity. Every mass on or near Earth feels a force of attraction toward Earth's center. Since Earth is so large, an arrow pointing toward its center doesn't change direction very much as you move around the room, or even around a good-sized city. So for a lot of our work with gravity, we will assume gravity is a force that has a constant direction, and we define that direction to be vertical down (that is, normal to Earth's surface).

We also observe, to a very good approximation, that the magnitude of the gravitational force experienced by an object is directly proportional to the mass of that object. If we call the magnitude of the gravity force F_g, then this means that $F_g = gm$. Note that we have labeled the constant of proportionality g, and in the SI system of units, it has a value very close to 9.8 m/s^2. We'll use this numerical value for g in all of our examples in this book. Notice that g is a positive number.

The force of gravity experienced by every mass on Earth is the familiar force that we call *weight.* Weight cannot be the same thing as mass because they have different dimensions. Weight must always be given in force units (newtons) while mass is measured in kilograms. It is understandable, though, why some people confuse mass and weight. Here on Earth, an object's weight is always directly proportional to its mass. Something with twice as much mass as another object will also weigh twice as much. But in physics problems, substituting mass for weight (or vice versa) is a recipe for disaster.

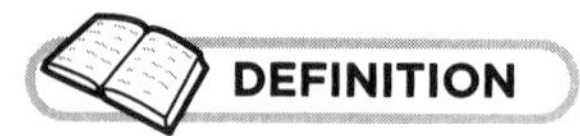

Weight is the force of gravity acting on a mass. Since it is a force, it is a vector quantity. The same mass can have a different weight depending on where in the universe it is located. On Earth, the magnitude of the weight force is mg for any mass m.

Instead of writing the weight formula $F_g = gm$, most books write it as $F_g = mg$. The first way emphasizes the role of g as the constant of proportionality between m and F_g. The second way of writing it looks a lot like Newton's second law. In this form, it makes sense to call g the "acceleration due to gravity." Don't confuse g with gravity itself, though. Gravity is a force, and g does not have the dimensions of a force.

Writing $F_g = mg$ does not mean that the mass m is actually accelerating at 9.8 m/s^2 in some direction. We will often use this expression to calculate the weight force acting on a known mass (on Earth). The weight force is always there, but it may not be the only force acting.

Free-Body Diagrams

A lot of physics problems can be solved by using $\boldsymbol{F}_{net} = m\boldsymbol{a}$ correctly. We already know how to work out the vector on the right side of this expression, multiplying the acceleration vector by the scalar mass. Now let's look at how to deal with the left side, the sum of all forces that are acting.

First of all, if there is only one force, then it is really easy. Whatever that single force is, $\boldsymbol{F}_{net}$ is exactly equal to it. Knowing the mass and seeing how it accelerates is an effective way to measure the force. If you observe a mass whose velocity changes, just measure the velocity at the start and end of a certain time interval. Construct the vector $\Delta\boldsymbol{v}$ as we always do, and divide it by the time interval. The mass multiplied by that average acceleration will give you the average force during that time (both the magnitude and direction).

If instead you have one force and you want to know how a given mass will accelerate, you just divide the force by the mass. You can also use the second law to find an unknown mass from a given applied force and acceleration. Again, if there is only one force, the acceleration vector has to be in the same direction as the force. To calculate the mass, you would only need to divide the magnitude of the force by the magnitude of the acceleration.

More often, however, there are multiple forces acting on a single mass. It might be tempting to calculate an acceleration vector for each of the forces, and then somehow say the motion is some combination of all those accelerations. Avoid this temptation! There may be many forces acting simultaneously, but a mass has only one acceleration at any given time.

In working with Newton's second law, we should apply it as written. As previously demonstrated, the left side of the equation is the vector sum of all forces acting on the mass m. A helpful tool for correctly finding this sum is a certain kind of visual aid called a *free-body diagram.*

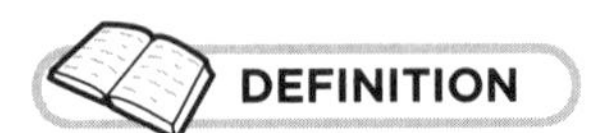

A **free-body diagram** is a sketch showing all the forces that act on a single mass.

Figure 4.2 illustrates two ways to make a free-body diagram. The purpose of a free-body diagram is not to portray everything you know about a situation. Its specific purpose is to help you add all of the forces correctly, so it should only show what is necessary to do that. It should feature only one mass, and the mass can be represented by a simple circle or box or even a dot. That mass is the mass to which you want to apply Newton's second law.

A free-body diagram doesn't have to be drawn to scale, but try to use your knowledge of the situation to point the arrows in approximately the right directions, and even use the lengths of the arrows to roughly show magnitudes, if possible. The important thing is to draw and label *all* of the forces that act on the mass in question. In the center pane of Figure 4.2, the combined mass of the boy and sled is represented as a box. The three forces are shown as arrows, with the tail of each arrow located at the point where the force acts on the mass. $\boldsymbol{F}_p$ labels the pulling force of the rope, $\boldsymbol{F}_N$ is the normal force, and $\boldsymbol{F}_g$ is the force of gravity, the weight pulling the mass straight down. At left is a small xy-coordinate system to remind us which direction is which, in case we want to separate these forces into their components.

In the lower pane of Figure 4.2, the mass is just shown as a dot. The three forces have the same labels. In cases like this where the physical size of the mass doesn't matter, this method of drawing the diagram can be easier and clearer. In this case, for illustration purposes, we have resolved the forces into their x and y components. The components are shown in a lighter shade so you don't get confused and think these are additional forces. The same three forces are shown in both of the free-body diagrams, and you can choose either way you prefer to draw them.

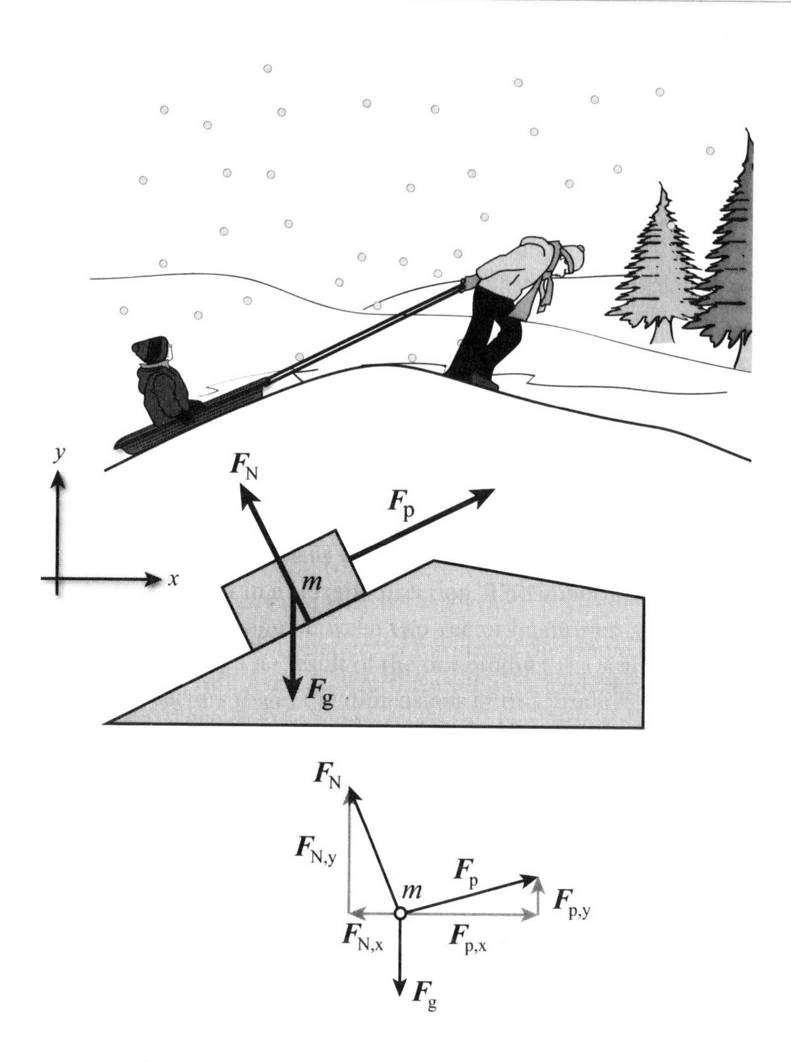

Figure 4.2

Here, a girl is using a rope to pull a boy on a sled (combined mass = m*) up a small slippery hill. The center and lower panels are examples of corresponding free-body diagrams.*

Once you have a good free-body diagram, it is relatively easy to write out the left side of Newton's second law. Just make sure all of the forces you drew are taken into account. Since the best way to add vectors is to add their components, there will often be two or three equations, one for each spatial dimension in the problem. To help you decompose the force vectors into components, you may want to include a sketch of your coordinate system with the free-body diagram, but do it off to the side a little, so you don't confuse the axes of the coordinate system with additional forces. To show you how all of this works in practice, we will tackle a few examples in the next chapter.

CONNECTIONS

It seems like we encounter countless forces every day. You may therefore find it surprising that when you boil it all down, there are only four fundamental forces in nature. These are the gravitational force and the electromagnetic force, which we will deal with in this book, as well as the more exotic strong and weak nuclear forces.

In more complicated situations, you may have more than one mass to worry about. In such cases, you should make a separate free-body diagram for each mass you are interested in. Some of the examples in the next chapter will show how free-body diagrams can help solve problems dealing with Newton's second law.

The Least You Need to Know

- Forces are the physical quantities that are needed to cause any change in an object's motion.
- Forces are vector quantities, so they have both magnitude and direction.
- The exact way that forces change motion is summarized in Newton's three laws of motion.
- You will encounter a wide variety of forces in physics, including many that require physical contact, and some (like gravity) that don't.
- The free-body diagram is a useful tool for applying Newton's second law.

CHAPTER

5

Everyday Applications of Mechanics

In This Chapter

- Motion under the influence of a constant force
- Equilibrium, when forces balance
- Multiple forces in two dimensions
- Friction, springs, and other forces

Aristotle once said, "For the things we have to learn before we can do them, we learn by doing them." In this spirit, before we assume you've fully grasped the basic concepts of the previous three chapters, we're going to consider a few applications of them. First, we will do some for you, and then we'll give you a few to do yourself.

We will begin with several examples designed to show how Newton's laws of motion (mostly the second law) are applied to real situations. We'll start simple, and gradually add more forces and complexity. By the end of this chapter, you'll have a much firmer grasp on the abstract ideas we introduced in Chapters 2, 3, and 4.

The bottom line is that all you have to do to master problems like this is remember that direction matters, that forces sum by vector addition, and that acceleration is not the same as velocity (but is rather the rate of *change* of velocity).

Single Constant Force

We already know how to deal with the very simple case of only a single force acting on one mass. With only one force, that force is exactly the net force that appears in Newton's second law. We can clearly see that a mass under the influence of one single force will accelerate in the same direction as that force. The magnitude of the acceleration is simply the magnitude of the force divided by the mass.

So how much force is required to get a hockey puck moving at 36 m/s in 1.8 seconds? We are assuming the puck was just sitting on the ice motionless to begin with. An NHL regulation puck has a mass of about 170 grams, which we know is the same as 0.17 kilograms. We're also assuming a very smooth ice rink that is also perfectly level, so the direction we are shooting the puck in is purely horizontal. Since all of the motion is going to be horizontal, we only need to apply a horizontal force to cause that motion. Other forces may be acting vertically, but we'll ignore those for now and address them in the next section instead.

Going from zero to 36 m/s in 1.8 seconds means that the magnitude of the required acceleration is 36 m/s ÷ 1.8 s = 20 m/s^2. Multiply that by the mass, and you quickly see that we need 3.4 newtons of force, assuming it is constant for the whole 1.8 second time period. Easy, huh?

Now let's think about the other end of the rink. Say this puck was shot straight toward the goal (and stays in contact with the ice). After we stop applying the 3.4 newtons of force, it continues to slide in a straight line at a roughly constant speed of 36 m/s. Now the goalie wants to stop it with his glove. He reaches his gloved hand down to the ice surface, but it is only 40 centimeters in front of the goal. How much force must he apply to prevent the puck from crossing the goal line? Obviously, the goalie has to apply a force in the direction opposite to the velocity of the puck, so that the resulting acceleration will reduce that velocity to zero in time. What isn't so obvious at first is how much time he has in which to do it.

But you do know how far (at most) the puck can travel in that unknown time: 40 centimeters or 0.4 meter. Recall our discussions of one-dimensional motion back in Chapter 2. If we assume that once it makes contact, the goalie's glove applies a constant force to slow the puck, then we know the acceleration is also constant during that time. Since the speed of the puck goes from 36 m/s down to zero, its average speed during that time must be (36 m/s + 0 m/s) ÷ 2 = 18 m/s (that came from averaging the starting speed and the ending speed). Divide the distance by the average speed and you will find that the time interval must be only 0.4 m ÷ 18 m/s = 0.0222 s (rounding off).

Finally, we have enough information to use Newton's second law to find the force that the goalie must apply. We get the average acceleration from the one-dimensional definition. If we take the direction the puck was moving as the positive direction, then the average acceleration is (0 m/s – 36 m/s) ÷ 0.0222 s = -1,620 m/s^2. Then, we get the required force by multiplying the acceleration by the mass: -275 newtons. The minus sign tells us that the stopping force had to be directed opposite to the direction the puck was traveling as it approached the goal line.

Note what the physics shows. It took a much larger force to stop the puck than to get it going. Why? Because the same change in velocity had to happen a lot quicker, which means a much larger acceleration. Also, we hope you took note of the importance of using the correct units.

If you are dealing with forces in units of newtons, it is essential that all your times be in seconds, your distances in meters, and your masses in kilograms. If you faithfully stick with these standard units, your acceleration units will be correct, and newtons will come out automatically.

Equilibrium

Let's look at another simple case. Let's say you're in a room, and on the windowsill you spot a small potted plant just sitting there, without moving. Like any other object, the potted plant has mass. Therefore, we know for sure that there is a downward force on the object due to gravity (its weight) with a magnitude equal to $F_g = mg$. If we observe that the pot is not moving, then we also know that the acceleration of the object is equal to zero. Therefore, the weight force can't be the only force acting on the potted plant.

If the object's acceleration is zero, Newton's laws tell us that the total force on it must also be zero. We need at least one other force besides gravity to act on the pot, so that the vector sum of forces can be zero. In this case, the obvious source for the additional force is the contact with the windowsill. That contact results in a normal force, which on the standard, horizontal windowsill will be directed straight up. The observation that the potted plant has zero acceleration leads us to conclude that the windowsill exerts a force on it that is equal in magnitude to the weight, but directed upward.

Mathematically, it would look like this: Let's call the normal force $\boldsymbol{F}_N$, and take the vertical upward direction to be positive. In this example, we only need to consider one dimension, so we can replace the vector by its magnitude and explicitly account for direction by using positive and negative values. In the vertical direction, Newton's second law would be $F_N - F_g = F_N - mg = ma = 0$. We put a minus sign on F_g because we know that the weight force is always directed down, and we have assigned that as the negative direction. We always consider g to be a positive number, so that there is no hidden minus sign there. In this case, with a obviously equal to zero, solving for the normal force gives $F_N = mg$. If the potted plant has a mass of 2.1 kilograms, then the normal force exerted upward on it must be equal to $mg = 2.1 \text{ kg} \times 9.8 \text{ m/s}^2 = 20.6 \text{ N}$. This is numerically equal to the weight of the potted plant, but don't forget, the weight force is down and the normal force on the pot is directed upward. The vector sum of the weight force and F_N is zero, and so is the acceleration.

We figured all of this out using simple observations and Newton's second law of motion. If we invoke the third law, we can figure out another force, the force exerted by the pot on the windowsill. The interaction between the pot and the windowsill is due to the direct contact. The windowsill exerts a force of 20.6 newtons upward on the pot, so the pot must exert a force on the windowsill equal to 20.6 newtons downward. (The latter acts on the *windowsill*, not on the *pot*, which is why we didn't include it when applying Newton's second law to the pot.)

The term we use for this general case, when a mass experiences a net force equal to zero, is *equilibrium*. "Static" equilibrium describes the case where there is no motion as a result, but zero net force also allows for constant non-zero velocity. The latter case, with constant velocity but still zero net force, is referred to as "dynamic" equilibrium.

DEFINITION

Equilibrium describes any state where the total or net force on an object is zero.

The hockey puck in our previous example started out in static equilibrium before we decided to accelerate it. When we looked at the effect of the force pushing it toward the goal (horizontally) we conveniently ignored any forces which might have been acting in the vertical direction. Now we see why this was okay. In the vertical direction, there were two other forces acting, the weight of the puck down and the normal force from contact with the ice pushing up. Just like in the case of the potted plant, these two forces add up to zero, so they have no effect on the motion.

Those forces were still there when we applied the horizontal force, but they had no effect on the horizontal motion. This is a consequence of the way that vectors work, specifically the way that vectors can be split into their components. For the usual perpendicular x- and y-axes, the x component of a force vector can only affect acceleration in the x direction, and the same goes for the y component. You can analyze motion separately for the two directions. Newton's second law, being a vector equation, is like two separate equations in two dimensions, one for the x components and one for the y components. This idea will be especially useful when we take up projectile motion in Chapter 6.

CONNECTIONS

If you ever eat too much at Thanksgiving and feel the need to lose the extra weight, you may decide to do a bit of extra exercise. But it turns out there is a much faster way to shed that extra weight—just take your bathroom scale to the moon! The bathroom scale measures the force exerted by the floor on your feet when you stand on it as Earth's gravity pulls you down. If you stand still, that force happens to be exactly equal (in magnitude) to your weight. Since the force of gravity on the moon is less than that on Earth, the reading on your bathroom scale will also be lower. Sadly, this does not mean that your overall mass has decreased, so you'll still need to hit the treadmill to burn off all the turkey.

Now consider a heavy lamp hung from the ceiling and then attached to a wall by a wire, as shown in Figure 5.1. Say the mass of the lamp is 25 kilograms. The wire on the left exerts a tension force on the lamp equal to 125 newtons, and it is 20 degrees from horizontal. If the lamp is stationary, how much tension force is in the wire to the ceiling? You should be able to see that there are three forces acting on this mass, two tensions and gravity. As in any other equilibrium situation, we know that the total of all forces acting on the lamp must be zero. Let's use our standard coordinate system, where the positive *y* direction is vertical and up. If the net force is zero, we also know that the *x* components of all forces must add to zero, and so must the *y* components. We'll call the force due to the left wire $\mathbf{F}_1$ and the force due to the upper wire $\mathbf{F}_2$.

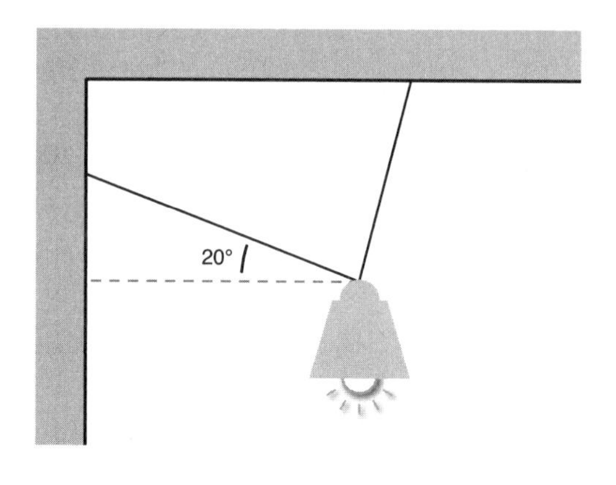

Figure 5.1

The wire attached to the wall has a tension force of 125 N acting on a lamp, and it is angled at 20° with respect to the horizontal.

Let's look at the horizontal direction first. The force of gravity has an *x* component which is equal to zero, so with no acceleration we immediately see that $F_{1x} + F_{2x} = 0$. From the data given, we know that $F_{1x} = -(125\ \text{N}) \times \cos(20°) = -117.5\ \text{N}$. Therefore, $F_{2x} = +117.5\ \text{N}$. In the *y* direction, all three forces contribute. The sum of the forces in the *y* direction looks like this: $F_{1y} + F_{2y} - 245\ \text{N} = 0$. Using trigonometry and the known magnitude of F_1, we see:

$$F_1 \sin(20°) + F_{2y} = 245 \qquad \textbf{Equation 5.1}$$

$$42.75 + F_{2y} = 245$$

$$F_{2y} = 202.25\ \text{N}$$

Using the fact that $F_2^2 = F_{2x}^2 + F_{2y}^2$, we can easily calculate that the magnitude of the tension force in the second wire. We find that it must be about 234 newtons.

Inclined Plane

Now let's consider a two-dimensional example in a nonequilibrium situation. Take a very slick, flat board of some kind and tilt it up by some angle. We generally reference the angle of inclination in such cases to the horizontal. So let's say the board is tilted by 30 degrees from horizontal, and some object with a mass of 5 kilograms is placed on it. What happens?

In this case, we will say that the board is so slippery that it can only exert a contact force on the mass that is perpendicular to its surface (i.e., there is no friction). Thus there are only two forces acting on the mass, the normal force and the weight. But the mass can't be in equilibrium, because there is no way that these two force vectors can add up to zero.

Given the geometry of the situation, it is very helpful to depart from our conventional orientation of coordinate axes. This is because we already have some intuition that the mass will slide along the tilted plane, but that it will not spontaneously jump off of it. There may be motion and acceleration parallel to the surface, but the velocity perpendicular to the surface will be zero at the start and it will not change. Therefore, let's tilt our coordinate axes so that the positive y direction is not vertical, but rather perpendicular to the surface, while the positive x direction points along the surface downward, as shown in Figure 5.2.

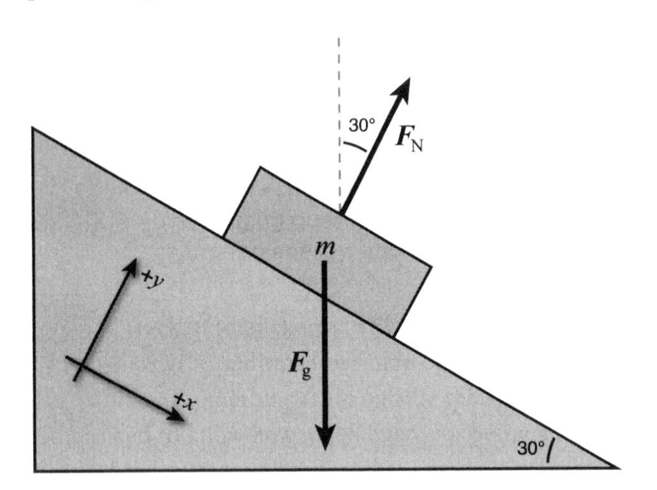

Figure 5.2

Mass m *is located on a perfectly smooth plane tilted at 30° with respect to the horizontal.*

Now the problem is easy. To predict the motion of the mass, we only need to look at the new x direction. The normal force has no component in this direction, so it can't have any effect on the motion. We can therefore forget about it entirely! Only the component of the weight force that is parallel to the surface will cause the mass to accelerate down the ramp. If the ramp is 30 degrees from horizontal, then $F_{g,x} = mg \times \sin(30°) = 24.5$ N. Now, Newton's second law in the x direction says $F_{g,x} = ma_x$, so $a_x = 4.9$ m/s^2. This means that however it starts out, the mass will have an acceleration down the slope equal to 4.9 m/s^2. If you release it from rest, it will immediately start sliding down (as you expect) at an ever-increasing speed. If instead you start by giving it some speed up the slope, it will slow down, reverse direction, and then go faster and faster down the ramp.

You might notice in this last example that you will get the same acceleration no matter what the mass of the object is. This is because the weight force is proportional to the mass, and the acceleration is inversely proportional to the mass. As long as the angle stays the same, any mass you use will have the same acceleration on a frictionless inclined plane.

TRY IT YOURSELF

5. For the frictionless plane inclined by 30 degrees (Figure 5.2) we were able to find the acceleration of the mass without knowing the strength of the normal force. Now use Newton's second law to calculate the magnitude of the normal force acting on the 5 kg mass as it slides down the plane.

Force in Circular Motion

Most of the situations we will look at in this book will involve constant forces; that is, forces which do not change in time. At the introductory level, you might encounter situations where forces change, but they will probably change suddenly, such as when they start or stop.

TRICKS AND HACKS

If you encounter a situation with abrupt changes in force, your best bet is usually to just separate the problem into different time periods, with forces constant during each time period, even if they are different in the separate periods. We did this already with the hockey puck example. While the puck was stationary on the ice, the net force was zero (static equilibrium). Then for 1.8 seconds we applied a constant force, say by contact with a hockey stick, to accelerate it. Then it slid down the ice at a constant velocity for a while, again with zero net force on it (dynamic equilibrium). Finally, in that last 0.0222 seconds, the goalie's glove was slowing it to a stop with a different constant force.

One case that we can easily handle where the force is not constant is uniform circular motion, which we described in Chapter 3. It was there that we encountered centripetal acceleration, an acceleration that is not constant, but which has constantly changing direction and constant magnitude (so long as the speed of the moving object is also constant). This is certainly a non-zero acceleration, so there must be some net force. Like the acceleration in this case, the total force will have a constant magnitude, but its direction will be constantly changing.

Since we know that the magnitude of the acceleration is $a_c = v^2/R$, we can calculate the magnitude of the net force once we know the mass. And even though the direction is constantly changing, we can still determine the direction using Newton's second law. Since the acceleration is directed toward the center of the circular path, so is the total force.

Let's consider an example of this. Take a string that is about a half-meter long and tie it to a small rock (or any other small mass that is handy). Swing the rock around so that it moves in a circle. Orient it so that the circular path is completely horizontal, and try for a constant speed of the rock as it moves around. You should eventually be able to keep your hand almost perfectly still, but the rock will keep spinning. The string will have an angle to it, as shown in Figure 5.3.

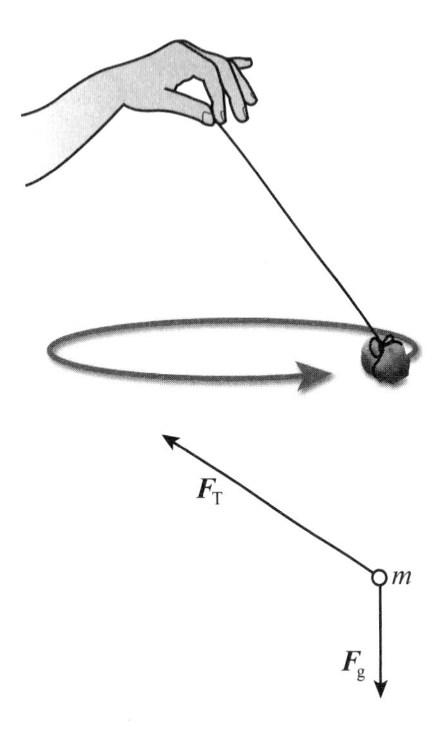

Figure 5.3

In this situation, a stone of mass m *is swinging on a string in a horizontal circle.*

The free-body diagram shows that there are only two forces acting on the rock, the weight force F_g and the tension in the string F_T. This is not an equilibrium situation, because these two forces obviously don't add up to zero, and the acceleration is not zero either.

Let's put some numbers in, and then try to figure out how much tension must be in the string (that is, what the magnitude of the force $\boldsymbol{F}_T$ must be). Let's say the radius of the circle is 32 centimeters, and let the mass of the rock be 0.5 kilograms. In some problems, you might be given the speed directly, but in this case, we watch the rock for a while and we see that it takes 1.24 seconds to go around each full circle.

To calculate the force, we're going to need the exact acceleration. For that we need the radius of the circle and the speed of the rock. The radius was given, but how do we get the speed? Just use the distance traveled and the time it takes. A full circle has a circumference of $2\pi R$ if R is the radius of the circle. Therefore, the speed of this rock is $2\pi \times 0.32$ m $\div$ 1.24 s = 1.62 m/s. That makes the acceleration $v^2/R = 8.22$ m/s^2.

Now that we know the acceleration, we use Newton's second law to deduce that the net force must have a magnitude equal to $ma_c = 4.11$ N, and that the force must point toward the center of the circle at all times (horizontally to the left in our figure). But note that this is the net force, the sum of all the forces acting on the rock. It is not the tension force that we are looking for, but the sum of the tension and the weight.

Since we know that $\boldsymbol{F}_{net} = \boldsymbol{F}_g + \boldsymbol{F}_T$, and we also know the weight force $\boldsymbol{F}_g$, we should be able to find the vector $\boldsymbol{F}_T$ that we are looking for. To do this, we again look at the horizontal and vertical components separately. In the y direction, $F_{net,y} = -mg + F_{T,y}$. Since F_{net} is completely horizontal, $F_{net,y} = 0$ and $F_{T,y} = mg = 4.9$ N. In the x direction, at the instant in time shown, we know that $F_{net,x} = F_{T,x}$ because the weight force does not have any x component. Thus $F_{T,x} = -4.11$ N, the full value of the net force we calculated above. We square the two components, add them, and take the square root (as in the first part of Equation 3.3) to discover that the magnitude of the tension force must be $F_T = 6.4$ N.

TRY IT YOURSELF

6. For the same case described above and illustrated in Figure 5.3, what angle does the string make with respect to the horizontal? Assume the same numerical values, $m = 0.5$ kg, radius $R = 32$ cm, and speed $v = 1.62$ m/s.

Friction Between Surfaces

It's about time that we stop imagining only slippery surfaces and ignoring friction forces. These are really quite common in the real world, so we need to learn how to handle them. For now, we will confine ourselves to friction that occurs when surfaces are in contact with each other, and we will only look at flat surfaces.

RED ALERT!

When we use the word "flat," we do not mean that the surface is perfectly horizontal. It could be oriented at any angle. Flat just means that the surface is not too bumpy or curved. Scientific texts may use the word "level" to indicate that a surface is horizontal; don't confuse this with flat.

Recall that a friction force can be seen as the parallel component of a force that comes from contact with a surface, as distinguished from the normal or perpendicular component. These directions are in reference to the direction in which the surface extends, whatever its actual orientation. We must distinguish two different situations in which friction acts: *static friction*, when there is no motion between the surfaces, and *kinetic friction*, when one surface is moving relative to the other.

Static friction is a force exerted by a surface in contact with an object, parallel to the surface, when the object is stationary at the point of contact. $F_s \leq \mu_s F_N$, where μ_s is the coefficient of static friction.

Kinetic friction is a force exerted by a surface in contact with an object, parallel to the surface, when the object is moving relative to the surface with which it is in contact. $F_k = \mu_k F_N$, where μ_k is the coefficient of kinetic friction.

It turns out that in both cases, the *magnitude* of a friction force is directly related to the magnitude of the normal force that is acting at the same surface. This leads to the use of dimensionless coefficients to relate the magnitudes of the friction forces to the normal force. The values of the coefficients are different depending on the details of the surfaces which are in contact (their smoothness, the presence of lubrication, etc.). Simple physics can't tell you what the coefficients should be; in real life you just have to measure the forces and calculate the ratios. In physics textbooks, on the other hand, the value of a friction coefficient may simply be a given if you need it to do a problem.

We can, however, give you a feel for typical magnitudes of friction coefficients. For one thing, the coefficient of static friction is almost always larger than that of kinetic friction, for the same surface. If you ever tried to slide a heavy box across a floor, you may have noticed this already. It takes more horizontal force to get the box moving from a standstill than it takes to keep it moving at a constant speed.

Typical values of the coefficients for unlubricated metal surfaces, for example, are around 0.5. The addition of a lubricating oil between the surfaces can easily reduce that to less than 0.15. Objects moving on a smooth ice surface typically have a tiny μ_k of less than 0.05, mainly because the normal force will cause a bit of the ice to melt and lubricate the contact area with liquid water. No matter what, a smaller μ means less friction, all else being equal. The largest friction coefficient you can expect to run into would have a value around 1, which would be typical for μ_s of rubber on concrete—where the rubber meets the road, so to speak.

The kinetic friction case is pretty straightforward. If an object is sliding on a surface, the magnitude of the friction force $F_k = \mu_k F_N$. If you know the normal force and the coefficient, then you know exactly how strong the friction force is. The direction of the force is always opposite to the direction of the relative motion, so friction acting alone will always tend to slow things down. Conversely, if you observe how fast something slows down while sliding, that acceleration combined with the mass will give you the friction force. If you also happen to know the normal force, then you can deduce the value of μ_k by taking the ratio. Again, the reason that μ_k has no dimensions is that it is the ratio of two quantities that have the same dimension (force).

RED ALERT!

Although we have used the equation $F_k = \mu_k F_N$, this does not mean that the friction force is in the same direction as the normal force. This is strictly a scalar equation, which tells us about magnitudes but says nothing about direction.

Static friction can be a little trickier to deal with. The formula given in the definition above only provides an upper limit to the magnitude of the static friction force. Static friction can be zero or any other value up to that maximum, depending on the situation. The fact that it is static means you know that the object in question has no acceleration, and therefore the sum of the forces on it must equal zero. The static friction force takes on whatever magnitude are needed to make sure that all the forces add up to zero.

TRICKS AND HACKS

A common physics exercise is to find some condition where the static friction is just about to be overcome, which means that the magnitude of the static friction is at its maximum. In these cases (and only these cases) you can say that the static friction force is equal to $\mu_s F_N$ in magnitude.

As an example, let's add some kinetic friction to the inclined plane situation we looked at above. Recall that on a frictionless surface that was inclined by 30 degrees, the 5-kilogram mass accelerates at 4.9 m/s^2 down the slope. Instead of being frictionless, we now imagine that the coefficient of kinetic friction of the surface is equal to 0.35. What acceleration would we get in that case?

Now we do need to calculate the normal force. Since there is no acceleration perpendicular to the surface (our y direction in that example), the normal force must be $mg\cos(30°) = 42.44$ N. This makes the magnitude of the friction force $0.35F_N = 14.85$ N. If the block is sliding down the incline, then this force is directed upward. You would subtract it from the downward force of 24.5 newtons that we calculated before (in the positive x direction, using the coordinate system from before). The result is a total force in the x direction equal to 9.65 newtons, and an acceleration down the slope of only 1.93 m/s^2.

Simple Springs

The force exerted by a spring depends on how much the spring is compressed or stretched. In real life, some springs can't be compressed at all, because their coils are tightly packed in their relaxed state. In this book we'll avoid any situations where the coils of a spring touch each other. Our somewhat idealized springs can be stretched or compressed to be longer or shorter than their natural length.

Some external force is needed in order to stretch or compress a spring. If the external force is removed, the spring returns to its natural length. If some external agent applies a force to stretch a spring, then Newton's third law says that the spring applies an equal and opposite force pulling on that outside agent. It is that force exerted *by* the spring that we will consider.

For the simple springs we will consider, the magnitude of the spring force is directly proportional to how much the spring is stretched or compressed. We can write a simple expression for this if we set up an x-axis along the length of the spring, as indicated in Figure 5.4.

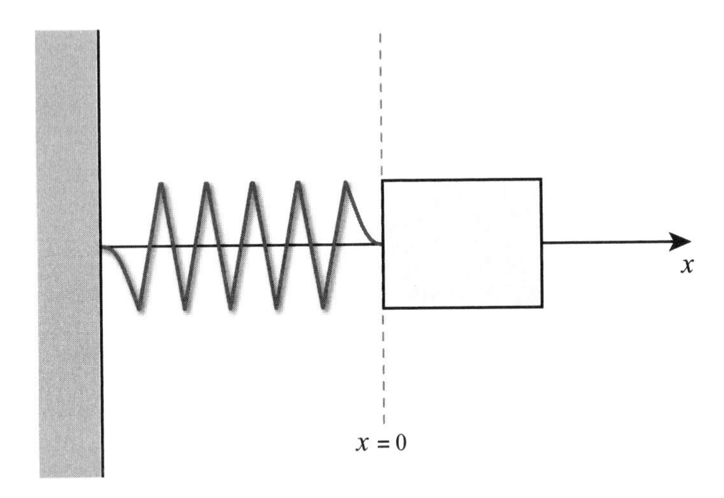

Figure 5.4

In this situation, a mass is connected to a stationary wall by a spring, and the spring is neither stretched nor compressed.

Imagine the left end of the spring is fixed in place, and let x be the location of the other end. We must locate the origin of this coordinate axis at the location where the spring end would be when no force is applied (i.e., when the spring is in its natural equilibrium state). In that case, the force applied by the spring will be $F_s = -kx$ in the x direction. The positive constant k is known as the spring constant. This constant is different for different springs, depending on their size and what they are made of. The larger the value of k, the more the spring opposes your attempt to expand or contract it, the "stronger" we say the spring is.

Notice how this force works. The force exerted by a spring is always directed back toward the "natural" position. The spring force must be zero when $x = 0$ in order to use the simple expression above. This expression is also known as "Hooke's Law." Like our strings and ropes, we will ignore the mass of our simple springs for simplicity.

TRY IT YOURSELF

7. A spring hangs vertically from a hook on the ceiling. When you attach a 7 kg mass to the bottom of the spring, you notice that it stretches by 0.15 m. What is the spring constant k for this spring?

The Least You Need to Know

- Newton's second law is the powerful vector equation that can be applied to a wide variety of situations that relate forces and motion.
- Equilibrium is a physical state that adds another condition to the analysis, the requirement that all force vectors add up to zero.
- Friction describes the parallel component of any force exerted by a flat surface.
- Frictional forces come in two different flavors, static and kinetic, which are treated differently and have different coefficients associated with them.
- An external force is needed to stretch or compress a spring from its natural length, and that force is proportional to how much the stretched or compressed length differs from the natural length of the spring.

CHAPTER

6

Gravity

In This Chapter

- The ever-present weight force near Earth's surface
- Trajectories of objects in free fall
- Horizontal range and maximum height
- Gravity beyond Earth's surface and orbits

In this chapter, we will take a closer look at the force of gravity, which we have already seen is responsible for the weight force we feel every day of our lives. When you kick a soccer ball or shoot an arrow, this force is mainly responsible for shaping the path taken by that projectile. Half of this chapter will be taken up with analyzing such paths.

But physics is not limited to the surface of our modest planet. The same genius we met in Chapter 4, Isaac Newton, realized that the same gravity that keeps us attached to Earth is felt by massive objects all over the universe. So the second part of this chapter will be an introduction to universal gravity, and the motions of bodies beyond our astronomical neighborhood.

Gravity on Earth

As we've already discussed, the force of gravity near the surface of Earth is the familiar force we call weight. For a given mass, to a very good approximation, this force is constant no matter where you are on Earth's surface. And we mean constant in the vector sense; the force due to gravity is constant in both magnitude and direction, in that it always points straight down. What's more, the strength of the force is directly proportional to the mass of the object feeling the force, as in $F_g = mg$.

It is worth mentioning that the magnitude of the acceleration due to gravity (g) is not a "fundamental" constant. If we were living on a planet of a different size, the local value of g would be different from 9.8 m/s^2. The reason for this will be obvious later in this chapter, when we take a closer look at the true nature of gravity. Even on this planet, g varies ever so slightly at different locations, and is noticeably different on high mountains or very deep wells. But you have to travel pretty far before g will change by even a small amount, so in any practical motion problem, it is not bad to consider the gravity force as constant.

According to Newton's second law, constant net force means constant acceleration, in the same direction as the net force. If all an object's motion is in line with the force (in either direction), it will stay that way, and one dimension is all you need to describe all of the motion. In the case of gravity, that means vertical motion problems are straightforward and can be solved using the methods we described way back in Chapter 2. That includes dropping objects (initial velocity equal to zero) or starting objects with some velocity directly up or down (like rocket's blasting off).

On the other hand, if there is any component of the velocity that is not vertical, the motion is a little more complicated, even if the only force acting is vertical gravity. This is a pretty common occurrence, for example when a baseball is batted, so we'll look at such motion in some detail.

Projectile Motion

For our purposes, *projectile motion* is any motion experienced by an object near Earth's surface, during which time gravity is the only force acting on it. This means we will be ignoring all effects of the air, which does exert some force on any real-world projectile. For objects with low density, like feathers, balloons, and ping-pong balls, this will be a rather poor approximation of their actual motion. But for denser objects that don't have wings, our results will be a lot more realistic. This situation is often called *free fall.*

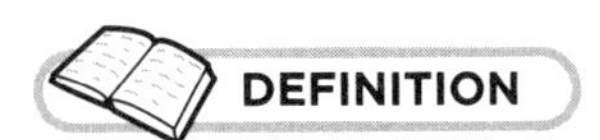

Free fall describes the state in which a massive object is moving solely under the influence of gravity near the Earth's surface. The specific kind of motion that occurs in such a case is called **projectile motion**.

The key to understanding projectile motion is to think of it as a combination of two independent motions in two different directions: horizontal and vertical. As usual in coordinate systems, the two directions are at right angles to each other. Physically, what happens to the horizontal components of position and velocity vectors is completely different and independent from what happens to their vertical components. These components can and should be analyzed separately.

TRICKS AND HACKS

Strictly speaking, the period of free fall is the time during which gravity is the only force acting on a projectile. It extends from the last contact with the launcher (pitcher's hand, kicker's foot, Civil War cannon, etc.) until the moment of impact when some contact force stops or otherwise changes the motion. Of course, some other force(s) will be needed to launch the projectiles in the first place and then stop them in the end. We can forget about all those, though, since they act before and after the period of free-fall motion. All we need to know when dealing with projectile motion is an object's velocity vector immediately after launch (and the value of g, of course).

When the only force acting is vertical, then the most general motion will always be confined to a (two-dimensional) vertical plane. That's nice for us, because it makes it easy to draw pictures of the trajectories of objects in free fall. If at any time, the mass you are looking at has any component of its velocity that is not vertical, then that velocity vector defines a unique vertical plane. Put your xy coordinate system in that plane and you will have the tools you need to describe its motion.

The gravity force vector points straight down, in what we usually call the "negative y" direction. That means that projectiles experience no force whatsoever in the x direction. As we know, zero force means zero acceleration, which means no change in velocity. Whatever velocity the object has in the horizontal (x) direction remains constant the whole time it is in free fall.

The same is not true for the vertical (y) component of the velocity. In this direction, there is a constant force, and so v_y is changing at a constant rate, getting more negative at a rate of 9.8 m/s per second. The result of these two different behaviors in x and y is a trajectory that is curved in a particular way. The name for this kind of geometric curve is the parabola.

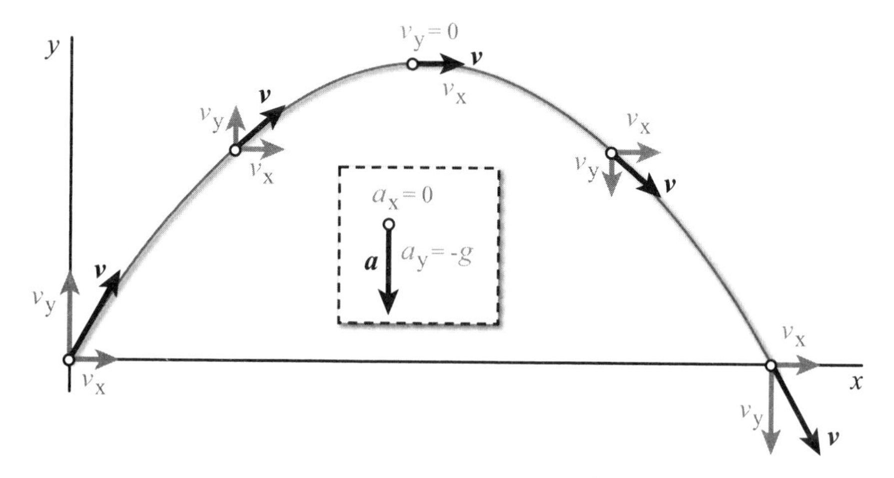

Figure 6.1

This is a general illustration of the path taken by a mass in free fall. The acceleration is straight down at all times, and the changing velocity vector (along with its x *and* y *components) is shown at several different times.*

The parabolas we are talking always open downward. They can be wider or narrower depending on how large the horizontal velocity is, compared with the acceleration due to gravity. And depending on the exact situation, only part of the parabola might be seen during the period of free fall, as you can see from the following figure.

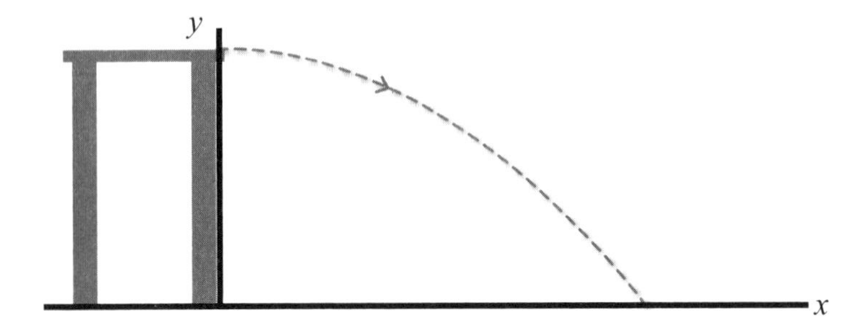

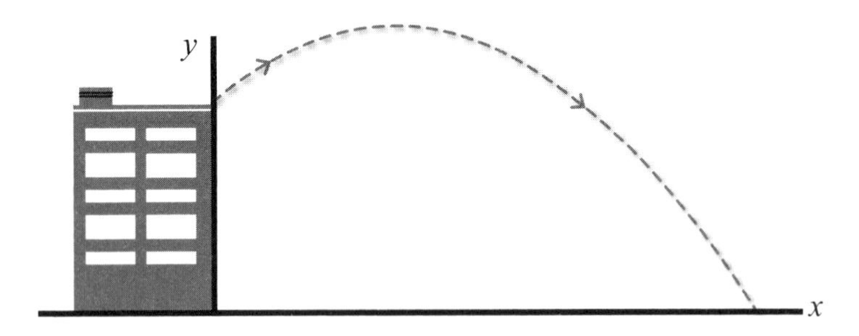

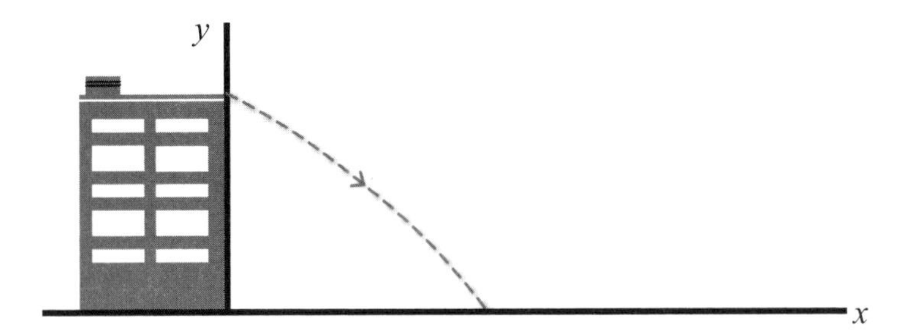

Figure 6.2

Three different examples of free-fall trajectories: a marble rolling off of a level table (top), a ball thrown upward from the top of a building (middle), and a ball thrown downward from a building (bottom).

Let's consider the first of these: a marble rolling off of a level table. The marble will roll off with a certain speed, say, 1.5 m/s. If the tabletop is 0.7 meters above the floor, how far from the edge of the table does the marble hit the floor?

We know how fast the marble is moving horizontally the whole time, because there is no horizontal force and v_x is constant. It is and will be 1.5 m/s until it hits the floor, when friction is bound to slow it down. To determine how far from the table's edge it will travel, we need to know how much time the marble was in free fall. But figuring out the time requires us to look at the vertical part of the motion.

In the vertical direction, it is just as if the marble were dropped from the 0.7 meter height. It has an initial velocity component $v_y = 0$, but as soon as it leaves the table, it immediately starts gaining negative vertical velocity. We don't actually need to use $\boldsymbol{F} = m\boldsymbol{a}$ in this case, because we already know $a_y = -9.8\ \text{m/s}^2$ and is constant. So we can go back to our formulas from Chapter 2 and use $\Delta y = a_y t^2/2$:

$$-0.7\ \text{m} = \frac{-gt^2}{2}$$

Equation 6.1

$$t = \sqrt{\frac{(1.4\ \text{m})}{g}} = 0.378\ \text{s}$$

This is exactly the elapsed time from when the marble leaves the table until it hits the floor. The interesting thing is that this is the same time no matter what is happening horizontally. To calculate this time, we only needed to know the vertical distance the marble must travel, plus the fact that the vertical velocity was initially zero. The horizontal motion does not affect the vertical motion, or vice versa.

Now that we know the time interval, we can return to the x dimension to determine the horizontal distance traveled. Since v_x is constant, the horizontal distance is simply the product of horizontal speed and the elapsed time. In other words: $\Delta x = v_x t$. Putting in the known values, the marble hits the floor 0.567 meters from the edge of the table.

Horizontal Range

A similar analysis will allow us to calculate something called the *horizontal range* in projectile motion. This is useful in so many cases that we will derive a single formula for it. Consider the general case where an object is launched from the ground with an initial speed of v_o at a direction making an angle θ with the horizontal. We want to know how far it will go in the x direction before it returns to the ground.

DEFINITION

The **horizontal range** is the horizontal distance traveled by a projectile during the time it takes to rise and then return to the same height from which it was launched.

As above, we know that the horizontal component of the velocity will be constant, so it will remain the same as its starting value the entire time. From trigonometry we see that the horizontal velocity starts out, and will remain, at $v_{ox} = v_o\cos(\theta)$. Just as with our falling marble, if we can figure out how long the object will be in flight, we would just multiply that time by this speed to get the horizontal range.

Once again, for determining the time, we look at the vertical motion. Now we do have some initial vertical velocity, $v_{oy} = v_o\sin(\theta)$, but instead of remaining constant, this velocity will decrease and turn negative due to the acceleration g in the negative y direction. Now, take a careful look at what happened to v_y in Figure 6.1. You should see that the symmetry of the parabola implies that if the object returns to its original height, the vertical velocity will reverse in sign but have the same magnitude it started with. The change in the vertical velocity during the flight we are considering is thus twice the original vertical velocity and in the negative direction.

Let t be the total time that the projectile is in flight. Since the definition of acceleration is the rate of change of velocity, we can write:

$$\frac{\Delta v}{t} = -g \quad \text{and} \quad \Delta v = -2v_o\sin(\theta)$$ **Equation 6.2a**

If we now insert the Δv from the right equation into the left equation, and do a tiny bit of algebra, we see that:

$$t = \frac{2v_o\sin(\theta)}{g}$$ **Equation 6.2b**

Then, the horizontal range, which we will call R, will be equal to the time we just calculated multiplied by the constant horizontal velocity:

$$R = v_o\cos(\theta)\frac{2v_o\sin(\theta)}{g}$$ **Equation 6.2c**

This can be simplified a bit by using a trigonometric substitution for the product of the sine and cosine of the same angle. In its most useful form:

$$R = \frac{v_o^2}{g}\sin(2\theta)$$ **Equation 6.2d**

So much for the general formula. Now, let's recall what all of these terms mean. The constant g we know, as long as we are near Earth's surface. The symbol v_o stands for the initial speed given to the object, which is launched at an angle of θ above the horizontal. If you put in specific values for a certain launch, this expression will tell you how far the object will travel before it returns to the same height.

Notice that if $\theta = 0°$, then R would be zero. Given the properties of the sine function, the same is true for $\theta = 90°$. This should also make physical sense. If you throw an object with an initial direction that is 90 degrees from horizontal (i.e., straight upward), it doesn't move horizontally at all. You can also see from this formula that the maximum range is reached at an angle that is halfway between 0 and 90 degrees. Aspiring quarterbacks take note: this means that if you throw as hard as you can, a ball thrown at 45 degrees will travel farther than if you threw it at any other angle.

Maximum Height

Given an initial speed and launch angle, there is another quantity that is especially useful for projectile motion: the maximum height achieved. Since this is so important, we'll derive a general expression for this quantity just as we did for the horizontal range.

If the motion is purely vertical and we start an object moving upward, then (as we noted earlier) the velocity will decrease and eventually turn from positive to negative. This implies that the velocity is zero at just the instant it changes sign. That exact instant is also the time at which the object reaches its maximum height. If the trajectory is not exactly vertical, no matter. Since horizontal motion is independent of vertical motion, the maximum height still occurs at the time the vertical velocity shrinks to zero before reversing sign. (Of course, the overall speed of the projectile in such a case is not zero at that time, given that the horizontal component of the velocity is not zero.)

Once again, we'll find the distance traveled by multiplying the average velocity by the elapsed time. On the part of the flight from launch to maximum height, the vertical component of the velocity goes from $v_0\sin(\theta)$ to zero. Since the velocity changes at a constant rate, the average vertical velocity is halfway between the starting and ending values, i.e., $\frac{1}{2}v_0\sin(\theta)$. Now, how much time did it take? The vertical velocity changed (in the negative direction) by the amount $v_0\sin(\theta)$, and the rate of change is $-g$. Therefore the required time is the ratio of these two, or $t = \frac{v_0\sin(\theta)}{g}$. (You may notice that this is equal to half the time of the whole trajectory we calculated for the horizontal range, which makes sense given the symmetry of the projectile's motion.) We can now determine the maximum height by simply multiplying the average vertical velocity by this time. If we call this height y_{max}, then we have:

$$y_{max} = \frac{v_0^2\sin^2(\theta)}{2g} \qquad \textbf{Equation 6.3}$$

Curiously, in both of these derivations, the results do not involve the mass of the projectile. If you launch any two projectiles with the same initial speed and angle, they will follow the same trajectory, even if their masses are different. Now imagine one projectile is a box, and the other one is a ball inside the box. Those two objects will follow the same path, even if they don't touch each other at all. We'll come back to this idea later in this chapter.

Universal Gravitation

We perceive gravity as a constant force because we live on a solid planet that is a whole lot larger than we are. The true nature of gravity is a bit more complicated, but its effects are still relatively easy to describe.

The truth is that gravity is a force that acts between any two masses located anywhere in the universe. Our old friend Isaac Newton was the first human to realize this, and he was able to write down a simple mathematical description of all the properties of gravity. Keeping in mind that force is a vector, let's look at the magnitude and direction of this "universal" gravitation separately.

The direction of the gravitational force between masses is always attractive. For relatively small masses, we can think of all the mass being located at a single point (say, the approximate center of each mass). To keep things simple, imagine two masses some distance apart. Each mass attracts the other, and the force exerted by one mass on the other is directed exactly toward the first mass. If we consider each mass to be concentrated at a single point, then the gravitational force experienced by one mass due to the other is directed precisely on a line drawn between the two points.

If the two objects have appreciable size and are relatively close together, then determining the direction of the attractive force is a little more complicated. For example, imagine a pea that's a few centimeters away from a pumpkin. In what direction is the gravitational force attracting the pea to the pumpkin? Hypothetically, you could divide the pumpkin into lots of smaller pieces, determine the attractive force between the pea and each of these small pumpkin pieces, and then add up the attractive forces for all the individual pieces to find the net attractive force. Of course, since all of those forces would be acting in slightly different directions, the net force of attraction would be a vector sum of all of those forces.

We're going to avoid this mess, mainly to spare you unnecessary grief and a whole bunch of calculus. It turns out that we can do pretty well even for relatively large masses if we continue to assume that all of their mass is located at a single point somewhere near their geometric center.

CONNECTIONS

When Isaac Newton was considering this very problem—how to calculate the gravitational force due to a large mass like Earth—he lacked the mathematics that he needed. He therefore took the time to develop some new mathematical tools, which turned out to be the foundation for what we now call calculus. In essence, this entire branch of mathematics was a mere byproduct of Newton's work on universal gravitation.

Universal gravity is mostly interesting when the distances between masses are relatively large anyway, and in such cases a vector pointing toward the mass doesn't vary in direction very much as long as it points somewhere within the mass. Even better, Newton proved that masses that are shaped like spheres (planets and stars, for example) behave *exactly* as if their entire masses were all located at their centers (as long as the second mass is located outside the sphere).

Now we are ready to reveal the expression for the magnitude of the attractive force acting between any two masses in the universe. Let's call the two masses simply m_1 and m_2, and imagine that they are separated by a distance r. Then the magnitude of the attractive gravitational force between them is:

$$F_g = \frac{Gm_1m_2}{r^2}$$ **Equation 6.4**

The G in this formula is a constant, and this constant really is fundamental. As far as we know, this constant has the same value everywhere in the universe and for all time! That value is $G = 6.67 \times 10^{-11}$ Nm^2/kg^2, a pretty small number. If we present the units of this constant in this way, it is easy to see that Equation 6.4 gives the force in newtons, if the masses and the distance are in standard SI units.

RED ALERT!

Although many books use the letter r in the formula for the strength of gravity, you should not assume that this r is the radius of a circular path or spherical object. In determining the force of gravity, this r must be the distance between the two objects, and if the objects have appreciable size, r has to be the distance between the centers of the objects.

Notice that this formula for F_g gives the same result whether you call a particular mass 1 or mass 2. Reverse the labels, and the force still has the same strength. Separately, we know the force is attractive, so one mass feels a force that is the same strength but in the opposite direction as the force felt by the other mass. Thus this picture of gravity is completely compatible with Newton's third law of motion.

Another amazing thing about this formula is what it tells us about how far the gravitational force reaches. There is no limit on how large r can be. Two masses will still attract each other by the gravitational force even if they are extremely far apart. Even huge galaxies separated by 10^{23} meters are attracted to each other by gravity. But we also see that the force does get weaker as the separation increases, and that it's not just linear. If you double the distance of separation, the strength of the force decreases by a factor of four, not two.

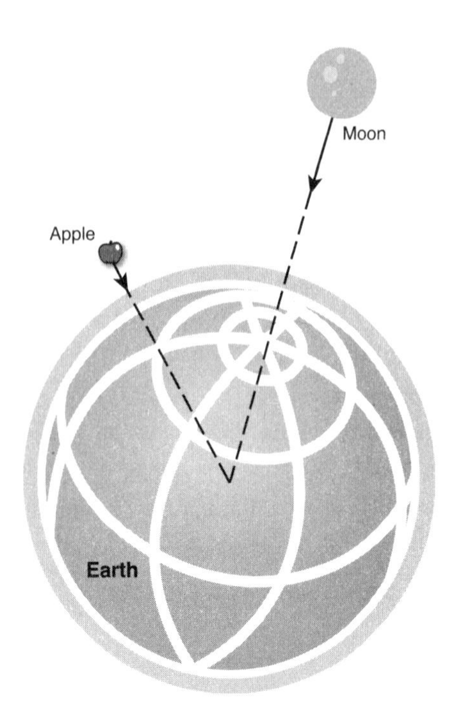

Figure 6.3

Newton's theory of universal gravitation says that the force that attracts a falling apple to Earth is the same kind of force as that attracting the moon, even though the latter is much further from Earth.

Let's see how this new, improved expression for the gravity force compares with the way we have been calculating weight. Say your mass is $m_1 = 100$ kg. What is your weight on Earth? That is, what is the gravitational force of attraction to the huge mass of planet you are standing on? We want to use the universal law of gravity, so all we need is Earth's mass (m_2) and how far you are from its center (r). Earth's mass is known to be 5.98×10^{24} kg. And, if you are standing near sea level (and we assume Earth is a perfect sphere) then we can simply use Earth's radius for r (6.38×10^6 meters). Plug all this into Equation 6.4, and the result will be 980 newtons. Our old formula $F_g = mg$ with $g = 9.8$ m/s^2 gives us exactly the same result!

Measuring the value of g is pretty easy, and was done hundreds of years ago by measuring the time it took objects to fall various distances. It was not so easy, however, to know the mass of Earth. Equating these two different expressions for the force of gravity was actually one of the first ways scientists used to deduce the mass of our planet.

The important insight is that it is not wrong to say your weight is mg, but that is only true as long as your distance from Earth's center is approximately 6.38 million meters. Even if you take a trip on a commercial jet airline, you won't be changing your distance from Earth's center by very much. But if you could go out above the atmosphere into space, then the force of attraction between you and Earth would definitely decrease, and mg would no longer be a valid way to calculate your weight.

Motion of Satellites and Planets

When we do start to consider distances on such a scale, we find that not only does the magnitude of the gravitational force change. We see that the shapes of trajectories are different, too. No longer are the straight line and the parabola the only choices for the paths taken by moving objects influenced by gravity. Small masses move around larger objects in paths that we call orbits. These paths repeat themselves, so that the smaller mass follows the same path over and over around the larger mass, repeating the motion in equal amounts of time. The time required to traverse a complete orbit once is called the orbital period, and is usually designated by T.

Recall that in order to talk about gravity at all, we need at least two masses. From now on, for simplicity, we will adopt a coordinate system that has its origin at the center of the larger of the two masses. There are some complications that occur if the two masses are nearly equal, so we will assume that the larger mass is much greater than the smaller mass, and that only the smaller mass is moving (i.e., the big mass is anchored to the origin of our coordinate system). This situation covers a lot of real situations very nicely, such as satellites in orbit around Earth, and all of the planets orbiting the much more massive sun.

If the smaller mass has a large velocity, it is also possible for it to escape from the gravitational pull of the larger object altogether. In such cases, there is no repeating orbit. Newton's laws allow us to predict the trajectories in these cases as well, but in this book, we will concentrate on repeating or so-called "closed" orbits.

One example of a closed orbit would simply be a circle. With a large mass that is stationary at the origin, a small mass would undergo uniform circular motion if the gravitational attraction were just enough to cause the centripetal acceleration we introduced in Chapter 3. Recall that this acceleration was equal to v^2/r. If we say that the smaller mass is equal to m, and the larger mass is M, then Newton's second law for a circular orbit would look like this:

$$G\frac{Mm}{r^2} = m\frac{v^2}{r}$$ **Equation 6.5a**

Now multiply both sides of this equation by r, divide by the smaller mass m, and we have:

$$\frac{GM}{r} = v^2$$ **Equation 6.5b**

This result tells us that if a small mass (m) moves in a circular path around a large mass (M), then its speed (v) is determined by the radius of the circle (r) and the mass of the central body. (Note that in this case, r is still the distance of separation between the centers of the two masses.) If everything is in balance, the force of gravity will be just enough to provide the centripetal acceleration, and the mass m will keep moving on that same circular path forever.

Let's try to relate the speed of this moving mass to the orbital period (T) we introduced above. We know that the distance traveled by m during one period is exactly $2\pi r$ (the circumference of the circle). Hence, using the definition of speed, we see that $v = 2\pi r/T$. If we square this result, put it into the right side of Equation 6.5b, and then do a little simplification, we find:

$$T^2 = \frac{4\pi^2}{GM} r^3$$ **Equation 6.6**

This is a perfectly general result for circular orbits that result from gravitational attraction. Remember your standard units, especially that the time period must be measured in seconds. This result tells us that for a given central body of mass M, the size of the circular orbit determines its period, or vice versa. There are lots of satellites orbiting around Earth in circular paths, and they all obey this relationship. Even the orbits of the planets in our solar system are approximately circular, so this relationship gives a rough idea for how long it takes all the different planets to complete their orbits.

TRY IT YOURSELF

8. The moon is in a nearly circular orbit around Earth. Using Earth's mass (5.98×10^{24} kg) and the observed period of the moon's orbit (27.3 days), calculate the distance between Earth's center and the center of the moon.
9. If the moon has a mass of 7.35×10^{22} kg, what is the force of attraction between Earth and the moon?

Note that once again, Equation 6.6 tells us something about a trajectory for a body of mass m that does not depend on the value of m. Any small object orbiting in a circle around a larger mass M will have the same period and speed as long as it is orbiting at the same distance—regardless of its mass.

It turns out that the circle is not the only shape that a closed orbit can take. Years of careful observations of planetary orbits revealed that the most general shape for closed orbits is a shape called the ellipse. An ellipse is sort of a squashed circle, longer in one direction than it is in the perpendicular direction. When an object of mass m moves in an elliptical path around a larger mass M, the larger mass is not located at the center of the ellipse. It is off center in the direction of the longer dimension, at a location called a focus of the ellipse. Once in each elliptical orbit, there is a point where the mass m makes its closest approach to M. On the other side of the orbit is the point where the separation is maximum.

The rules that govern the motion of planets (and other masses orbiting around the sun) are often called Kepler's laws (after Johannes Kepler, who first derived them between 1609 and 1619). We won't go into all of the interesting applications of Kepler's laws, but we will say that they follow directly from Newton's laws of motion and the universal law of gravity.

CONNECTIONS

Neptune is the only planet that can't be seen with the naked eye. It therefore evaded discovery until long after the other known planets had been identified. In fact, astronomers initially inferred its presence indirectly, when observing that Uranus was perturbed slightly from the path predicted by the laws we've presented in the chapter. By studying the motion of Uranus, astronomers mathematically deduced the potential location of an undiscovered planet. Turning their telescopes in that direction, they caught Neptune's first glimpse in 1846.

It's also worth mentioning that a form of Equation 6.6 (also called the law of periods) works for elliptical orbits, too. As we mentioned, an ellipse has one direction in which its size is longest. In measuring the size of an ellipse, it is customary to use half of this longest dimension, called the *semi-major axis*. If we designate that length as a, then the law of periods for an ellipse is:

$$T^2 = \frac{4\pi^2}{GM} a^3$$ **Equation 6.7**

A circle is just the limiting case of an ellipse, for which the short dimension becomes equal to the longer dimension, and a becomes the radius of the circle.

When Weightlessness Isn't Weightless

You may have heard or seen that astronauts in orbiting spacecraft experience a state of "weightlessness." This is a bit of a misnomer, and now that you know a little physics, you can see why.

Astronauts in orbit about Earth are certainly subject to its force of gravity. If there was no gravity acting, then Newton's laws tell us the astronauts and their spacecraft would move in a straight line. Gravity is needed to bend that path into a closed orbit. The reason that they seem to float around in their spacecraft with no weight is that they are accelerating toward the center of Earth at the same rate as the spacecraft surrounding them.

"Free fall" is a better description of this state than weightlessness. It is a consequence of the fact we mentioned above that objects with different masses will all accelerate the same way. This means that if they start together, they will follow the same paths or orbits. No force needs to be exerted on the astronaut by the walls or floor of his or her ship in order to keep him or her moving on the same trajectory. It seems like there is no gravity, because the occupant of the spacecraft can't see any effects of gravity. But the gravitational force is definitely there. We can never escape it, no matter how hard we try.

The Least You Need to Know

- Gravity is a universal force of attraction between massive objects anywhere in the universe.
- The strength of gravity is proportional to each of the two masses attracting each other, and inversely proportional to the square of the distance separating them.
- On Earth, free-fall trajectories are shaped like parabolas, and we can calculate their properties based on initial speed and launch angle.
- For larger separation distances, free fall leads to closed orbits shaped like circles or ellipses, with periods that vary in a predictable way.

CHAPTER

7

Work and Energy

In This Chapter

- The importance of energy
- Kinetic energy and its relationship to work and force
- An introduction to potential energy
- The concept of power
- What physicists mean by the conservation of energy

This chapter brings us to one of the most important quantities in all of physics: energy. The concept of energy is crucial not only in physics but in every other physical science. Without energy (or more precisely, transfers of energy) nothing at all would ever happen.

Despite its importance, energy as a general idea can be tricky. You usually can't see, taste, or feel energy itself. In addition, there are a lot of different kinds of energy that seem to be unrelated at first. When a science teacher starts talking about energy, it can therefore seem pretty mysterious. It is customary in most physics courses to start by describing a few specific kinds of energy that can be easily defined, and hope that students will gradually build up a sense of the general concept. That's what we'll do in this book, too, but we can't resist a glance at the big picture first.

What Is Energy?

Before getting into what energy *is*, it's worth stressing a couple of things that energy is *not*. The precise physics concept of energy is only vaguely similar to the way the word "energy" is used in everyday terms. When someone says they "have no energy" first thing in the morning, or describes someone else as "energetic," they aren't talking about the exact physical quantity we have in mind here.

It is also important to distinguish energy from another quantity we have already talked about: force. Energy and force are two different physical quantities (like distance and time) and are measured in completely different units.

So if energy is not a feeling or a force, what is it? Any general definition will try to make use of the fact we noted above: energy is that essential thing that is required to make anything happen. We've already seen that a force is needed to change the motion of an object, but force alone is not sufficient. Recall our discussion of equilibrium. It is possible for forces to be present without changing the motion of an object, i.e., when the forces add up to zero. Many textbooks define energy as the ability to do work on an object. That definition doesn't help very much until we know the precise meaning of work, so we'll get to that in the next section.

For now, let's establish that energy is not a substance, but a quantity of something that can be well defined in a variety of situations. It comes in many forms, and all the interesting things that happen in the universe happen when energy changes form or when it's transferred from one place to another. Energy is a scalar quantity; it doesn't have any direction in space like the vector quantities we identified earlier. It's only the amount of energy that matters.

Forces and Work

In order to define work, let's review a simple application of Newton's second law of motion. Think of an object with a mass of 5 kilograms, which is initially motionless in some ideal location where no forces whatsoever are acting on it. (This means you shouldn't be thinking of any place on Earth, where we always have the force of gravity, among others, to worry about.) Newton's laws tell us that as long as no forces act on our object, there will be no change in its motion, so it will just remain at rest in the same place.

Now, somehow, we apply a constant force of exactly 10 newtons to that 5 kilogram object. Call the time at which the force starts zero seconds on the clock ($t = 0$). Starting at this time, the force is always the same, a constant 10 newtons in magnitude and always in the same direction. What happens to the object?

You already have the tools you need to answer that question. Since $\boldsymbol{F} = m\boldsymbol{a}$, that 5 kilogram mass is going to accelerate, in the same direction as the force. The magnitude of the acceleration will be $a = F/m$, in our case exactly 2 m/s^2. That means for every second after $t = 0$, the object gains 2 m/s of speed, as long as the applied force has not changed. As you'll recall, constant force and constant mass imply constant acceleration.

When a force acts on an object that moves, we say that the force is doing something that physicists call *work*. Once again, the precise physics meaning of the word "work" is not the same as common usage. The force in our example is acting not only over some period of time, but also over some distance (since the object on which it's acting is moving the whole time). The work done by a force is precisely defined as a product of that force and the distance over which the object moves.

Work is the scalar product of force and displacement, $W = \boldsymbol{F} \cdot \boldsymbol{d}$.

The kind of multiplication we need here is the scalar product, which is a particular way to multiply two vectors together and give a scalar result. (Sometimes the scalar product is called the "dot product" because of the symbol used to show it.) In general, the scalar product is found by multiplying three factors: the magnitudes of the two vectors and the cosine of the angle between those two vectors. If we let θ represent the angle between $\boldsymbol{F}$ and $\boldsymbol{d}$, then in this case $W = Fd \cos(\theta)$.

Instead of the displacement vector $\boldsymbol{d}$, we could have used $\Delta \boldsymbol{r}$ again, since we are talking about a distance traveled in a certain direction. The important thing to remember is that the scalar product is an operation between two vector quantities, so it is not the same as multiplying a vector by a scalar. The result of the dot product is a scalar, and you also multiply the dimensions of the two vectors to get the dimensions of this product.

In a simple case of a single force acting on an object that wasn't moving to begin with, the displacement ($\boldsymbol{d}$) and the force that caused it ($\boldsymbol{F}$) are both in the same direction, and the angle θ is zero. This means the scalar product is the same as the regular product, and for such a case we can ignore complications that occur when the directions are different. In this case, the work is simply the product of the magnitudes Fd.

To calculate the displacement, we need to know how long the force acts on the object. In our example, let's say this is one minute. Since the acceleration here is 2 m/s^2, you immediately know that after one minute of this, the speed of the object has changed from zero to 120 m/s. But how far did it travel in that minute?

That's easy, too, using the methods we worked out in Chapter 2. All of the motion is along a single straight line. Since the speed grows smoothly from zero to 120 m/s, the average speed during that time was 60 m/s. Traveling at an average of 60 m/s for 60 seconds means the total displacement was 60 s × 60 m/s = 3,600 m.

So for our case, the work done by the 10 newtons of force is 10 N × 3,600 m = 36,000 N m (newton-meters). It turns out that work is just one of the many ways to transfer energy, so the units of work are the same as the units used to measure energy. Standard work units are N m. When used to measure energy, we call one N m a joule for short. So standard energy units are joules, abbreviated J.

Kinetic Energy

Now we have a particular kind of energy to contemplate. One of the easiest kinds of energy to visualize is the energy associated with a mass in motion. This kind of energy is called *kinetic energy*. A mass that is not moving has no kinetic energy whatsoever. In our example above, the force caused our mass to begin moving, and it moved faster and faster as time went on. Therefore, the force was continuously adding kinetic energy to the object.

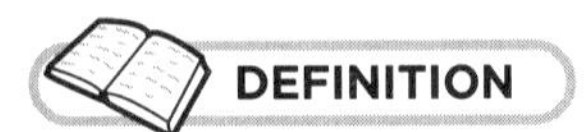

DEFINITION

Kinetic energy is the energy possessed by any mass that is in motion. For a mass equal to m moving at speed v, the amount of kinetic energy is exactly $\frac{1}{2}mv^2$.

After one minute of forcing, we found that the speed of our object became 120 m/s. At that instant in time, the mass had acquired 36,000 joules of kinetic energy, because $\frac{1}{2}mv^2 = 36{,}000$ J for $v = 120$ m/s and $m = 5$ kg. If we stopped applying the force at 60 seconds, the object would continue to move and the amount of its kinetic energy would stay the same, because the speed would stop changing and remain constant. You should be able to convince yourself that the units of kinetic energy are joules (kg m^2/s^2).

TRICKS AND HACKS

Note that kinetic energy can never be negative or less than zero. If you ever do a calculation that results in a mass having negative kinetic energy, you'd better look for your mistake!

The energy that was added to the object came from the action of the force, and it was added gradually, not all at once. In fact, the energy added was exactly equal to the work done by the force, as defined above. This idea is often called the "work-kinetic energy theorem" in textbooks. Any single force acting on an object in motion can change its kinetic energy by an amount equal to the work done, as previously defined. In our example, the 10 newtons of force added 10 joules of kinetic energy to the object for every meter traveled, slowly at first, then faster and faster.

In other cases, an object might be moving in a direction that is *not* the same as the direction of a force that is being applied. This often happens in the real world when more than one force is acting on the same object at the same time. (For example, picture a soccer ball being kicked to the left or right as it was rolling straight downhill.) As long as you know the magnitude and direction of each force, and the distance and direction of the object's motion, you can calculate the work done by that force using the scalar product.

Notice that in general, the work done can be positive, negative, or zero depending on the directions of the two vectors, force and displacement. The maximum positive value for the work happens only when the two are in the same direction, like our example above. In every other case, the work is less than the simple product of the force magnitude and distance. Even if an object moves while a strong force is acting on it, the work can be zero, if the direction of the force is 90 degrees away from the direction of the motion. If that is the case, this particular force would not change the speed of the object at all.

RED ALERT!

There is another way for a force to act on an object and do no work. That is, if the object doesn't move at all, i.e., zero displacement. This is obvious from the definition, but a little counterintuitive. If you are holding a big barbell one meter off the ground but not moving it, it still *feels* like you are doing work. It certainly requires force to hold it up. But in physics, as long as there is no motion, there is no work being done.

What about negative work? We've seen that there is no such thing as negative kinetic energy, but the same is not true for work. If a force acts on an object that is moving opposite to the direction of the force, then the work $W = \mathbf{F} \cdot \mathbf{d}$ will be less than zero. Doesn't this violate the work-kinetic energy theorem somehow?

Fortunately, the answer is no. The key to understanding the work-kinetic energy theorem is to realize that the work equals the *change* in kinetic energy. In our simple example above, our force did positive work on an object even after it was already moving, thereby adding kinetic energy to what was already there. But if a force acts to stop an object that was already in motion, then that force reduces the speed of the body and decreases the kinetic energy. The result is *negative* work!

The classic example of negative work in the real world is applying the brakes on a moving car or bicycle. If your bike is coasting along nicely to the east at, say, 25 m/s, applying the brakes is effectively applying a force on your bike directed to the west. The amount of work done by that "westward" braking force (as you continue to move east) is negative, gradually reducing your kinetic energy and thus your speed.

Potential Energy

Sometimes when a force is applied to a body in motion, the velocity doesn't change, so there is no change in kinetic energy either. As we saw previously, this can happen if the force is at right angles to the displacement. But it can also happen if the force is in the same direction as the displacement, so long as there is more than one force at work.

Take the example of lifting your textbook straight upward. Let's say the mass of your textbook is exactly 3 kilograms. Wherever you happen to be on Earth, the force of gravity on that book will be about 29.4 newtons straight down, right? If you're careful, you can slowly lift that book straight up and keep the speed of the book constant while you are lifting. In this case, you are applying a force to the book that is also 29.4 newtons in magnitude, but straight up. Start at the floor and lift it for 2 meters, and you clearly did 58.8 joules of work ($\boldsymbol{F} \cdot \boldsymbol{d} = 29.4 \times 2 = 58.8$). But, since the speed of the book was not changing as you lifted it, what happened with the work-kinetic energy theorem?

It still applies, but we have to consider the effects of *all* forces acting simultaneously (in this case, two). The total, or net, change in kinetic energy of an object is equal to the sum of the works done by all the forces. While you were doing positive work by forcing the book in the same direction (up) as its motion, gravity was doing negative work by acting in the opposite direction (down). As long as your lifting force was exactly the same as the weight force, the negative work done by gravity exactly cancelled all of your positive work.

If your lifting force happened to be a little greater than the weight, then the book would be accelerating as it rises. You can use the work-kinetic energy theorem to find the kinetic energy after 2 meters in two different ways. Either calculate the work done by each of the two forces separately and add the resulting scalars (one will be negative, so it will look like subtraction) or you can get the total work by using just the net force in the $W = \boldsymbol{F} \cdot \boldsymbol{d}$ calculation. The net force, just like before, is the vector sum of all the forces acting (two in our case).

TRICKS AND HACKS

Whenever there are two ways to do the same problem, you can do it both ways to check your work. This is always a good idea until you get very comfortable with any particular type of calculation.

The important point is that work is a way of transferring energy. In the case where you were lifting the book at constant speed, you were clearly doing positive work. Yet, your work did nothing to increase the kinetic energy. So where was it going? We say that it must have gone into some other form of energy.

This other form is something we call *gravitational potential energy*. Gravity is a good example to work with because it has practically the same value at all locations near Earth's surface, as we saw in earlier chapters. Whenever you do work against the force of gravity, you are essentially storing energy that later could be converted to kinetic energy (and therefore motion) without any further effort on your part.

You lift a book slowly from the floor and then hold it there. Since it is not moving, there is no kinetic energy. But if you let it go, that ever-present gravitational force will do positive work and increase the kinetic energy of the book (at least until it hits the floor). The work you did in lifting the book went to increasing the potential energy. Then later, when you released the book, that same energy was finally converted to kinetic energy of the falling book.

It is extremely useful to think of work and potential energy this way. You can always associate with gravity a potential energy that is determined by the vertical height above some reference point. It is customary to use the variable y for the vertical height, and to make the positive y-direction correspond to up. Then, since the magnitude of the gravitational force (weight) is exactly mg, the gravitational potential energy is simply given by mgy for an object whose mass is m.

Gravitational potential energy is given by the formula $U_g = mgy$, where m is the mass of an object, g is the acceleration due to gravity, and y is the vertical height above a reference point. Gravitational potential energy may be negative if the mass happens to be lower than the reference point.

When you do problems with gravitational potential energy, notice that the reference height you choose (where $y = 0$) is arbitrary. This is because only *changes* in potential energy really matter. As long as you make sure "up" corresponds to the positive y direction (and y is vertical), then it doesn't matter where you say $y = 0$. The gravitational potential energy will be zero at your reference height where $y = 0$. Just be sure to stick with whatever choice you make in a single problem.

There are other forms of potential energy besides gravitational. Different textbooks may use different symbols for potential energy, but here we will use U in general, and a subscript (like g for gravity) to indicate which specific kind of potential energy we are talking about.

Another common form of potential energy is that associated with springs. As we saw in Chapter 5, an external force is required to stretch or compress a spring to any length that is longer or shorter than its natural length. That means that such a force can do work on a spring, without producing any kinetic energy. Once again, we say potential energy is stored in a spring that is thus compressed or stretched. This kind of potential energy is often called *elastic potential energy*, because it can apply to other things besides springs (rubber bands, bending rods, etc.). Nevertheless, we will use the symbol U_s for this spring-type potential energy.

Elastic (or spring) potential energy is given by the formula $U_s = \frac{1}{2}kx^2$, where k is the spring constant (see Chapter 5) and x is the amount that the spring is stretched or compressed from its natural length.

You can think of x as the position coordinate of one end of the spring, along the axis of the spring. As we learned in Chapter 5, the origin of this coordinate axis (where $x = 0$) must be located at the point where the spring is neither stretched nor compressed, and the spring force is zero. Unlike the case of gravitational potential energy, the point where $x = 0$ is not at all arbitrary. The potential energy stored in the spring is zero only when the spring is at its natural length (that is, where $x = 0$). The variable x can be positive (stretched) or negative (compressed), but U_s will be a positive number in either case since we are dealing with the square of x.

Hopefully, it is clear that all kinds of potential energy are measured in joules just like any other kind of energy. And like any other energy, potential energy is always a scalar quantity. One of the reasons that energy is so useful in solving mechanics problems is that it is a scalar quantity (not a vector). When doing calculations with kinetic and potential energy, we don't have to worry about trigonometry or vectors or such, as we do with forces. We just have to keep track of which way energy is flowing, and add or subtract the appropriate scalar quantities. We'll do some examples of this in the last section of this chapter.

Power

As mentioned before, energy is an extremely important idea in all of physics. Besides being the necessary ingredient to drive any physical process, keeping track of energy actually makes it easier to predict the outcomes of many physical situations and figure out what is going to happen.

In real physical situations, however, energy may not be the only limiting factor. Think about carrying a heavy suitcase up a flight of stairs. You will be supplying, mostly with your legs, the force that is needed to raise that suitcase against gravity. You probably have enough energy to move that suitcase from the ground floor to the fifth floor, but how fast could you do it? The change in potential energy will be the same no matter what speed you climb the stairs, but the faster you go the more difficult the task will seem. There is certainly some maximum speed at which you can increase the potential energy of that suitcase (and yourself). The rate at which energy is transferred is also an important physical quantity. In general, we call this rate *power*.

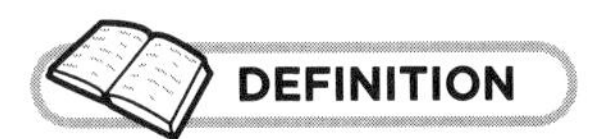

Power is the general term for the rate of energy transfer or conversion: $P = \Delta E/\Delta t$. Just as there are many different forms of energy, there are many kinds of power. If energy is being transferred by a force doing work, the power is the rate of doing work. The dimensions are energy per unit time, and the standard unit for power is the watt (W), with 1 W = 1 J/s.

If work is being done or energy is being produced or converted from one form to another at a constant rate, then you can simply take the amount of energy produced or converted in a certain period of time, divide by that time interval, and you will get the average power. If the rate of energy transfer is not constant, then our earlier comments about instantaneous power apply.

CONNECTIONS

Next time you pay the electric bill, you may notice that you are paying for electricity in units of kilowatt-hours, or kWh. Is this a unit of power or energy? The answer is obvious once you realize that one kWh is the product of 1 kilowatt times 1 hour. In other words, it is a power multiplied by a time. From the definition, we see that this product will have dimensions of energy. So the kWh is a nonstandard but widely used unit for measuring energy.

TRY IT YOURSELF

10. A kilowatt-hour (kWh) is the amount of energy transferred in one hour, if the rate of energy transfer is exactly 1,000 watts. How many joules are in 1 kWh?

Conservation of Energy

When scientists talk about the conservation of energy, they don't mean the same thing as environmentalists do when they urge us to conserve energy. In physics, conservation of energy is a fundamental and unbreakable law of the universe, so there is no point in urging anyone to follow it. It will happen no matter what!

In simple terms, the law of conservation of energy states that energy is neither created nor destroyed. Ever. Physical processes can only transfer energy from one body to another or from one place to another. Energy can also change forms (as we saw above when potential energy was converted to kinetic energy). To handle this mathematically, we would focus on the total energy of a "system" of objects.

The system idea is really necessary to make good use of the concept of potential energy. The idea of a system is just to imagine a boundary around the objects that are relevant, separating the objects you are interested in from the rest of the world. It is most useful if the system is "isolated," which just means that there is no energy flowing into or out of the system, during the period of time you are interested in. If that is the case, then details about the configuration of the system may change, but the total energy will not.

The total energy of the system (which we will denote E) will be the sum of its kinetic energy and any kinds of potential energy that are relevant. For parts of the system that aren't moving, there is no kinetic energy. If something happens, think about the total energy of the system before and after that happening, and then compare the two states of the system at the two different times. For an isolated system that goes from state 1 to state 2, $E_1 = E_2$. In different contexts you may see these states labeled as "initial" and "final," or "before" and "after," but it doesn't matter. As long as you have a clear idea of which states you are comparing, and you account for all of the energies, this simple equation can be used. It is not limited to two states, either.

CONNECTIONS

You've just learned that energy can be neither created nor destroyed, but merely converted between various forms. It turns out, though, that there is a minor caveat to this statement. In 1905, Albert Einstein deduced that—under the right circumstances—energy can in fact be converted into matter using his famous formula $E = mc^2$ (where E is an object's energy, m is its mass, and c is the well-known speed of light). To avoid violating energy conservation, this discovery forced physicists to rethink their understanding of matter—specifically the notion that an object's mass is an absolute and unchanging quantity. For the scope of this book, however, mass will never change to energy, and the two will be separately conserved.

Keeping track of total energy will also prove useful for nonisolated systems. In cases like that, you just have the additional wrinkle of a process or processes that can change the energy of a system, like if something from outside the system can do work on a mass in the system. You'll either need to know how much energy is being added or taken out, or your measurements of two states of the system will reveal that energy is missing or has been added.

Examples

Now that we've covered the general ideas behind energy and energy transfers, let's look at a couple of examples from the real world in which energy changes forms.

Throwing a Baseball onto the Roof

Let's say your two-story house has a flat roof, and that roof is exactly 9.5 meters off the ground. While standing on the ground, you throw a ball up so that it lands on the roof. As the ball leaves your hand, it has a speed of 21 m/s. At that same moment, it is 1.8 meters above the ground. How fast is the ball moving just as it lands on the roof?

We could do this problem using the methods of the previous chapter for analyzing trajectories. We would ignore the drag force caused by moving through the air and just assume that the ball was in free fall from the time it left your hand until landing on the roof. Only the known force of gravity would act during that time. If we knew the angle at which you threw the ball, we could figure out not only how fast it was moving when it landed, but exactly where on the roof it would land.

But if we don't care where it lands (maybe we're only wondering about potential damage to the shingles), then energy conservation gives us a much easier way to answer this question. Let's choose ground level to be the zero of our y scale for the purposes of measuring gravitational potential energy. The first state we want to look at is the ball and Earth at the moment the ball leaves your hand. The ball has potential energy $U_g = mg \times 1.8$ m and kinetic energy $K = \frac{1}{2}m(21 \text{ m/s})^2$. The second state is just before the ball hits the roof, when it has $U_g = mg \times 9.5$ m and $K = \frac{1}{2}mv^2$. This v is the speed we are looking for.

Why do we include Earth in our system? Because if we only included the ball, we would not have an isolated system. Through the force of gravity, Earth can do work on the ball as it is moving from my hand to the roof. If we include Earth, then we can make use of gravitational potential energy and write $E_1 = E_2$ as:

$$mg(1.8 \text{ m}) + \frac{1}{2}m(21 \text{ m/s})^2 = mg(9.5 \text{ m}) + \frac{1}{2}mv^2 \qquad \textbf{Equation 7.1}$$

The m in Equation 7.1 is, of course, the mass of the ball. Since it appears in every term, we can divide it out and solve for v whether or not we know the actual mass. This means our answer will be correct for any mass (as long as ignoring air drag is justified). The numerical result, 17 m/s, is accurate no matter what angle the ball was thrown at, so long as the ball actually makes it to the roof.

TRY IT YOURSELF

11. Standing right next to the house, what is the minimum speed you would have to give the ball to just barely make it to the roof, assuming the same heights as in our example?

Stopping the Runaway Caboose

Here's an example using spring potential energy. Say our old friend the little red caboose is coasting along the level track with negligible friction. Its mass is 4,000 kilograms and its speed is 25 m/s. At the end of the track is a bumper attached to a long spring which is fixed at the other end (Figure 7.1). The spring has a spring constant $k = 70{,}000$ N/m. What is the maximum distance that the spring will compress when the caboose hits it, before the caboose rebounds back the way it came?

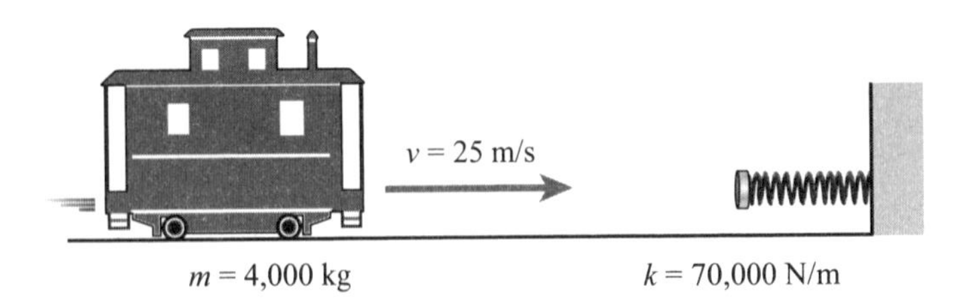

Figure 7.1

A caboose is moving to the right on a level track when it strikes the spring-mounted bumper.

Since the track is level, we don't have any gravitational potential energy to worry about this time. Assuming well-engineered wheels, we can safely ignore friction, too. We can make use of spring potential, though, so we should include the spring and the caboose in our isolated system. (The two vertical forces on the caboose cancel out, so the outside world does no work and transfers no energy into or out of this system.)

Before the caboose hits the bumper, it has kinetic energy and no potential energy is stored in the spring. At the moment of maximum spring compression, the caboose's velocity is changing direction, so for that instant the speed is zero. Thus between the two states, there has been a complete conversion of kinetic energy into spring potential energy.

Call the maximum compression distance that we are looking for x. Since spring potential energy is a function of x, we can easily write down an equation that we can solve for x. One way to do it would be to calculate the initial kinetic energy, which turns out to be 1.25 million joules (1.25×10^6 J). Set that equal to $\frac{1}{2}kx^2$, which is the maximum spring potential energy, and you can verify that $x = 6.0$ m is the maximum distance the spring will compress.

Vertical Spring Launch

Finally, consider a spring fixed at the bottom of a vertical tube, with spring constant $k = 900$ N/m. A 0.3 kilogram ball is placed in the tube, on top of the spring. Now compress the spring by 15 centimeters and release it, launching the ball straight up. What maximum height will the ball reach above its initial location (prerelease), if air drag is again neglected?

The initial state is the compressed spring, and no kinetic energy. The system now includes Earth and the spring as well as the ball. The final state we want to look at is the ball at its maximum height, which is an instant when it again has no kinetic energy. If we set $y = 0$ for the location of the mass when the spring is compressed, then equating the total energies in these two states would look like this:

$$\frac{1}{2}kx^2 = mgy_{max}$$

Equation 7.2

The values for k, m, and g are known or given, and y_{max} is the maximum height we are looking for. Note that the initial kinetic energy and the initial gravitational potential energy are both zero, so these do not figure into the left side of the equation. Similarly, the final kinetic energy and final spring potential energy are both zero, so these do not appear on the right side. If you insert $x = -0.15$ m (the given amount that the spring was compressed before launching), you will see that a little over 10 joules of energy was stored in the spring. That energy was converted to kinetic after the ball was released, but then that kinetic energy all got converted to gravitational potential energy by the time the ball reached maximum height. That y_{max} would be about 3.44 meters.

The Least You Need to Know

- Energy is required for just about every physical process you can imagine.
- The energy of a mass in motion is called kinetic energy, and is easy to calculate for a given mass at any speed.
- Potential energy is energy that is stored in one form or another, usually in the way objects are located or configured.
- Energy can't be created from nothing, nor can it ever be destroyed; it can only change form or move from one place to another.

CHAPTER

8

Linear Momentum

In This Chapter

- Momentum, another conserved quantity
- Analyzing all kinds of collisions
- The concept of the center of mass

In the last chapter, we introduced a very important conserved quantity in physics: energy. Here we introduce another, called momentum. Unlike the scalar energy, momentum is a vector, so besides having different dimensions, conservation of momentum is more restrictive than conservation of energy since momentum is conserved in all dimensions. For the physicist, that is not necessarily a bad thing. Applying the law of conservation of momentum actually gives you more information to help you predict the outcome of various situations.

We will show you how to apply this important new concept to solve a number of problems that deal with colliding bodies. As we'll explain, this is not only useful when playing billiards or bocce ball, but it also helps us study the inner working of atoms, nuclei, and even their constituent particles.

What Is Momentum?

There are two kinds of momentum in physics, and this chapter is devoted to the first, called *linear momentum.* In the next chapter, when we start discussing rotation, we will encounter the second type, called angular momentum. But for now, we will continue our focus on straight-line motion. We will also continue thinking about masses whose size does not matter, where we don't lose anything by assuming the mass is located at a single point. We will use the letter $\boldsymbol{p}$ to represent the linear momentum vector.

Linear momentum is a vector quantity possessed by any massive object in motion. It is the product of the mass and its velocity, $\boldsymbol{p} = m\boldsymbol{v}$.

Linear momentum is used more often than angular momentum, so if in this book or some other source you see a reference to "momentum" with no modifier, it probably means the linear variety. There is no special unit for momentum; it is simply measured in the composite units of kg m/s. Why is this quantity interesting? It turns out that Newton's second law of motion was originally stated using momentum. Recall that acceleration is the change in velocity divided by time. The right side of the second law ($m\boldsymbol{a}$) is therefore the scalar mass times the change in velocity, divided by time. If we assume that mass doesn't change, it is easy to see that the mass times the change in velocity is equal to the change in linear momentum, $m(\boldsymbol{v}_2 - \boldsymbol{v}_1) = m\boldsymbol{v}_2 - m\boldsymbol{v}_1 = \boldsymbol{p}_2 - \boldsymbol{p}_1$. The upshot is that we can write Newton's second law in an alternate form:

$$\boldsymbol{F}_{\text{net}} = \Delta \boldsymbol{p}/\Delta t$$ **Equation 8.1**

So another way of defining force is to say that force is the thing that is needed to change the linear momentum of a particle. As long as no force acts on an object, the momentum of the object will not change. If there is some net force, then the momentum of the object will change in the same direction as that total force.

It is useful to try to quantify just how much the linear momentum will change. If only one force acts, and it is constant during the specified interval of time, then it is a simple matter to multiply both sides of Equation 8.1 by the time interval Δt. Even if the force is not constant, the resulting equation would be true if we consider the average value of the force during that time. Thus for any given time interval, we can write $\Delta \boldsymbol{p} = \boldsymbol{F}_{\text{avg}}\Delta t$, where $\boldsymbol{F}_{\text{avg}}$ is understood as the force acting on the object averaged over the time Δt. In physics, we call this quantity *impulse.*

Impulse is defined as the change in linear momentum caused by a force. It is a vector quantity with the same dimensions as momentum.

We can see from the definition that impulse also has dimensions of force times time. You should be able to verify that one newton-second is the same as the standard momentum unit of kg m/s, by breaking the derived unit of the newton into its more fundamental parts. Impulse is most often used when a force acts suddenly and for a brief time. In such a case, the time interval Δt completely contains the force; the force is zero both before the start and after the end of Δt. The details of how the force changes *during* that time period may be complicated and unknown, but if you are only interested in the effect on the motion, this is a convenient way to talk about what happens.

An example of such a force might be the brief contact between a rubber ball and the floor when the ball bounces. Let's say we have a ball with a mass of 0.25 kilograms. Say it hits the floor moving at a speed of 9 m/s, at an angle of 70 degrees from horizontal, as shown in Figure 8.1. It immediately rebounds upward, at an angle of 67.5 degrees, with a speed of 8 m/s. What impulse did the ball receive from the floor?

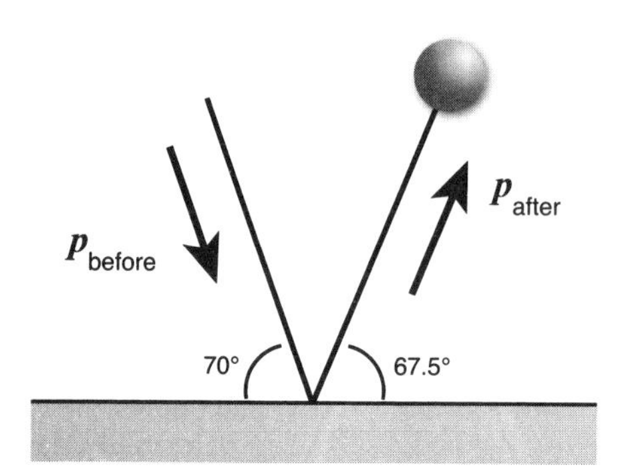

Figure 8.1

The example of a ball coming down from the left, bouncing off a level floor and rebounding up to the right.

Impulse is simply the change in the vector momentum, so we just calculate the momentum after the bounce and subtract the momentum before. We'll use components to do the subtraction, as we must do with vectors, so before we go any further we need to define a coordinate system. We'll take our customary directions for x and y, so the initial momentum has a positive x component and a negative y component.

Look at the horizontal (x) direction first. Before impact, $p_x = m(9 \text{ m/s})\cos(70°) = 0.77$ kg m/s. After the bounce, $p_x = m(8 \text{ m/s})\cos(67.5°) = 0.77$ kg m/s. Note that this is the same as the x component before the bounce. Subtracting the final x component from the initial x component, we find that this is equal to zero. Thus the change in momentum (and thus the impulse) has no x component.

In the vertical direction, we would have:

$$m(8 \text{ m/s})\sin(67.5°) - [-m(9 \text{ m/s})\sin(70°)] = 4.0 \text{ kg m/s} \qquad \textbf{Equation 8.2}$$

Since the vertical component of the momentum changes from negative to positive during the bounce, the change is in the positive direction (note where we subtracted the negative y component from before the bounce). Thus the magnitude of the impulse is 4 kg m/s, and in this case it is all in the vertical up direction.

We note that Newton's third law is still fully in effect. If two objects interact by briefly exerting a force on each other, they will also give each other an impulse. The impulse imparted to one will be equal but in the opposite direction to the impulse imparted to the other (since the time interval is the same from either particle's point of view). In our bouncing ball example, the floor gave an upward impulse of 4 kg m/s to the rubber ball. We should assume then, that the ball gave the same impulse to the floor, directed downward.

However, the floor is attached to Earth, which we have seen previously has a truly tremendous mass compared to our 0.25 kilogram ball. Since momentum is equal to mass times velocity, it is completely impossible to detect a change in Earth's velocity that results from giving it an extra 4 kg m/s of downward momentum. When two objects that have similar masses interact, then it will be a lot easier to see how they change each other's momenta, due to these reciprocal impulses.

What Good Is Momentum?

If we look back at Equation 8.1, we see that in any situation where the net force is zero, the momentum doesn't change in time. The true usefulness of momentum comes into play when we consider systems of multiple interacting bodies exerting forces on one another. Any system of moving bodies has a total linear momentum, which is equal to the vector sum of the momenta of all the moving bodies. If we have an isolated system, meaning no forces can act on the bodies from outside, then the total momentum of the system cannot change. This is again a situation we describe using the concept of a conserved quantity. In this case, the conserved quantity is the total linear momentum. Bodies within the system can interact with each other, and their individual momenta may change, but the total will remain the same. That is the law of conservation of momentum in a nutshell.

Let's consider a simple example. A circus cannon, normally used to launch a stunt clown through the air, has a big spring inside it. It is initially compressed, and then released to launch its payload. Instead of a clown, let's imagine instead that it is used to launch a heavy bowling ball, and that it is aimed horizontally instead of up toward the roof of the circus tent.

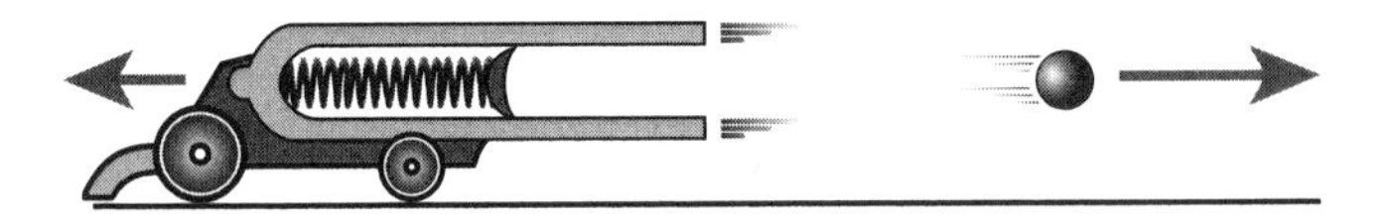

Figure 8.2

The circus cannon is initially motionless (top). After launching the bowling ball, it recoils to the left as the ball moves to the right, conserving horizontal linear momentum (bottom).

We also imagine that the cannon is mounted on wheels that roll freely. The cannon, loaded with the bowling ball, is initially motionless on a smooth level floor, but the effect of this elaborate setup is that there are no horizontal forces that are able to act on the system made up of the cannon and cannonball. The possible force exerted by the spring on the ball is internal to the system, and has no effect on the total momentum of the system.

Let the mass of the empty circus cannon be $M = 100$ kg, and the mass of the bowling ball be $m = 12$ kg. With no horizontal forces acting, the horizontal component of the total momentum of the two-body system cannot change. We say that the horizontal momentum is conserved in this case. Now the cannon launches the ball in the positive x direction so that it gets a velocity of 30 m/s relative to the ground. What happens to the cannon? It will have to recoil in the negative x direction in order to preserve the total momentum of the system (which has to stay zero, since that's what it was before the launch). Let v_x be the final horizontal velocity of the cannon. Momentum conservation then requires that $Mv_x + m(30 \text{ m/s}) = 0$. With the given masses, you can solve this with simple algebra to find $v_x = -3.6$ m/s.

Thus one of the uses of momentum conservation is to explain the familiar phenomenon of recoil. Momentum is helpful in many other situations as well, and most of them are similar to this in that total momentum is conserved. Don't forget that momentum conservation is applicable in situations where objects within a system may interact with each other, but no "external" forces act on the objects from outside the system.

Collisions

The interactions for which conservation of linear momentum is most useful are called collisions. The term *collision* in physics refers to a very specific kind of interaction between objects. It most often occurs when objects make brief contact with each other (although technically, actual physical contact is not required). Our strategy will be to concentrate on the situation immediately before and just after the very brief collision, and demand that initial and final linear momentum of the two-body system remain the same.

TRICKS AND HACKS

Conservation of momentum can only be applied in the absence of external forces. This condition is actually pretty strict, since we know that gravity, for example, is always acting on objects we deal with here on Earth. Fortunately, during most collisions, the interaction force must be brief and relatively strong. Recall that the impulse is a product of the average force and the time interval. By requiring the time interval to be very short, we are necessarily saying that the average force must be very large. During the actual collision, the force of interaction between the objects is so large that we can ignore other external forces (like gravity) in comparison and assume that linear momentum is truly conserved.

For the remainder of this chapter, we will only consider collisions between two bodies. In each case, our system will be made up of those two bodies, and for the purposes of analyzing the collision, external forces will be negligibly small. We won't need any details of the brief force by which the two bodies interact. We will simply use the fact that the total linear momentum of the two bodies must be the same before and after the collision.

Elastic vs. Inelastic

Although linear momentum is conserved during all collisions, kinetic energy is a different matter. In fact, the fate of the total kinetic energy provides us with a way to classify different kinds of collisions. In general, some kinetic energy will be converted to other forms of energy during a collision, and the result is that the objects after collision will have a total kinetic energy that is less than what they had before. When there is any appreciable change in the total kinetic energy, we call such a collision *inelastic*. This is the same as saying kinetic energy is not conserved during an inelastic collision.

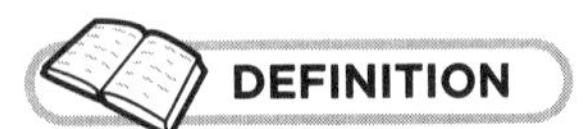

An **inelastic collision** is one in which the total kinetic energy after the collision is different (usually less) than it was before the collision.

An **elastic collision** is one in which the total kinetic energy is conserved.

In some collisions, such as between billiard balls and other hard objects, the change in total kinetic energy is negligible, and we can say that the kinetic energy was conserved. Such collisions are classified as *elastic* collisions. Many important interactions of particles at the atomic and molecular level are perfectly elastic, so it is useful to understand this distinction. On the other hand, you should never assume that any particular collision is elastic.

RED ALERT!

It is understandable that students often assume kinetic energy is conserved in all collisions, given the importance of the law of conservation of energy we talked about in the last chapter. The reality is, however, that there are many other forms of energy that kinetic energy can turn into during a real collision (e.g., sound, heat, or work done to dent or deform an object). If you are only considering kinetic energy, you can't assume it is conserved unless that is a given piece of information. Of course, it remains true that energy as a whole is conserved during all collisions.

The class of inelastic collisions is much larger and wider than the class of elastic collisions. If kinetic energy is conserved in addition to momentum, then you have a very restrictive set of conditions that must be satisfied before and after the collision. However, if the collision is inelastic, it means you don't know how much kinetic energy was lost; it could be a little or a lot. The only restriction is that linear momentum must be conserved, which usually prevents you from losing all of the kinetic energy. This point will become clearer when we look at some concrete examples.

One-Dimensional Collisions

The simplest situation to imagine is a case where two objects only move along a single straight line. They could be confined by something like railroad tracks or it could just be dumb luck, as when two billiard balls strike each other head-on. With only one dimension to worry about, it is pretty easy to apply the conservation of momentum condition.

Let's imagine two big beads on a straight, rigid, horizontal, and very slippery wire. We will let bead 1 move toward bead 2, which is initially at rest. Let's say the mass of bead 1 (m_1) is 2 kilograms and that of bead 2 (m_2) is 3 kilograms. If the first bead has an initial velocity of 7 m/s, what happens after the collision?

You might guess that the second bead will start moving due to an impulse delivered by the first bead, and that would be correct. That means that m_1 will have to slow down, so that the total momentum remains unchanged. The total momentum will be 2 kg × 7 m/s = 14 kg m/s, both before and after the collision. So far, so good.

However, after the collision you have two masses which have unknown velocities. There are a lot of different velocity combinations that will give a sum of 14 kg m/s for the momentum, so you don't have enough information to predict exactly what either velocity would be after the collision.

So whoever poses the problem would have to give you more data. They might tell you that the 3-kilogram bead moves at 5 m/s right after the collision. Then you could confidently predict the final velocity of the 2-kilogram bead. By requiring the total momentum to remain unchanged, you could write:

$$14 \text{ kg m/s} = (2 \text{ kg})(v_1) + (3 \text{ kg})(5 \text{ m/s}) \qquad \textbf{Equation 8.3}$$

Here, the variable v_1 stands for the final velocity of the 2-kilogram bead. Thus momentum conservation requires $v_1 = -0.5$ m/s, or in other words, the first bead bounces backward at 0.5 m/s after hitting the more massive bead that was stationary.

Is this collision elastic or inelastic? In order to answer that question, we only have to calculate the kinetic energies before and after the collision and compare them. Before the collision, the only kinetic energy to be found was in the motion of bead 1, so the kinetic energy would be $\frac{1}{2} m_1 v^2 = 49$ J. After the collision, both particles have kinetic energy, and the total kinetic energy is:

$$\frac{1}{2}(2 \text{ kg})(-0.5 \text{ m/s})^2 + \frac{1}{2}(3 \text{ kg})(5 \text{ m/s})^2 \qquad \textbf{Equation 8.4}$$

$$0.25 \text{ J} + 37.5 \text{ J} = 37.75 \text{ J}$$

This is clearly less than the 49 joules of kinetic energy that we started with, so it was definitely an inelastic collision.

It was not necessary that the second mass be stationary before the collision. If both beads were moving toward each other, they still would have collided and total momentum would still have been conserved. You would just have to include both masses in calculating the initial momentum. Notice that in one dimension, the sign of the velocity is important for the momentum, but not for the kinetic energy. Kinetic energy is always positive, which is taken care of by squaring the velocity. When we work in more than one dimension, only the magnitude of the velocity (the speed) is used to calculate the kinetic energy.

Let's go back to the original bead example, but take away the information about the final velocity of the 3-kilogram bead. Instead, let's say that the two beads are coated with molasses, and stick together after they collide. It turns out that this additional fact gives us enough information to completely determine what happens after the collision (given that we know the initial speed of the first bead). Now the momentum conservation condition would look like this:

$$14 \text{ kg m/s} = (2 \text{ kg})(v_f) + (3 \text{ kg})(v_f) \qquad \textbf{Equation 8.5}$$

Here, we are using v_f for the final velocity shared by both beads. Since the beads are stuck together after the collision, this is the only unknown quantity. A quick bit of algebra yields a final velocity of 2.8 m/s for the object which now has a total mass of 5 kg, and it is all moving in the positive direction.

It turns out that whenever the two bodies are stuck together following the collision, they have the smallest amount of kinetic energy allowed by momentum conservation. Such two-body collisions, even in two dimensions, are sometimes called "perfectly inelastic." The minimum final kinetic energy occurs when the two masses after the collision share the same velocity, no matter the reason. This applies when the motion is in more than one dimension as well.

TRY IT YOURSELF

12. How much kinetic energy was lost in the perfectly inelastic collision between the beads in the previous example?

Collisions in Two Dimensions

The same classifications and rules apply to collisions in two dimensions as we illustrated in one dimension. The only difference is that now the momentum vectors can all have two components. Note that the direction of a momentum vector for a certain mass is the same as the direction of its velocity vector, although momentum and velocity have different units. But don't forget that it is not velocity that is conserved, but rather the momentum.

TRICKS AND HACKS

If you are given a physical situation with velocities or momenta, and you have a choice on how to define your coordinate system, it is usually a good idea to align one of your coordinate axes with one of the momentum vector directions. Then at least one of your momenta will only have one component to worry about.

Let's consider an example with two hockey pucks on a level surface of ice. We'll say the ice is smooth enough that we have no frictional forces acting in a horizontal direction, and vertically the weights are perfectly offset by the normal forces. Then in the plane of the ice, we know that the total momentum of the two pucks will be conserved.

We'll say that these are not regulation hockey pucks, so that they can have different masses. Similar to the 1D example, we'll start with a 0.9 kilogram puck (m_2) at rest on the ice. We'll send a smaller puck, with $m_1 = 0.4$ kg toward the resting puck, with a speed of 16 m/s. A glancing collision causes puck 1 to change its speed and direction of motion, as shown in Figure 8.3.

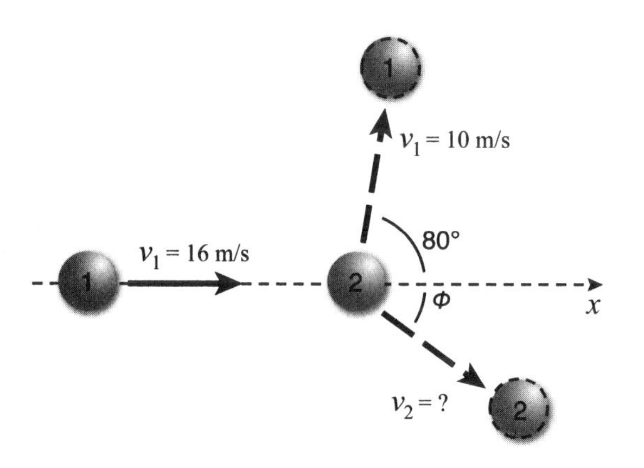

Figure 8.3

Puck 1 slides from left to right and strikes a second puck that was at rest. Dashed lines indicate the positions and velocities of the pucks following the collision.

If we know what happens to the first puck after the collision, then we can use momentum conservation to determine the velocity vector of the second puck after the collision. Whatever the actual direction of the first puck before impact, it makes sense to use that direction as our reference axis, which we call x. Then the direction perpendicular to that can be called y. Here's how it would work.

We would look at the momentum separately in the two orthogonal directions. If we look at the x direction first, we calculate that before the collision, the total momentum was 0.4 kg × 16 m/s = 6.4 kg m/s (only puck 1 was moving). After the collision, the x component of the first puck's momentum is p_{1x} = 0.4 kg × 10 m/s × cos(80°) = 0.695 kg m/s. Conservation of momentum says that 6.4 kg m/s = 0.695 kg m/s + p_{2x}, so we see that p_{2x} = 5.7 kg m/s. To get the x component of the velocity, we only have to divide this by the mass (0.9 kg), giving v_{2x} = 6.33 m/s.

For the y direction, we first note that the total momentum is zero. Therefore, $p_{1y} + p_{2y} = 0$ after the collision, and the unknown $p_{2y} = -p_{1y} = -0.4$ kg × 10 m/s × sin(80°) = -3.94 kg m/s. Once again, we would divide this by the second puck's mass to get $v_{2y} = -4.38$ m/s.

We can then apply trigonometry to determine the velocity vector for puck 2 that results from the collision. The puck that was originally motionless is now moving at 7.7 m/s at an angle of about 35 degrees with respect to the x-axis.

Motion of the "Center of Mass"

Whenever we talk about systems that contain multiple bodies, there is a particular location within that system that is especially interesting. That location is called the *center of mass* of the system. This point has several uses which will come into play later in this book, but this seems like a good time to introduce this idea for the simplest systems.

Center of mass refers to the weighted average location of all of the mass making up a system.

Suppose we have two objects with masses m_1 and m_2. In general, we will allow them to move freely in the xy-plane, but for the moment, imagine they are each located somewhere on the x-axis. The center of mass can be viewed as an "average" location of the two objects. No matter where the objects are, their center of mass is located somewhere between them along a line that joins their two centers. With both objects on the x-axis, we would have:

$$x_{cm} = \frac{m_1 x_1 + m_2 x_2}{m_1 + m_2} \qquad \textbf{Equation 8.6}$$

In words, we find the location of the center of mass by multiplying the mass of each object by its corresponding location, adding those products, and then dividing by the total mass. This recipe can be extended to any number of particles, and applies to any spatial coordinate.

It should be clear that the center of mass will be exactly halfway between two objects if their masses are equal. Otherwise, it will be located closer to the one that has more mass. Since the masses are constant, the distance from one object to the center of mass will always be the same fraction of the total separation distance, which should convince you that the center of mass is always on a line joining the two objects even in two- or three-dimensional space.

Let's now imagine that the two objects are moving. Think about Equation 8.5 for two successive times. Subtract the expression for x_{cm} at the earlier time from that at a later time. Each object will have a certain change in location Δx to go along with a short time Δt. Following a bit of tedious algebra, it should not be too surprising that the usual definition of velocity will yield:

$$v_{cm,x} = \frac{m_1 v_{1,x} + m_2 v_{2,x}}{m_1 + m_2}$$ **Equation 8.7**

$$(m_1 + m_2) v_{cm,x} = m_1 v_{1,x} + m_2 v_{2,x}$$

On the right-hand side of the second line of Equation 8.7, you recognize the sum total of the momenta in the *x* direction. For collisions or other circumstances where momentum is conserved, this sum is constant. Now look on the left. Since $(m_1 + m_2)$ is constant, the velocity component $v_{cm,x}$ must also be constant. This same analysis holds true for any component of the center of mass velocity.

CONNECTIONS

The best way to figure out how something works is to take it apart. There are some things in nature, though, that aren't so easy to disassemble. The smallest entities in the known universe—the so-called fundamental particles—are one important example. Rather than screwdrivers and wrenches, physicists have developed highly advanced tools to study the innards of these tiny entities. At particle colliders like CERN in Switzerland, physicists direct powerful beams of particles into special targets or even oncoming beams and then—using basic concepts like the conservation of energy and conservation of momentum—study the fragments to piece together an understanding of these tiny particles' structures.

This result tells us one way that the center of mass is significant. As long as momentum is conserved for a system of objects, it means that the center of mass of that system will have a constant velocity. It will be motionless (if the total momentum is zero) or moving in a straight line at a constant speed, regardless of what all the particles within the system are doing.

The Least You Need to Know

- For all collisions in which external interactions are negligible, linear momentum is strictly conserved.
- Collisions in which kinetic energy is also conserved are called elastic, while all others are inelastic.
- The momentum vector must have all of its components conserved separately in collisions in more than one dimension.
- When the momentum of a system of massive bodies is conserved, the center of mass moves with a constant velocity.

CHAPTER

9

Rotational Motion

With this chapter we will close out our survey of basic mechanics. Rather than trajectories in Cartesian coordinates, we will now consider rotations about an axis. A lot of our analysis will look similar to previous work, though there will be some important differences.

For the first time we will have to pay attention to the sizes of objects. So far we've gotten away with the approximation that all of the objects we are tracking are relatively small, and that the way their mass is distributed is irrelevant. This is equivalent to ignoring any possible rotation of the mass, either because the orientation of the object doesn't change, or because it doesn't have any effect on the forces or motion. Obviously, this is not a very good approximation for many real-world objects, which can spin and roll as well as move through space. We shall ignore rotation no longer.

Our strategy will be similar to how we began our study of motion in the first place. We will begin by simply describing rotation in all of its aspects. Then we will discuss how to account for whatever might cause a change in rotation. We're going to pack a lot into this one chapter, so be forewarned. But by the end of this whirlwind tour, you should be able to appreciate a nice analogy between these new ideas and what we have learned in the previous seven chapters.

In This Chapter

- Describing rotational motion
- Torque and its importance
- The moment of inertia
- A new kind of momentum

Rotation About a Fixed Axis

Rotation is everywhere. You see it when you toss a Frisbee or a football, when dancers twirl, or when you watch a carousel at the amusement park. As you surely learned in elementary school, even the planet you are living on spins on its axis. Some of this rotation, like the poorly thrown football, is too complicated for us to describe well in this book. But we'll at least make a good start.

Rotation is only relevant for systems with many particles in them, or bodies whose mass is distributed over some appreciable volume. We'll focus on the latter, which we sometimes call "extended" bodies. In order to keep things manageable, we will only consider rigid extended bodies in our initial discussion of rotation. This just means that whatever object we are considering, it keeps its shape. All of the pieces of mass that make up a rigid extended body maintain the same relative distances between them. So you should be thinking of completely solid objects, not blobs of Jell-O or rotating cans of soda pop.

The next restriction we impose is that the axis of rotation must remain stationary. The axis is the (imaginary) line about which the whole body rotates. The word itself might bring to mind the axle of a wheel, and an axis is similar to that. While an axle has some definite size and may or may not rotate when the wheel does, the axis is a purely geometric line. We will think of this line as fixed in the rotating body, and we won't allow that line to change either its location or its orientation in space.

All that said, we will not require that rotating objects be round, nor that the axis of rotation be located in the center of the object. A lot of rotating objects that we encounter out in the world are round and centered, but many are not; the analysis we present here will work in either case.

Rotation really means a change in orientation (of the rotating body, not the axis) so we must specify a way to describe this orientation at any particular time. The best way to do this is to picture the rotating body by looking straight down along the axis, as in the following figure. Figure 9.1 illustrates the orientation of the body at one instant in time.

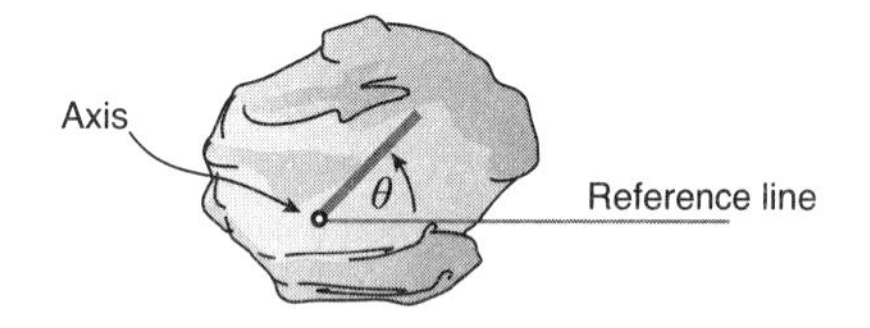

Figure 9.1

An extended body, viewed along its axis of rotation. The thicker line is attached to the body.

To describe rotation, we need some kind of mark on the body, or at least an imaginary line that is attached to the body but that also goes through the axis of rotation, at a right angle. We also need a line passing through the axis that does not rotate with the body, but which is fixed in our stationary frame of reference. You can think of both of these lines as defining a plane that is perpendicular to the axis. Only with both of these lines can we uniquely define the orientation, or angular position, of the body in question. Since these two lines meet at our stationary axis, we can define the *angular position* of the body simply as the angle between the two lines. That angle will change if the body actually does rotate.

The **angular position** of a body is defined relative to a fixed reference line in space. It is the angle between a line fixed to the body and that reference line, in a plane perpendicular to the axis of rotation.

Let's call the angular position θ. As we did with linear coordinates, we must decide what orientation corresponds to $\theta = 0$, and which direction will be considered to be positive. In this way, θ behaves just like a single position coordinate in one linear dimension (like x back in Chapter 2). From a given viewpoint, we usually call counterclockwise rotation the positive direction ($\theta > 0$), making clockwise rotation negative ($\theta < 0$). The orientation we call $\theta = 0$ occurs when the two lines in Figure 9.1 coincide.

One difference between the angular coordinate (like θ) and a linear coordinate (like x) comes about because the angle of rotation eventually repeats itself, instead of potentially increasing forever. After you have rotated by 360 degrees for example, you will be back at the same orientation that we called 0 degrees. The other difference, of course, is that angles are not measured in the same units as distances or lengths; they have a different dimension.

You may be used to measuring angles in degrees, where a full rotation is divided into 360 degrees. But physics works a lot better if we use a more natural unit for measuring angles called the *radian*.

The **radian** is a natural unit for measuring angles, based on the length along a circular arc and abbreviated "rad." A full circle corresponds to 2π radians, which is also 360 degrees. This means 1 rad = 57.3°.

To gain a solid understanding of what a radian is, it helps to think about the geometric concept of arc length of a circle. An arc is defined as some portion of a circle. If you have a certain arc, you can draw radius lines to the center of the circle from each end of the arc. The angle between those lines is the unique angle that defines the arc. The larger the angle, the longer the arc. If we define the arc length Δs to be the distance along the curved path of the arc, then it is easy to see that Δs is directly proportional to the angle (call it $\Delta\theta$). If we measure the angle in radians, then the constant of proportionality is just the radius of the circle:

$$\Delta s = r(\Delta\theta) \qquad \textbf{Equation 9.1}$$

The larger the radius, the longer Δs will be for the same angle range $\Delta\theta$. One radian is actually the angle for which the arc length is exactly equal to the radius of the same curve. In a sense, the angle in radians is defined as the ratio of these two lengths, so it is essentially a dimensionless unit.

Now we can state formally what we mean by rotation. A body rotates when its angular position θ changes with time. We can even define an angular velocity to keep track of how fast and in which direction the object is rotating. We'll use the Greek lowercase letter omega (ω) for *angular velocity.*

DEFINITION

Angular velocity is the rate at which the angular position of a rotating body changes. It is the rate of rotation about a specific axis, with the algebraic sign representing the direction of the rotation. The standard units are rad/s.

$$\omega_{avg} = \frac{\Delta\theta}{\Delta t}$$

Angular acceleration is the rate at which the angular velocity of a rotating body changes, with a sign that indicates whether ω is getting more positive or more negative. The standard units are rad/s^2. $\alpha_{avg} = \frac{\Delta\omega}{\Delta t}$

The value of ω may or may not be constant. If it is changing, we define the instantaneous angular velocity in the usual way: the value of ω_{avg} as the time interval Δt gets really small.

If the angular velocity is changing, we are often interested in the rate at which it is changing, as well as the direction. The rate of change of angular velocity is defined the way you would expect, and is called *angular acceleration.* The Greek letter α is used most often for angular acceleration.

Like the angular velocity, α can be positive or negative. In the case of ω, the sign tells you which direction the object is rotating, clockwise or counterclockwise. Conversely, the sign of α tells you in which direction ω is changing. If ω is positive and increasing, α is positive. But if ω is positive and decreasing, α is negative—even while ω is still positive. All this works out nicely if you remember the way we defined Δ.

Mathematically, you can think of θ just like a distance traveled, even though the object is not actually going anywhere. If you know that something is rotating at a constant rate ω for a certain time, then you know how much its orientation θ has changed during that time. For the case where the angular acceleration is constant, you can use the same tricks and formulas that we introduced back in Chapter 2 for motion in one dimension.

TRY IT YOURSELF

13. A rotating object makes 20 complete revolutions in one minute. What is its average angular velocity, ω_{avg}?
14. A wheel is rotating initially at +22 rad/s. If it then has an angular acceleration of α = -0.3 rad/s^2, how long will it be before it stops rotating?
15. In Exercise 14, how much did the object rotate while it was slowing down?

Motion of a Part of the Body

Let's take a closer look at one little piece of the rotating object. This will help us as we try to incorporate concepts like force and energy into this new realm of rotation. Let's label this piece *P*, and let's say that it is located at a distance *r* from the axis of rotation. *P* is an actual chunk of the rigid body, not too big but not just a tiny point, and it has a certain mass. Since *P* is part of a rigid body, rotating around an axis that is fixed, the distance *r* doesn't change. The resulting motion of just that little piece is a circle of radius *r* centered on the axis. If the angular velocity ω happens to be constant, then *P* is undergoing uniform circular motion, as we described in Chapter 3.

We know that *P* is therefore accelerating toward the axis at $a_c = v^2/r$ (in m/s^2), a centripetal acceleration. But what is *v*? The actual velocity of *P* would be in a direction tangent to the circle, which is also perpendicular to the radius line. If we want to know the magnitude of that velocity, or the speed at which that piece moves around the circle, just imagine a short time interval Δt, and think about the arc length traveled by the piece during that time. Divide both sides of Equation 9.1 by the time interval Δt, and you will see that $\Delta s/\Delta t = v = r\omega$.

This tells us that the linear speed of any piece of a rotating body depends on both the angular speed of the object as a whole as well as how far that piece is from the axis. Also, note that by substituting $r\omega$ for *v* in the old formula for centripetal acceleration, we arrive at an alternate formula for this acceleration. Any part of the body that is a distance *r* from the axis is accelerating toward that axis at a rate of $a_c = r\omega^2$.

CONNECTIONS

We know that $F = ma$, so for the piece *P* of the rotating rigid body, there must be a force acting on *P* that points toward the axis. Where does this force come from? It comes from the structure of the body itself. When a rigid body spins, internal forces act on all of the pieces of the body to hold it together. If it spins very, very fast, it is possible for these forces to be insufficient and overwhelmed, and the object will disintegrate.

Recall that we have been considering the case where the angular velocity of the object is not changing. Now, let's do a similar analysis for the case when it is changing. Since $v = r\omega$, any change in ω will result in a corresponding change in v. If we once again think about the rate of this change, we can calculate the linear acceleration experienced by any piece of a rotating rigid object located at a fixed distance r: $a_t = \Delta v/\Delta t = r\Delta\omega/\Delta t = r\alpha$. We call this acceleration a_t because it is in the same direction as the motion of the piece, tangent to the circle it is moving on. This component of acceleration is separate and in addition to the centripetal acceleration. Recall that the centripetal acceleration points toward the center of the circle and has a magnitude $a_c = v^2/r = r\omega^2$ even if the velocity is constant. If the velocity is not constant, it means that $\alpha \neq 0$, and both a_t and a_c are at work. The total acceleration in that case would be the vector sum of $\boldsymbol{a}_c$ and $\boldsymbol{a}_t$.

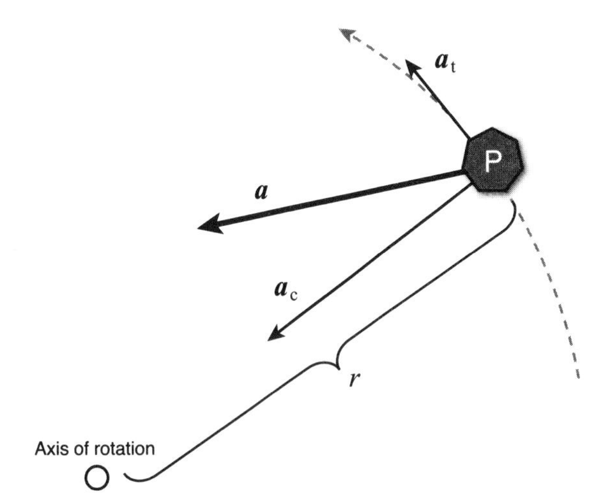

Figure 9.2

P is part of a rotating rigid body a distance r from the axis of rotation. Here we show the accelerations that affect P in the case of positive ω and positive angular acceleration α.

Torque

We're now equipped to describe the motion of a rotating object. But how do you actually get an object to start rotating in the first place? You might expect that a force would be involved. But now, not just any force will do. Imagine your bike is sitting upside-down on the ground. The front wheel is not moving. If you apply a force that is directed straight toward the axle, or directly away from it, you will not cause any rotation. To cause any rotation, the direction of the force must be at some angle that is not pointing to or away from the axis.

The Role of Torque

Physicists quantify this type of thing using a physical quantity called *torque*. Torque measures how effective you are at causing something to rotate (or to change its angular velocity if it already is rotating). Torque requires the presence of a force, and as a vector quantity, both the strength of the force and its direction matter. For torque, there is one more quantity that matters, and that is the location on the body where the force is actually applied. We didn't need to consider this when we were ignoring the size of objects, but now it is critical.

To get a better idea of what we're talking about, take a look at Figure 9.3. The wrench is effectively attached to the head of a bolt because the opening fits the shape of the head exactly. The bolt can rotate about its center, an axis that is perpendicular to the page. Let's imagine this is a pretty big wrench, with a long handle. You can grab the handle of the wrench anywhere along its length, and push or pull it in any direction. We want to translate that force into a torque.

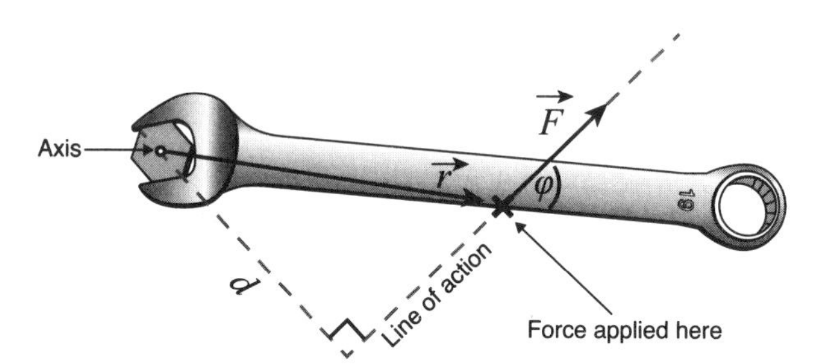

Figure 9.3

A force applied to a wrench produces a certain torque on a bolt.

How to Calculate Torque

A good way to determine how much torque is applied is to define something called a *lever arm*, which we will label d. The lever arm can be determined by first drawing a line (called the line of action) through the force vector. The location of this line depends on the direction of the force and where it is applied, but not on the magnitude of the force. Somewhere along the line of action, there will be a place where a line drawn from the axis to the line of action makes a right angle with the line of action. The length of that new line is the lever arm. You can also think of d as the minimum distance between the axis of rotation and the line of action of the force.

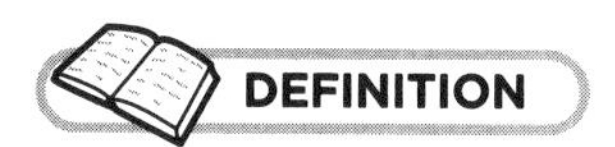

Torque is a physical quantity that corresponds to a tendency to cause rotation. When a force of magnitude F acts with a lever arm d, the magnitude of the torque is $\tau = Fd$. Standard units of torque are N m.

The **lever arm** associated with a force is the closest distance between the line along which the force acts and the axis of rotation.

The magnitude of the torque is simply the product of the force and the lever arm: $\tau = Fd$. To calculate the lever arm d, you can make use of the angle made between the vectors $\boldsymbol{F}$ and $\boldsymbol{r}$, where $\boldsymbol{r}$ is a position vector that points from the axis to the point where the force $\boldsymbol{F}$ is physically applied. If you define the angle between these two vectors as φ, then the length of the lever arm is $d = r\sin(\varphi)$. Thus, we see that the magnitude of the torque is $\tau = rF\sin(\varphi)$.

TRICKS AND HACKS

If you have worked with vector multiplication before, you may recognize that torque looks like the magnitude of the "cross product" between $\boldsymbol{r}$ and $\boldsymbol{F}$. The result of taking the cross product of two vectors is another vector. It is indeed possible and useful to define torque as a vector in three dimensional space, such that $\boldsymbol{\tau} = \boldsymbol{r} \times \boldsymbol{F}$, but then the direction of the torque vector has a different meaning. By sticking with a fixed axis of rotation, the only "directions" we need to worry about can be specified by using positive or negative signs for the sense of the rotation.

To determine the sign of the torque, we will use the same convention we used for θ and ω. If the torque would tend to rotate something counterclockwise, we call it positive; otherwise, the torque is considered to be negative. The sign is especially important when more than one torque acts on the same object, in which case the total torque is the sum of all the acting torques. These could have different signs if they act to rotate the object in different directions, so be sure to keep track of all the pluses and minuses.

The torque is proportional to the magnitude of the applied force $\boldsymbol{F}$, but that force alone is not enough to determine the torque. The same size force can result in different torques depending on where it is applied (which impacts d), and at what angle. To increase the torque from a given force, you would try to apply it to the object as far from the axis as possible. Also, for a given force and application point, the maximum torque occurs when the applied force $\boldsymbol{F}$ is perpendicular to the position vector $\boldsymbol{r}$. Any other angle will reduce the lever arm d and hence the torque.

This explains why doorknobs are normally mounted as far from the hinges as possible. You could still pull a door open if the knob or handle was close to the hinge, but it would require a lot more force to give the same resulting rotation of the door.

TRY IT YOURSELF

16. Let's say we apply a force to a wrench that is directed perpendicular to the handle of the wrench, at a distance of 16 cm from the center of the bolt. If the magnitude of that force is 41 N, what is the resulting torque on the bolt?

RED ALERT!

The standard units for torque look the same as the units we used for work and energy, but this is just a coincidence. Torque and energy have totally different dimensions, so try not to confuse them. Usually, the context will make it very clear which one you are talking about. We will do our part by never using joules to indicate the value of a torque.

Moment of Inertia

We now know that the effect of torque is to cause a change in angular velocity (i.e., rotation). How can we quantify this effect? Let's assume there is one single torque, and we'll see what this does to our small piece *P* of a rigid body. Let's now specify the mass of that piece, call it m_i. If we just imagine applying a force on *P* that is perpendicular to its position vector ***r***, and if we only look at the tangential direction, then we know from Newton's second law that:

$$F = m_i a_t \qquad \textbf{Equation 9.2}$$

As we saw above, for this one piece *P*, $a_t = r\alpha$. Let's multiply both sides of Equation 9.2 by *r*, and substitute for a_t, which gives us $Fr = m_i r^2 \alpha$. Now we note that *Fr* will be the torque due to the force *F*, because the lever arm relative to the axis will be exactly *r*. If we make that substitution, then we have an expression relating torque to the angular acceleration of the little mass m_i located at *P*:

$$\tau = m_i r^2 \alpha \qquad \textbf{Equation 9.3}$$

We can extend this to the effect of torque on the whole body by adding up all of the little pieces with their individual masses, until we get back to the body we started with. Since the body is rigid, every piece has the same angular acceleration α. Each piece, however, could be at a different distance from the axis. So now we will think of *i* as a numerical label, or index. If we divided our object into 100 pieces, then *i* stands for a number between 1 and 100. Each piece is numbered; it has its own mass m_i and now we say that it is located at a distance r_i from the axis of rotation.

With this understanding, the right side of Equation 9.3 becomes $\Sigma m_i r_i^2 \alpha$, which in our case means the sum of the quantity $(m_i r_i^2 \alpha)$ for all 100 terms, one for each piece of our rigid body. Then the effect of applying that torque τ to a rigid body would be:

$$\tau = \left(\Sigma m_i r_i^2\right)\alpha \qquad \textbf{Equation 9.4}$$

The factor in the parentheses is a new and important property of the object. It is different for different rigid bodies. It depends on the total mass of the body, but also on how that mass is distributed relative to the axis of rotation, so it depends on where the axis is located in the body. We call this quantity the *moment of inertia.*

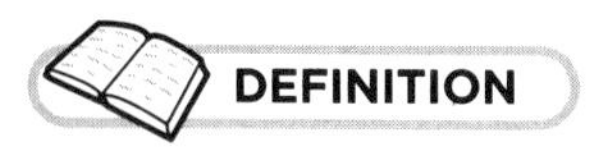

The **moment of inertia** is a property of a rigid body that measures its resistance to rotation about a specific axis. If we divide the object into discrete small pieces, each with mass m_i and located a distance r_i from the axis, then $I = \Sigma m_i r_i^2$. The standard units for the moment of inertia are kg m^2.

The moment of inertia will always be a positive quantity for any object. However, a given object does not have only a single moment of inertia; the value of I depends on where the axis of rotation is located. Two identically shaped objects with the same total mass don't necessarily have the same moment of inertia. If for one object more of the mass is located farther from the axis, then it will have a greater I.

As a concrete example, let's take an ordinary number 2 pencil. It is a rigid object with a definite mass, but you can choose to rotate it around many different axes. If you rotate the pencil about an axis that is aligned with its length, with the axis right down the center of the pencil, then the moment of inertia will be very small. This is because all of the mass of the pencil is located within a few millimeters of the chosen axis. On the other hand, you could choose to rotate the pencil end over end, around an axis that is again centered in the pencil, but at right angles to its length. In that case the same pencil will have a larger I than before, because now some of the mass is centimeters away from the axis. You can increase I even more by rotating it about an axis that goes through one end.

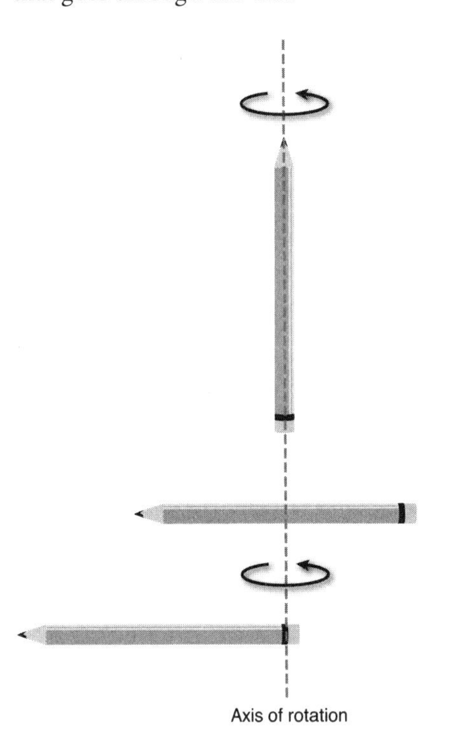

Figure 9.4

The moment of inertia of a pencil depends on what axis it is rotating around. I *is smallest for the case at the top, and largest for the case on the bottom.*

The nice thing about the moment of inertia is that it gives us a simple way to represent the effect of torque on a rigid body. As long as the body is rigid and can only rotate around a fixed axis, I will be constant, and there is a direct relationship between the angular acceleration and the applied torque. That relationship is simply $\tau = I\alpha$. In this formula, the τ on the left should be understood as the total or net torque applied to the object. This allows us to handle cases where more than one torque is applied at the same time (keeping in mind the convention for the sign of a torque). You should note a strong resemblance between this expression and Newton's second law of motion.

TRY IT YOURSELF

17. A light string is wound around the rim of a heavy wheel (radius = 12 cm) that is mounted on a frictionless bearing. Initially, it is not rotating. You pull on the string with a steady force of 3 N for exactly 3 seconds. The angular velocity after 3 seconds is 13 rad/s. What is the moment of inertia of the wheel?

Torque and Static Equilibrium

In Chapter 5, we looked at the special case where the net force on an object was zero. We noted that because forces are vectors, multiple non-zero forces in different directions could add up to zero. When this is the case, Newton's laws tell us that a mass with zero net force will have no acceleration, a state we call equilibrium.

The most obvious example of equilibrium is when an object is not moving (static equilibrium). When the net force acting on a mass at rest was zero, that fact was enough to assure you that the mass would remain at rest and be in static equilibrium. Now that we are considering extended bodies, however, that is no longer sufficient to guarantee no motion.

It is possible for two equal and opposite forces to act on a body and have a non-zero net torque. This would happen if the forces acted at different points, and did not have the same line of action. If these were the only two forces acting on the body, the net force would still be zero, but there would be some torque, and therefore there would also be some rotation. To achieve true static equilibrium for an extended body, we need to have both zero net force and zero net torque.

Energy in Rotation

As we saw above, every piece of a rotating body is moving at some linear velocity (at any given moment). Since every piece also has mass, we would assume that a rotating body must therefore have kinetic energy. It turns out that the moment of inertia also gives us a convenient way to specify how much total kinetic energy should be associated with rotation.

It is pretty easy to determine the total kinetic energy of a rotating body if we go back to our strategy of dividing the object into many small pieces. Given a certain angular velocity ω, we know that an individual piece P has a linear velocity $v_i = r_i\omega$, so its kinetic energy would be $\frac{1}{2}m_i(r_i\omega)^2$. If we now sum up the total kinetic energy of all the little pieces, we would get $K_r = \frac{1}{2}(\Sigma m_i r_i^2)\omega^2$. (We use the subscript on K_r to distinguish this rotational kinetic energy from the usual kinetic energy associated with translational motion.) Lo and behold, we see the moment of inertia making another appearance, so that $K_r = \frac{1}{2}I\omega^2$.

So rotation of a body with mass is another way to embody energy. Since torque is the thing that changes angular velocity ω, it must be able to do work in the same sense that a force acting over a distance does work. Recall how work in linear motion was able to change the kinetic energy of an object. Similarly now, torque can change the rotational kinetic energy of an object.

We can define the work done by a torque as $W = \tau(\Delta\theta)$. This work would have our usual energy units. If a constant τ acts over a certain change of orientation $\Delta\theta$, a definite amount of energy is transferred and we would expect the rotational kinetic energy to change by exactly that amount, $\tau(\Delta\theta)$. Note that if either τ or $\Delta\theta$ is negative, it means that the torque is opposed to the direction of the rotation, and the K_r would decrease.

TRY IT YOURSELF

18. How much kinetic energy was added to the wheel in the previous exercise? How much power were you pulling with?

Angular Momentum and Its Conservation

There is one more quantity to define that is useful for figuring out rotational motion, and that is something called *angular momentum.*

DEFINITION

The **angular momentum** of a rigid rotating body is given by $L = I\omega$. The standard units for angular momentum are kg m^2/s.

Angular momentum is a very important quantity in physics, and it has ramifications far beyond the scope of this book. We have only defined it for rotating rigid bodies, but it can be defined for other types of motion as well. It is fundamentally a vector quantity (as is torque), but by restricting ourselves to a fixed axis of rotation, we can again handle the direction of angular momentum with a simple algebraic sign. Since I can only be positive, L will have the same sign as ω.

Recall our rotational analog of Newton's second law: $\tau = I\alpha$. Consider constant torque, in which case we know that $\alpha = \Delta\omega/\Delta t$. Substituting this in to the previous expression gives us $\tau = \frac{I\Delta\omega}{\Delta t} = \frac{\Delta L}{\Delta t}$. The last part follows because, for a fixed axis, I can't change, so from the definition of angular momentum, $\Delta L = I(\Delta\omega)$. Therefore we can also think of torque as the thing required to change the angular momentum of a body. If no torque is applied, then there is no way to change the angular momentum, and L would be another conserved quantity.

This is most useful for objects that are not completely rigid and thus able to change their moment of inertia. The human body is a good example of this, and gymnasts, dancers, divers, and other athletes make use of conservation of angular momentum regularly. Next time you watch figure skating, pay attention to the spins. The figure skater spinning about a vertical axis through the tip of an ice skate can change her angular velocity by changing the positions of her arms and legs. Drawing her arms and legs inwards toward the axis decreases her I, which requires a corresponding increase in ω to conserve angular momentum. When she wishes to slow her spin, she extends her arms and/or legs, and slower she goes.

The Great Analogy Between Rotation and Linear Motion

We can summarize much of this chapter by noting that almost everything we learned about motion in one dimension has an analog in rotational motion. We have already seen that there are analogous definitions of velocity and acceleration, where the angular position θ plays the role of the linear dimension x. The analogy extends to things like energy and momentum, if we just substitute the moment of inertia where we previously used the mass. And for changes in angular motion, the newly introduced quantity torque plays the role of force. The whole story is nicely summed up in the following table.

Linear and Rotational Forms of the Most Common Quantities in Mechanics

Quantity	Linear Motion	Rotation
position coordinate	x	θ
velocity	$v = \Delta x/\Delta t$	$\omega = \Delta\theta/\Delta t$
acceleration	$a = \Delta v/\Delta t$	$\alpha = \Delta\omega/\Delta t$
momentum	$p = mv$	$L = I\omega$
Newton's second law	$F = ma = \Delta p/\Delta t$	$\tau = I\alpha = \Delta L/\Delta t$
kinetic energy	$K = \frac{1}{2}mv^2$	$K_r = \frac{1}{2}I\omega^2$
work	$W = \boldsymbol{F}\cdot(\Delta\boldsymbol{r})$	$W = \tau(\Delta\theta)$

The Least You Need to Know

- When an object can rotate about a fixed axis, we describe its orientation using an angle measured in radians.
- Rotation can be described by angular velocity and angular acceleration, defined the same way as the corresponding linear quantities.
- The moment of inertia measures the resistance of an object to rotation, by taking account of how mass is distributed relative to the axis of rotation.
- Torque quantifies how effective a force will be in causing or changing the rotation of a rigid body.
- Rotating bodies also have energy and angular momentum, which are intimately related to the moment of inertia.

PART

2

Waves and Fluids

Now that we have a handle on how things move, we're ready to branch out a bit. We'll start by considering a special kind of motion that repeats itself at regular intervals. We call this kind of motion oscillation, and it includes all kinds of swings and vibrations. Unless you walk around with your eyes closed, you will find oscillations almost everywhere you look.

Under the right circumstances, this can lead to a phenomenon called wave motion, when oscillations in one place cause some nearby matter to oscillate, which causes other nearby matter to oscillate, and so on, and a disturbance moves through some continuous distribution of matter. It turns out that ordinary sound is an example of such wave propagation, so we will devote a whole chapter to the properties of sound.

Mechanical waves require some continuous distribution of matter on which to travel, so this is also the natural place to take a closer look at the properties of fluids. We will briefly look at static properties like pressure and buoyancy, as well as some simple characteristics of fluids in motion.

CHAPTER

10

Oscillating Motion

In this chapter, we will apply our knowledge of mechanics to a special kind of motion, simple harmonic oscillation. You may not recognize this term, but you frequently encounter this type of motion in your daily life. We will see what is required to produce and sustain this kind of motion, and learn about its key features. We will take a very general approach, because there are so many specific systems in the world which exhibit this kind of motion. Lots of these systems involve objects which distort elastically when forces are applied. These can usually be modeled well by thinking about simple springs that obey Hooke's Law, which we mentioned back in Chapter 5. Although we will strive to be general, we will restrict our analysis to oscillating motion that occurs in only one dimension.

One specific example we will look at is the simple pendulum, a mass that swings back and forth in a vertical plane under the influence of gravity. The regular repetition of the pendulum's motion has long played a role in timekeeping. We will see what is required for it to approximate simple harmonic motion, and what determines how fast the pendulum swings.

In This Chapter

- The basic parameters for oscillation
- Conditions required for simple harmonic motion
- How energy flows in simple harmonic motion
- The physics behind the pendulum

Period and Frequency

The general definition of *oscillating motion* is any motion that repeats itself at regular time intervals. It can be any specific motion at all, as long as that motion gets repeated. Oscillation is all around us, and even inside us. The contraction of your heart repeats a regular pattern, at regular intervals, at least while you are relaxed. Guitar strings vibrate, a clown bounces on a pogo stick, a dog wags its tail, etc.

For the kind of oscillations we want to discuss, we will assume that the time it takes to repeat the motion is the same every time around. That time is then a constant of the motion called the *period* of oscillation. (Sometimes oscillation is called periodic motion.) The period can be long or short, and should normally be measured in seconds. If you observe some oscillating motion and you want to know the period, just measure how long it takes for the motion to return to some starting state. That may be difficult to do in some cases, maybe because the motion is too quick. In such a case, you may be able to get a good measurement by timing some number of complete cycles. Then just divide the total time by the number of cycles that occurred.

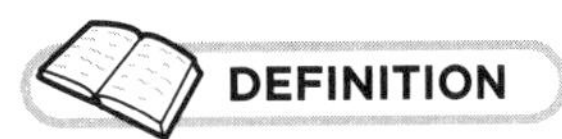

Oscillating motion is any motion that repeats itself at regular time intervals, typically in a back-and-forth fashion.

The **period** of oscillation T is the time it takes to complete one full cycle of the repeated motion. The standard unit for period is the second.

The **frequency** of an oscillator is the mathematical inverse of the period, $f = 1/T$. It represents the number of complete cycles of the motion which occur per unit of time. The standard unit for frequency is the cycle per second, or hertz (Hz).

The inverse of the period is another quantity that is often used to describe an oscillator. This quantity is called the *frequency*, labeled f. The frequency doesn't give you any information that you don't already know from the period; it's just another way to communicate the same information that is sometimes more convenient to use. It is used often enough that another unit has been created for it—cycles per second, or hertz (Hz). We should note that it is possible to have a frequency that is less than one hertz. That just means that a full cycle of the motion takes more than a second.

If you know the frequency of an oscillator, you can always calculate its period, and vice versa. Because of the inverse relationship, a very high frequency corresponds to a very short period of the motion, which means the motion itself is relatively fast. On the other hand, if the period is very long, that would be a case of low-frequency oscillation. In principle, there is no limit to how high or low the frequency can be unless there is some physical restriction on the motion.

CONNECTIONS

One branch of physics takes oscillating motion to an extreme, and in so doing may offer the most powerful interpretation of the universe yet derived. "String theory," as it is known, is actually an attempt to describe the four fundamental forces using a single, unified framework. It treats the smallest particles in the universe not as pointlike objects, but rather as tiny, one-dimensional entities called "strings." Depending on how the strings oscillate, they can take on different frequencies and therefore different energies, and even—thanks to the concept of mass-energy equivalence—the different masses observed in the laboratory.

Simple Harmonic Motion

Of the countless possibilities for oscillatory motion out there, there is one specific type that is particularly interesting for physicists (and physics teachers). Not only is it often seen in nature, it is simple enough to describe in some detail. We call this motion simple harmonic motion. It is a straightforward task to connect this type of motion to a physical cause, as we will now see.

There are just a few conditions required to allow simple harmonic motion to occur. The oscillating object must have mass, and the force acting on the mass must be a certain function of position in one dimension. That force must be zero at a location that will end up being the center of the motion. This location is also called the equilibrium position, because equilibrium is defined as a configuration where the net force is zero. At all other locations, the force must point toward that equilibrium position. Such a force is often called a "restoring force," because it acts to restore the system to equilibrium. What's more, away from the equilibrium position the force must be directly proportional to the distance from equilibrium. It is not necessary to have only a single force acting, but whatever the total force is, it must have these characteristics or the motion will not be simple harmonic motion.

In Chapter 5 we encountered a force that just so happens to meet these criteria, and that is the force exerted by a spring which has one end anchored, say, to a wall. If we line up our x-axis with the axis of the spring, and place the origin at the point where the spring has its natural length, then the force due to the spring is $F_s = -kx$. This force is zero when $x = 0$, and always points toward $x = 0$. Recall that k is a positive constant related to the strength of the spring. This means that whenever x is positive, the force points in the negative direction, whereas the force points in the positive direction whenever x is negative.

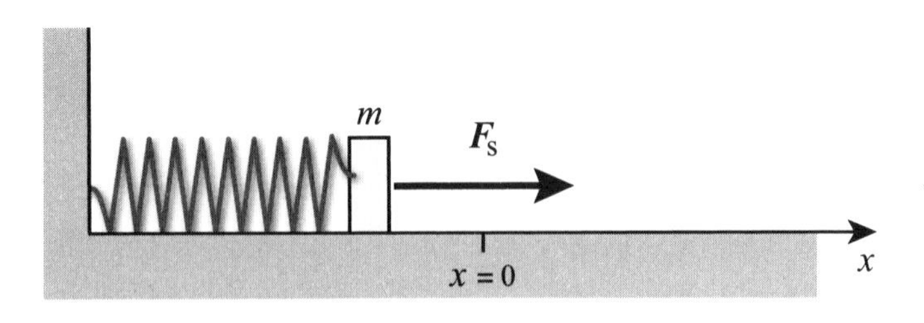

Figure 10.1

A mass moves on a frictionless surface attached to a spring, whose other end is fixed in place. That moment pictured here is some time after the mass was released. At this instant, the spring is compressed and the force exerted on the mass ($\mathbf{F}_s$) points in the positive direction.

Newton's second law and a little bit of math are all you need to derive a model for position vs. time that is consistent with a force like this. The result is either the sine or the cosine function from trigonometry. In general, it is possible to make either the sine or cosine fit a simple harmonic oscillator, but let's get a little more specific and pick the simplest form.

Let's return to our mass attached to an ideal spring, and assume it is free to move on a very smooth horizontal surface. The spring is also horizontal, fixed at the other end, and the mass only moves in a straight line along the axis of the spring. Imagine pulling the mass to a distance $x = A$ in the positive direction, and then releasing it. Start a clock so that the time $t = 0$ corresponds to the moment of release. The motion that results will be described by this equation: $x = A\cos(2\pi ft)$. If we use this form, f will automatically be the frequency of the resulting simple harmonic motion.

When we use a trigonometric function like this, the quantity in parentheses must be an angle, and we want to use our natural angle units, radians, to measure these angles. We assume that the time t is measured in seconds and f is in hertz (Hz). Since one hertz is one cycle per second, the product ft is then measured in cycles of the motion. The factor 2π must then represent the number of radians per cycle, to make the units work out.

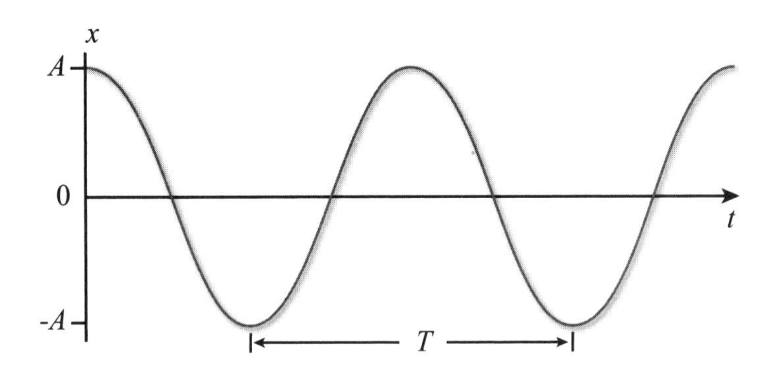

Figure 10.2

The position x as a function of time for simple harmonic motion, $x = A\cos(2\pi ft)$. The amplitude of the motion is A, while $T = 1/f$ is the period.

But what is this angle, $2\pi ft$? There is no geometric angle to see in the physical setup. All of the motion is along a single straight line. We call this quantity the *phase angle,* or sometimes just the phase of the motion. It is an abstraction, a purely mathematical quantity which allows us to model simple harmonic motion. For any given simple harmonic oscillator, the phase angle is a constant multiplied by the time, so it is a quantity that increases steadily as time goes on. As the argument of the cosine function, it determines exactly how x varies with time. The cosine function oscillates between -1 and 1, so x will vary between $-A$ and A (where A is another constant called the *amplitude* of the motion).

So 2π radians in the phase angle corresponds to one complete cycle of the motion. Starting at any time, a change in phase equal to 2π will bring you back to exactly the same location x. It is often useful (and will be for us in a few chapters) to know what the phase of the motion is at a given time or x value within the cycle.

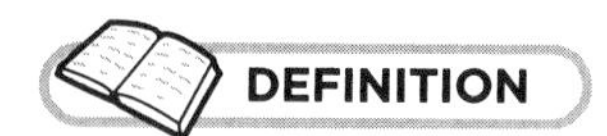

The **phase angle** in simple harmonic motion is the argument of the trigonometric function used to describe the position of the oscillating mass as a function of time. It is an imaginary angle equal to $2\pi ft$ and is measured in radians.

The **amplitude** of an oscillator is the maximum displacement from equilibrium.

The amplitude A is determined by how the motion is started. We started the mass at position $x = A$ at time $t = 0$. If you start the mass by simply releasing it (giving it no initial speed in either direction) then after one complete cycle of the motion, the mass will return to $x = A$, where it will stop momentarily and then reverse direction. It will never go to any larger values of x. Notice that this follows from the math and the quantities that we have defined. One full cycle takes a time equal to T; when $t = T$ the phase angle equals 2π radians. Halfway through the cycle, the phase will be π radians, and x will reach its most negative value, $-A$, before turning around and heading back toward zero.

For the same spring and mass, you can start anywhere you want (within reason) so A is determined by your choice. What determines the other constant, the frequency f? This parameter is different, and is determined by the strength of the spring and the mass of the oscillating object. The exact relationship turns out to be this:

$$f = \frac{1}{2\pi}\sqrt{\frac{k}{m}}$$ **Equation 10.1**

Recall that k is the force constant for the spring, and the stronger the spring the larger its value. We can see from Equation 10.1 that a very strong spring or a very small mass results in oscillations with a high frequency. Conversely, if the mass is very large or the spring is very weak, the oscillations will be slow, with a long period and low frequency. This should make sense to your physical intuition by now, just based on Newton's second law. The spring constant k also tells you how much force the spring exerts for a given displacement. A small force or a large mass means small acceleration, so it will take a while to get the mass moving toward equilibrium, then it will take a while to slow it down after it passes $x = 0$ so it can turn around and come back.

TRY IT YOURSELF

19. A 12 kg mass is attached to a spring with a spring constant of 800 N/m. If you pull the mass 21 cm from the equilibrium position and release it, what is the period of the resulting oscillation?

In order to understand simple harmonic motion, it's really important to understand the fundamentally different origins of f and A. It is also important to note that they are completely independent of one another. This is a bit of an idealization with respect to the real world, but in our model it is absolutely true. The frequency has no effect on the amplitude of the motion, and vice versa.

Energy in Simple Harmonic Motion

With no friction to worry about, we expect that the total mechanical energy of a simple harmonic oscillator will be conserved. This is reasonable because, in our example, the spring is able to store energy in the form of elastic potential energy. Let's take a closer look at what is going on during the motion.

Think back to when we first defined velocity and acceleration in one-dimensional motion (Chapter 2). Instantaneous velocity is the slope of the position vs. time graph at a given point in time. And by definition, the instantaneous acceleration is the slope of the velocity vs. time graph.

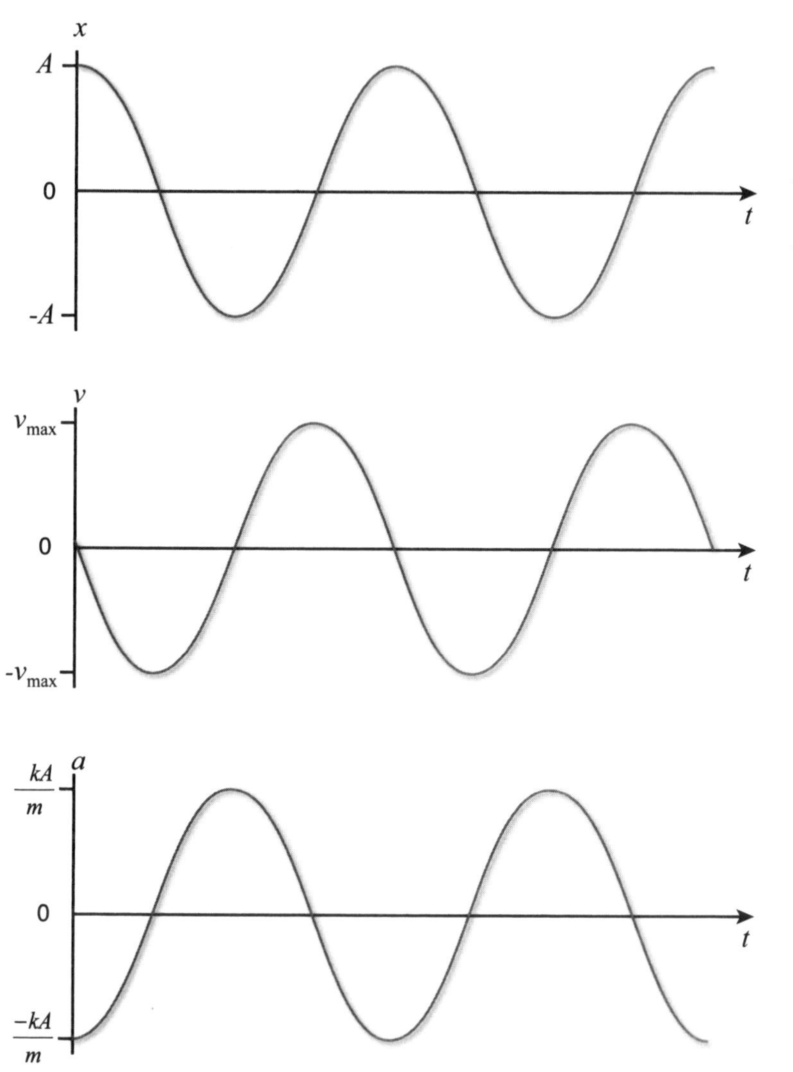

Figure 10.3

Position (top), velocity (middle), and acceleration (bottom) of a simple harmonic oscillator as functions of time. All three of these quantities look like sine or cosine functions with the same frequency.

Figure 10.3 shows all of these quantities for two full cycles of simple harmonic motion. These graphs show us many things, some of which confirm our qualitative observations. First of all, at the extremes of the motion when $x = \pm A$, the object is turning around and therefore changing its direction of motion. The velocity must be zero for just a moment whenever this happens. These points should also correspond to the maximum absolute value of the acceleration, since the force is proportional to how far x is from zero. Conversely, we would expect the acceleration to be zero at the center of the motion ($x = 0$, where the force is zero) and that the maximum speeds will also occur at $x = 0$.

You should also see from these figures that the acceleration is certainly not constant. None of our constant acceleration formulas will be of any use with simple harmonic motion. Fortunately, we don't need them. We already have a formula that gives us the exact position at every moment in time.

The maximum acceleration is easy to calculate, because of Newton's second law ($F = ma$) and Hooke's Law ($F = -kx$). Combining these and solving for the acceleration would give us $a = -\frac{k}{m}x$. The maximum positive acceleration will occur when x is at its most negative value, $-A$, so the maximum a will be kA/m. Thus we can say that the acceleration oscillates in a similar way to the position, but with an "amplitude" equal to kA/m.

In order to determine the maximum speed, let's look at energy conservation. We're going to work with the hypothesis that mechanical energy is conserved as our mass oscillates back and forth on the end of the spring. We know that there are two kinds of energy present: kinetic and potential. In Figure 10.3, we saw that the speed varies during the motion, so the kinetic energy must also be changing. We can still write it as $K = \frac{1}{2}mv^2$, but just need to remember that v is not a constant value. (Note that we have used uppercase K for kinetic energy, in order to distinguish it from the spring constant k.) What's more, we learned in Chapter 7 that a spring will store elastic potential energy any time it is stretched or compressed and that the spring potential energy is given by $U_s = \frac{1}{2}kx^2$. Here, too, the potential energy will vary with time.

So, as the mass travels back and forth, energy will transform between kinetic and potential energy. Importantly, though, the sum of these two remains constant since that is the total mechanical energy, which is conserved. How much total energy is there? To find out, consider that any time the position of the mass is at an extreme, at either $x = A$ or $x = -A$, we know that the velocity is briefly zero, so the kinetic energy must also be zero at these locations. Using the definition above, the spring potential energy is equal to $\frac{1}{2}kA^2$ at either of these locations. Since all of the energy is in the form of potential energy at these spots, the total energy must also be $E = \frac{1}{2}kA^2$.

Now consider a time when the mass passes through $x = 0$. At this location the spring is neither compressed nor extended, so the potential energy will be zero. All of the energy must therefore be in the form of kinetic energy. From Figure 10.3, we saw that this spot also corresponds to the maximum speed; therefore $\frac{1}{2}mv_{max}^2 = \frac{1}{2}kA^2$. From this we can use algebra to see that the maximum speed is $v_{max} = A\sqrt{k/m}$. The velocity oscillates between this value and its negative as the mass moves back and forth.

Figure 10.4 plots the kinetic energy K, the potential energy U_s, and the total energy $E = K + U_s$ as functions of the position x. Note that time is not visible on these graphs.

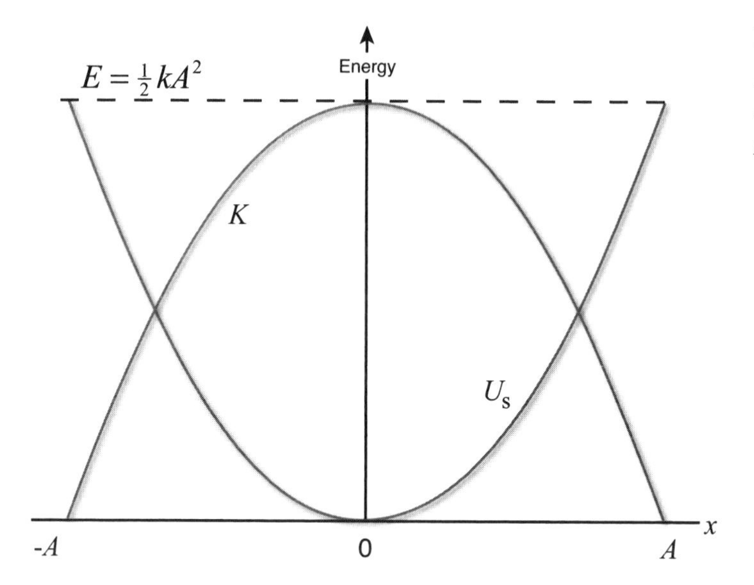

Figure 10.4

Kinetic, potential, and total energy for an object in simple harmonic motion, plotted against the position x.

TRY IT YOURSELF

20. A 12 kg mass is attached to a spring with a spring constant of 800 N/m. If you pull the mass 21 cm from the equilibrium position and release it, how fast will it be moving when it reaches $x = 0$?

In the real world, it is very difficult to set up an oscillator with no friction. So most any oscillation you observe will not continue with the same amplitude forever. The amplitude will gradually decrease along with the mechanical energy, until the oscillation ceases. How fast this happens depends on how much friction is present. This kind of motion can also be described mathematically, but it is quite a bit more complicated, so we will opt to move on to the next topic.

The Pendulum

Another classic example of oscillation occurs whenever a mass is suspended from a fixed point by a very light string that can't stretch. If the mass is displaced from hanging straight down and then released, it swings back and forth, repeating the same motion at regular time intervals. In this case, there is an obvious equilibrium location when the string is vertical and the mass is located directly below the support point. At that point there are no horizontal forces, and the mass could remain stationary. Does this situation also have a restoring force that is proportional to the displacement?

For this analysis, we don't want to worry about any rotational kinetic energy of the mass itself, so we will assume the size of the mass is very small, this time compared to the length of the string (L). We'll need a little trigonometry to see if we can make this motion fit our model of simple harmonic motion.

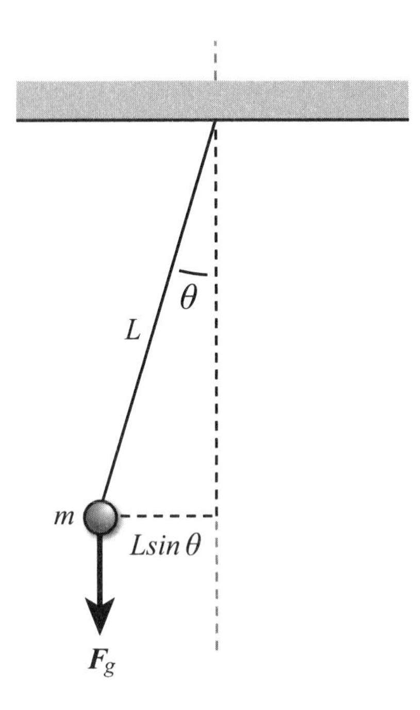

Figure 10.5

A simple pendulum is made of a small mass m *attached to a very light but unstretchable string of length* L.

Since we have assumed the string's length is fixed, the mass can only move along a path that is part of a circle. The trick is to think of the component of the weight force that is along this path (tangent). If you imagine the mass in motion, you will see that this force component always pushes it back toward equilibrium. As you can see from Figure 10.5, we now measure the angle of the string relative to the vertical, so that $\theta = 0$ also corresponds to the equilibrium configuration. If the mass is to the left (as in the figure), let's call those angles negative, so θ will be positive when it swings to the right of center.

Let's call the component of the weight force tangent to the path F_t. This is our candidate for a restoring force in this case. With our sign conventions, its magnitude is $F_t = -mg\sin(\theta)$, and its direction is always tangent to the circular path. (The positive direction is to the right for this force as well. In the figure, the negative θ means that $\sin(\theta)$ is also negative.) Here is where we employ a little trick. On the right side of this expression for F_t, we multiply and divide by the length of the string L:

$$F_t = -\left(\frac{mg}{L}\right)L\sin(\theta)$$

Equation 10.2

Now look at the last piece of this expression, $L\sin(\theta)$. This is exactly the horizontal distance of the mass relative to the equilibrium position, so we might as well call it x. Then Equation 10.2 looks just like Hooke's Law, if only (mg/L) plays the role of the spring constant k. Well, why not? The mass, the length of the string, and g are all constants. The important result is that we now have a restoring force whose strength is proportional to the horizontal distance from equilibrium. That's all we need for simple harmonic motion in the horizontal position x.

In order to figure out the frequency of this oscillation, all we need to do is go back to Equation 10.1. That expression tells us the frequency when you know the force constant. Just substitute mg/L for k, and you'll get:

$$f = \frac{1}{2\pi}\sqrt{\frac{g}{L}}$$

Equation 10.3

Often for the pendulum, we are more interested in the period of a complete back and forth swing. To find the period, all we need is the inverse of the frequency, so the period of a simple pendulum is:

$$T = 2\pi\sqrt{\frac{L}{g}}$$

Equation 10.4

We should confess that the tricks we used to come to this simple expression are only valid for a limited range of angles. But the expressions we derived for the period and frequency will be accurate enough as long as the angle never exceeds about 15 degrees on either side.

TRY IT YOURSELF

21. A simple pendulum can be a convenient way to measure the acceleration due to gravity, if the length of the string is known precisely. Let's say you take a pendulum, with length L equal to exactly half a meter, up to the top of a high mountain. You make the 250 g mass oscillate with a small angle amplitude. By careful timing, you observe that 10 oscillations take 14.3 seconds to complete. What is the value of g at this location?

The Least You Need to Know

- Oscillating motion is motion that repeats itself over a period of time T at a frequency $f = 1/T$.
- Simple harmonic motion is a special case of oscillating motion in one dimension, which requires a restoring force that is proportional to the displacement from center.
- For simple harmonic oscillators, the expressions for the position, velocity, and acceleration are sine and cosine functions of time.
- The frequency and period of a simple harmonic oscillator depends on the restoring force constant and the mass of the oscillator.
- A simple pendulum undergoes simple harmonic motion for small angle oscillations, and the period depends on the length of the string and the acceleration due to gravity (but not on the mass).

CHAPTER

11

Mechanical Waves

In This Chapter

- Definition and types of mechanical waves
- Conditions required for wave propagation
- What determines wave speed
- What happens when two waves combine

Waves are another example of a general physical phenomenon with a wide variety of specific realizations. In this chapter, we won't be discussing all of the different kinds of waves that are out there. Instead, we'll concentrate on just describing mechanical waves. Nevertheless, a lot of the descriptions and terms we will use for mechanical waves also apply to other kinds of waves, so this material will come in handy again in Part 5 of this book.

The best way to think about wave motion is to realize that what is moving is actually a pattern of some sort. The pattern may be a certain configuration of matter, or a mathematical function that describes how a property varies with position. It is not like the motion of a projectile, where some chunk of stuff moves from one place to another. The pattern that moves during wave motion is rather a kind of disturbance, a displacement away from a smooth uniform background state that exists when there is no wave.

This is one place where a book, with its words and static diagrams, is not the best medium with which to give a good account of the phenomena. Waves are inherently dynamic, always moving. If this topic is new to you, it might be wise to also explore some animated illustrations of wave motion, found on the websites listed in Appendix D. Here we will make sure to give you all the tools you'll need for analyzing mechanical wave motion. We'll also provide some useful keywords with which to search other resources—in particular, to better visualize the dynamic nature of waves.

Different Kinds of Waves

Among the wide variety of waves that exist in nature, there is one really major division we should be clear about right up front. The distinction comes from what makes up the background, what is actually being disturbed as the wave goes by. For a lot of the most familiar waves (e.g., ripples on a pond, waves on the ocean, seismic waves, or even sound) the wave is a disturbance in the arrangement of matter itself. We say that a *mechanical wave* is supported by a *medium*, some continuous distribution of mass. This medium could be a solid, liquid, or gas. But for this category of waves, the presence of some mass is essential. All of the examples we will study in this chapter and the next fall into the category of mechanical waves.

Mechanical waves are disturbances in some material medium that travel over time, such as ripples on a pond.

A **medium** is the background distribution of matter (e.g., water or air) that gets disturbed when a mechanical wave travels. The medium must possess mass and be connected in some way, so that one piece of the medium can influence adjacent pieces.

It is also possible for a certain kind of wave to propagate without any material medium at all. These waves are actually patterns in the strength of electric and magnetic fields, which can exist in space and are completely independent of mass. These are called electromagnetic waves, which manifest themselves as visible light, radio waves, microwaves, x-rays, etc. As you might imagine, these waves are fundamentally different from mechanical waves. We will cover electromagnetic waves in a later chapter.

Even if we confine ourselves to mechanical waves, there are still many kinds of waves to discuss. Mechanical waves can be distinguished by their frequencies, their shapes, their intensities, or by the kinds of media in which they travel. One very useful way to separate different kinds of waves is by the direction of the displacement relative to the direction the wave is traveling. If the mass gets displaced in the same direction as the wave is moving, we call it a *longitudinal wave.* On the other hand, if the mass moves at right angles to the direction of travel, that is called a transverse wave. *Transverse waves* on a one-dimensional medium, like a string, are the easiest kind of waves to visualize, so we will start our detailed study by looking at this example.

Longitudinal waves are mechanical waves in which the parts of the medium undergo small back-and-forth movements in a direction parallel to the travel of the wave.

Transverse waves are mechanical waves in which the motion of the medium is perpendicular to the direction in which the waves move.

It turns out that in certain media, such as the surface of a liquid, you can get waves that are a combination of longitudinal and transverse. In fact, the waves you have often seen on bodies of water, both large and small, are waves of this type. It is difficult enough to analyze transverse or longitudinal waves separately; we will not even try to analyze such complex combinations of wave types.

Transverse Traveling Waves

Imagine a rope on the smooth slippery floor of a very long hallway. The rope is attached to the wall at the far end, and you pull gently on the free end, so that the rope has a little tension in it. If you hold the rope still, it lies on the floor in a straight line from your hand to the point where it is attached to the far wall. Now imagine what happens if you quickly jerk your hand to the side and back. A "bump" will travel along the rope away from your hand. In this example, the rope is the medium on which the wave is traveling.

TRICKS AND HACKS

If you want to actually try some of these demos for yourself, one of the best things to use is an extra-long coiled cord from an old telephone handset. Sometimes you can find these at garage sales or thrift stores. They have a good combination of flexibility and mass per unit length.

A more scientific term for that bump is "pulse." The shape of the pulse depends on exactly how you moved your hand. The specific shape of the pulse is not so important, but that shape is what travels down the rope. As the disturbance passes by a certain spot on the rope, the piece of the rope at that spot moves up and down, away from the rope's equilibrium position, mimicking the motion of your hand or whatever generated the pulse in the first place. The motion of each individual piece of rope is at right angles to the undisturbed rope, while the pulse itself moves along the rope. This is why such a wave is called "transverse."

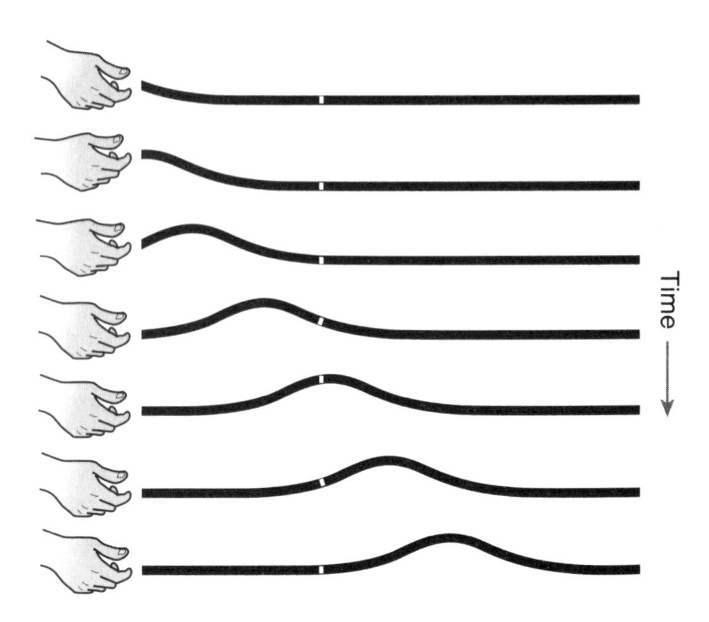

Figure 11.1

These "snapshots" of a rope at seven successive times show a pulse traveling along the rope due to a disturbance at one end. Note the motion of the white-tagged piece of rope as the pulse goes by.

Figure 11.1 shows the progression of a pulse as it travels along a rope due to motion of a hand at one end. At the top, the hand has just started moving up, and most of the rope is still at equilibrium. In the last four frames, the hand has stopped moving, but the pulse continues to travel to the right.

The reason that a pulse travels is that each "piece" of the rope is attached to other pieces on either side, so that each piece can exert a force on the piece next to it. Each piece also has its own equilibrium position, which is the position it takes when the rope is stretched along a straight line. When your hand moves the end of the rope away from equilibrium, the end pulls the next bit of the rope sideways, so that it will follow the position of the end a moment later. That piece causes the next piece to move, and so on.

We should note right away that this pulse does not come for free. If your hand is the thing that generates the pulse, you have to exert a force on the rope as you move it. When a force acts over a distance, we know that means work is being done. Work done on the end of the rope means that energy was transferred to the rope. This is a standard feature of wave phenomena. Waves of all kinds transport energy from one place to another without transporting matter. The pulse, once created, carries a certain amount of energy with it.

The rate at which the disturbance travels down the rope is what we call the *wave speed* (v). It is not the same as the transverse speed of any piece of the rope. What determines the speed of pulses on the rope? It is basically the delay in the response of one piece to the motion of the piece before it. To get a feel for this, we can again use Newton's second law of motion. The force of one piece on the next is basically what we call the tension in the rope. When one piece moves, that force will cause the adjacent piece of rope to accelerate. But that rate of acceleration will also depend on how much mass the piece of rope has (or, more precisely, mass per unit length). We would think that a greater tension would lead to a greater acceleration, which would lead us to expect faster wave motion. On the other hand, if the mass is greater, we would expect the wave to be slower.

DEFINITION

Wave speed is the speed at which the pattern of disturbance travels in a medium. It is not the actual speed of motion of any piece of the medium.

Let's use the Greek letter μ to stand for the mass per unit length of the rope. If the magnitude of the tension force in the rope is F_T, then the wave speed turns out to be:

$$v = \sqrt{\frac{F_T}{\mu}} \qquad \textbf{Equation 11.1}$$

While we're still talking about pulses, it will be useful to consider what happens to a pulse when it gets to the end of its rope. Whenever a pulse reaches the end of the rope, it will be reflected, meaning it turns around and heads back in the opposite direction. If the end of the rope is fixed, the reflected pulse will have the same shape but be inverted (that is, on the opposite side as the initial pulse). This is because of the reaction force exerted by the wall on the rope when the pulse tries to exert a force on the wall. If the end of the rope is loose, then the pulse gets reflected without being inverted. Either way, the pulse continues to carry the energy it was given initially.

TRY IT YOURSELF

22. 100 m of a certain rope has a mass of 12 kg. If that rope is stretched under a tension of 800 N, how long will it take a wave pulse to travel from one end to the other?

All of this analysis has assumed no friction forces are acting as the pulse moves. This is difficult to achieve in any real situation. If you try this on a smooth floor, some friction will oppose the transverse motion of the rope, and the pulse will die out as it travels. You can also try it with the rope suspended in the air, which will decrease the energy loss, but then you will have to deal with the fact that the rope is not straight, but sags under its own weight.

Sinusoidal Waves

Let's now consider a case where we generate more than a single pulse on the medium that is our rope. A most interesting kind of wave results when the hand holding one end of the rope moves like a simple harmonic oscillator, as discussed in Chapter 10. This oscillation would take constant effort, because it would be putting energy into the rope continuously.

We assume that this oscillation has a small enough amplitude that it doesn't change the tension in the rope very much. We do this so that we can consider the speed of the waves on the rope to be a constant. Then, since the disturbance propagates away from your hand at a constant rate, we know exactly what shape the rope will take. The moving pattern will look like a train of positive and negative pulses, all of which look exactly the same.

In order to get something we can describe mathematically, we need to make several simplifications, some of which we have already described. We still want to ignore the effects of any frictional forces. The wave speed must be constant all along the rope. The rope must have some constant mass per unit length, but we also want to ignore any effects of the force of gravity. The only force acting on the rope is the tension acting from one piece to the next. Finally, we don't want to deal with reflections, so we either assume that the rope is infinitely long, or that the energy of the wave is somehow absorbed suddenly and completely at the far end of the rope.

With all of this in mind, after the source has been oscillating for some time, there will be a pattern in the shape of the rope that looks exactly like a trigonometric sine function. This shape will continuously travel away from the source of the disturbance at the speed we gave previously (Equation 11.1), which you'll recall was determined by the tension in and the mass per length of the rope.

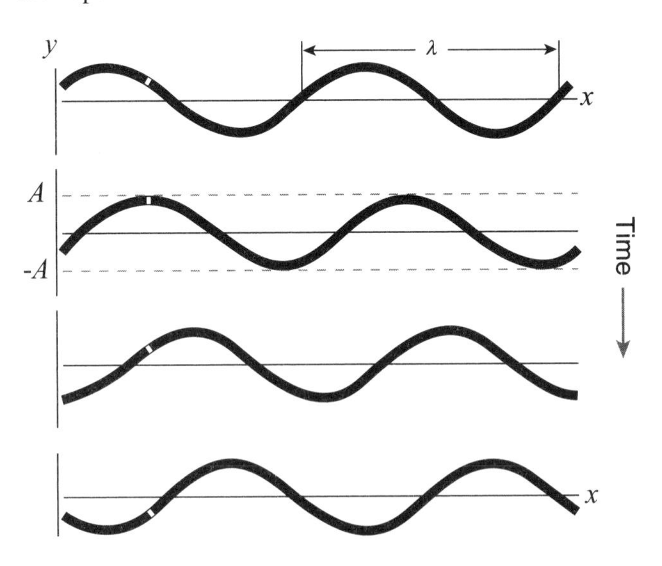

Figure 11.2

This series of snapshots shows the shape of a sinusoidal transverse wave in a rope at several successive time intervals. Note again that only the shape moves, not the whole rope.

A figure like this bears a strong resemblance to figures we used in the previous chapter to illustrate simple harmonic motion. Although sinusoidal waves and simple harmonic motion are related, their motions are completely different. The key is to take note of the axes on the plot (which you should always do for any graph you see). These figures plot two positions, y vs. x, while the graphs demonstrating simple harmonic motion had time on the horizontal axis.

Because the shape looks like a sine function, this type of wave is called a sinusoidal wave. As in Figure 11.2, we will place an x-axis along the length of the rope. At a right angle to the rope, the y-axis will then measure the transverse displacement of any piece of the rope. The equilibrium position of any piece will correspond to $y = 0$. As the wave travels down the rope, any given piece of the rope (at some location x) will oscillate through positive and negative values of y. The amplitude of this oscillation is labeled A, so at any point, y varies between A and $-A$ as time goes on.

With all of this information, we can now construct a mathematical formula that will describe the displacement y at any location x along the rope and at any time t:

$$y = A\sin\left[2\pi\left(\frac{x}{\lambda} - ft\right)\right]$$ **Equation 11.2**

We have already defined the three variables, x, y, and t. This expression gives y as a function of the other two variables, x and t. The other symbols are constants for a given wave. The amplitude of the transverse motion is labeled A, as stated above. The Greek letter λ (lambda) is the wavelength of the wave, as illustrated in Figure 11.2. It is the minimum distance over which the pattern repeats itself. It can also be described as the distance between any two adjacent peaks of the wave form, but it doesn't have to be measured at any specific location. Pick any point on the wave (frozen in time) and the *wavelength* is the distance to the nearest point where the wave is exactly the same.

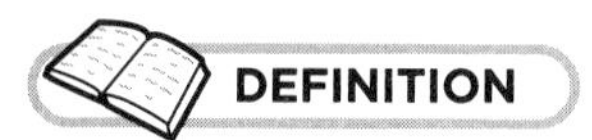

Wavelength is a distance between two successive peaks (or troughs) of any sinusoidal wave. A sinusoidal wave displaced by one wavelength along the direction of motion will look exactly the same as the original wave.

The last symbol we have to define is f. This is called the frequency of the wave, and it has the same definition as it had when we discussed oscillations. Recall that the wave in this case is generated by some simple harmonic motion at one end of the rope. This motion is periodic, with a frequency f (and therefore a period $= 1/f$). Every piece of the rope, at every location x, will undergo simple harmonic motion with exactly this same frequency and with an amplitude equal to A.

An important concept in wave dynamics is something known as the phase angle. This is just a fancy term for the quantity that appears in the square brackets in Equation 11.2. Since this is what feeds into the sine function, it must be an angle and it is measured in radians. But just as in the previous chapter, it is not a physical angle between any geometric lines. The form we have presented here is valid for a wave that moves in the positive x direction. You might as well think of this in our conventional way, where positive x is to the right in our figure.

It turns out that the minus sign in Equation 11.2 is critically related to the direction of the wave motion. Since it is the shape of the wave that is moving, we can look at what happens to the phase angle to see which way (and how fast) the wave will move. Since the wavelength λ is a constant, at any fixed time, the phase angle advances as you move to the right (increasing x). At a fixed location, the minus sign tells us that phase angle decreases steadily as time goes on. Put these ideas together, and you can see that to maintain a constant value of the phase, you have to move to the right as time goes on. If the negative sign were positive instead, then the wave would be moving to the left (in the negative x direction).

Since the pattern repeats in both space and time (x and t), we can write down another expression for the speed in terms of the constants that appear in Equation 11.2. Let's call the time it takes to repeat the motion of each piece the period, which we defined in Chapter 10 to be $T = 1/f$. The distance over which the wave repeats itself is λ, so the wave speed must be $v = \lambda/T = \lambda f$. This does not contradict what we said earlier about the wave speed. That speed is still determined by the tension and the mass divided by the length of the rope. But now we know that for a given physical situation, the wavelength and frequency are not independent of each other. With the speed determined by the physical characteristics, driving the rope at a certain frequency will result in waves with a definite wavelength. The amplitude, on the other hand, is independent of the frequency, wavelength, and wave speed.

TRY IT YOURSELF

23. Consider the same rope as in the previous exercise, 100 m long with a mass of 12 kg and 800 N of tension force. A sinusoidal wave with amplitude 2 cm and frequency 5 Hz is traveling along the rope. What is the wavelength of the waves?

Longitudinal Waves

Now let's look at the other kind of mechanical wave, where the oscillation of the medium is not transverse to the direction of the wave motion, but rather parallel to it. These so-called longitudinal waves can be either a single pulse, or a continuous pulse train, just like transverse waves. But we need a different sort of medium than a rope under tension in order to illustrate this case.

The classic example used most often to illustrate longitudinal waves is a long spring that can be either stretched or compressed. The equilibrium state is when all parts of this long spring are relaxed, neither stretched nor compressed. If one end is fixed, you can start a longitudinal pulse from the other end by giving one end a quick shove and then pulling it back to the equilibrium position. That will cause a local area of compression to propagate along the length of the spring.

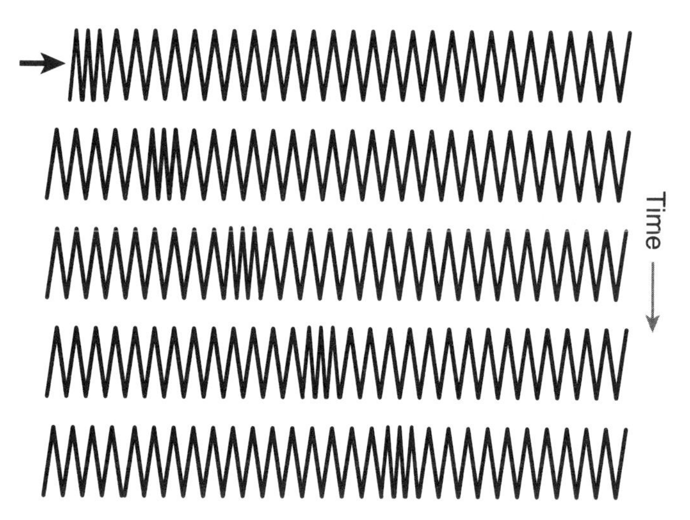

Figure 11.3

A quick shove on the spring at the left end causes a small region of compression to travel to the right.

A similar situation can be created with air (or any gas) confined to a tube, with a moveable piston on one end. Gases are compressible, in that they can be squeezed into a smaller volume if acted on by forces in the right way. An important quantity when considering this type of wave is density. We will explore this in great detail in Chapter 13, but for now it suffices to say that density is basically the total mass of gas particles per unit of volume. So, if a certain mass of gas is compressed to a smaller volume, it will have a greater density.

Consider now a long, gas-filled tube into one end of which we've installed an air-tight piston. In the equilibrium state, the density of the gas is the same everywhere along the tube. By applying a force to the piston, the piston is quickly moved a short distance into the tube and back. Right in front of the piston in the tube, the gas is bunched up, or squeezed a little bit, so that there is a small region of higher density. From that time on, the gas in the denser region wants to expand, so it pushes and compresses the gas next to it. As a result, that small region of higher density moves down the tube.

No portion of the gas moves very far down the tube. Instead, each portion of the gas moves back and forth a small amount as the disturbance goes by, in the same direction and amount as the piston moved, just delayed by some amount of time. These movements are qualitatively similar to the movements of the spring we considered a moment ago, and a region where the coils of the spring are closer together is analogous to a region of higher-than-average gas density.

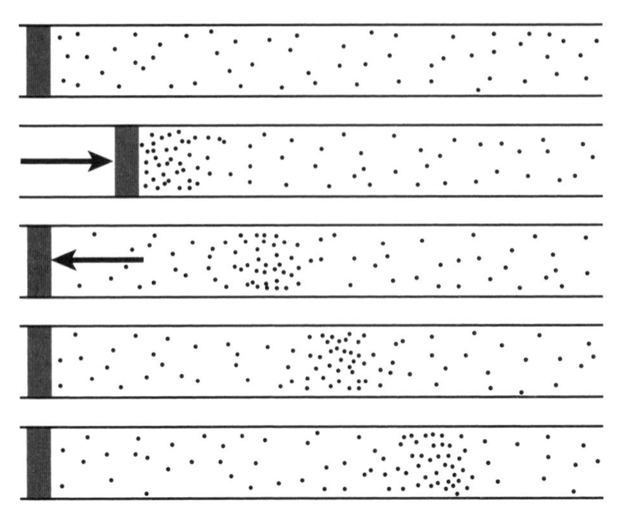

Figure 11.4

A quick shove on a piston into a gas-filled tube causes a small region of increased density to travel to the right.

These pulses are the same as transverse pulses on a rope in many respects. The main difference is the direction of the small movements of the medium. Pieces of rope move perpendicular to the wave motion, while pieces of spring or portions of gas move parallel to the wave. This is intimately connected with the forces between adjacent parts of the respective media. Some kind of force is necessary for the transmission of any wave. In the case of the rope, the tension is able to exert a force that is perpendicular to the rope. In the tube, one portion of the gas can only push on the gas next to it, so it can only move it in a direction parallel to the wave. Since there can't be any transverse force from one portion of the gas to another, there can't be any transverse displacement. In the case of a spring, one piece may be able to act on the next in different directions, but the main direction of the spring force is longitudinal.

If we again align an x-axis along the spring or the length of the gas-filled tube, we can see the similarities between longitudinal and transverse waves very easily. Imagine making a plot of density as a function of x. In the case of the traveling longitudinal pulse, snapshots of this plot would look exactly like Figure 11.1. With density plotted on a vertical axis, the pulse of high density moves along the medium with a certain wave speed.

By repeatedly shaking the end of the spring, or making the piston move back and forth continuously, we can create a repeating pattern of high and low density that travels along either medium. If the spring shaking or piston moving is simple harmonic motion, then the density profile also looks like a sinusoidal function (except that it is centered on the average density instead of about zero, since density cannot be negative). Just as with transverse waves, the density function will have a wavelength, the oscillations will occur with the same frequency as the driving force, and the frequency and wavelength will be related by the wave speed in the same way.

Superposition and Interference

As we saw in our sinusoidal examples, the moving shape of a wave can be described by a function of both position and time. Let's go back to thinking about transverse waves, where the function quantifies the transverse displacement. For sinusoidal waves, the transverse displacement goes positive and negative relative to the equilibrium position (where equilibrium corresponds to zero displacement).

One really cool thing about waves is that it's easy to determine what happens when more than one wave appears on the same medium at the same time. The transverse displacements due to the different waves simply add at all places and at all times. Physicists give this simple behavior a fancy name: the *principle of superposition.*

The **principle of superposition** tells us that the net effect of multiple waves on the same medium is determined by the simple sum of the displacements from equilibrium.

One consequence of this is that two pulses moving in opposite directions on the same medium will appear to move right through each other completely unchanged. If the transverse pulses are in the same direction, it will look like they fuse into one big pulse for a moment when they come together, but they will just continue to move apart after that, the speed of each pulse is determined by the tension and mass per unit length of the rope. Even more interesting, when the pulses are equal in magnitude but on opposite sides of the rope, they will cancel each other out during the moment that they overlap, but then will miraculously reappear and move apart as if nothing happened to either pulse.

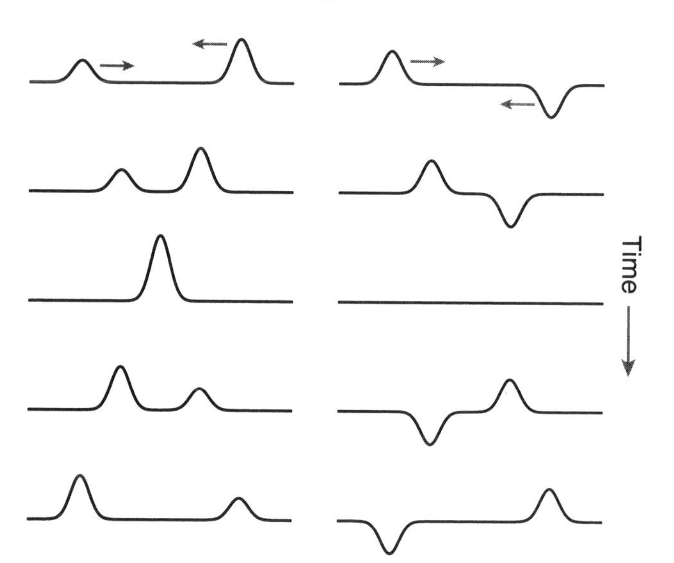

Figure 11.5

Here are two examples of transverse pulses moving through each other. Two positive pulses of different magnitudes (left), and two pulses of opposite sign but equal magnitudes (right).

To really drive this home, we consider two sinusoidal waves with the same amplitude moving in the same direction on one section of rope. In order to see all of the possibilities, we must allow there to be an adjustable offset between one wave and the other that is caused by starting one wave earlier than the other. (This would also correspond to a phase difference between the waves.) If the offset is small, then the peaks of one wave will be nearly aligned with those of the other. Since the transverse displacements add, the resulting wave will have nearly twice the amplitude of either wave that we started with. If the offset were exactly zero, then the combined wave would have exactly twice the amplitude of the individual waves. Whenever the combined wave is larger than the component waves, we have what is known as *constructive interference.*

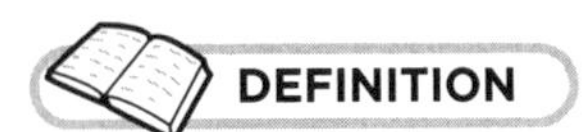

Constructive interference occurs when the superposition of two waves on the same medium results in a wave with a larger magnitude than either of the waves being added.

Destructive interference occurs when the superposition of two waves results in a wave with a smaller magnitude than the waves being added.

On the other hand, the offset could be about half a wavelength, such that positive peaks of one wave nearly line up with the negative peaks of the other. The result in this case would be a wave that is very small. If the offset between the waves is exactly half a wavelength, the combined wave has exactly zero amplitude. Any time the combined wave is smaller than the component waves, we have what is known as *destructive interference.* A whole range of resulting amplitudes is possible, from zero to twice the original amplitude, depending on how great the offset between the waves to be added is.

Standing Waves

There is one more important example of interference that is worth mentioning. It occurs naturally in media that have fixed boundaries (which leads to reflections, as we discussed earlier). Because of these reflections, it is easy to get waves that are moving in both directions on the same medium. So that's the setup we will use to illustrate this case: waves traveling in opposite directions on the same medium (unlike the example in the preceding paragraphs, where the waves were both traveling in the same direction).

We'll keep it simple, and consider two sinusoidal, transverse waves with the same amplitude and the same wavelength, moving slowly in opposite directions on the same rope. At some time, the peaks exactly line up, and the result of the superposition of the two waves will be a wavelike pattern that has twice the amplitude of the individual waves. A short time later, however, as the waves continue to move, the positive peaks of one wave will line up with the negative peaks of the other. At that time, the amplitude will sum to zero; the whole rope will be at $y = 0$.

It turns out that in this scenario, there can be some locations on the rope that never move at all. This happens whenever the length of the rope is at least as long as a wavelength. These locations, called *nodes*, are evenly spaced along the rope, half a wavelength apart. At all other locations the rope oscillates with an amplitude that depends on the position x. The net result is a pattern of oscillation that doesn't move along the rope. It "stands" in place, and hence it is called a standing wave.

A **node** in a standing wave pattern is a location where destructive interference results in no motion.

The specific standing waves that can occur on a medium depend on what we call boundary conditions. In the case of a rope, it matters how far apart the ends of the rope are, and whether the ends are fixed or loose. Let's say the length of the rope is L and that both ends are fixed in place and unable to move. In that case, the ends themselves would have to be nodes. This leads to a series of possible wavelengths, sometimes referred to as "resonant wavelengths," for which a whole number of half waves fit exactly between the two fixed ends. The longest of these would have $\lambda = 2L$ (only two nodes, one at either end). Shorter resonant wavelengths would have more nodes evenly spaced between the rope ends, and each successively shorter wavelength would have one more node. Moreover, for each resonant wavelength there is a certain frequency with which the rope vibrates up and down; these are called "resonant frequencies" of the system.

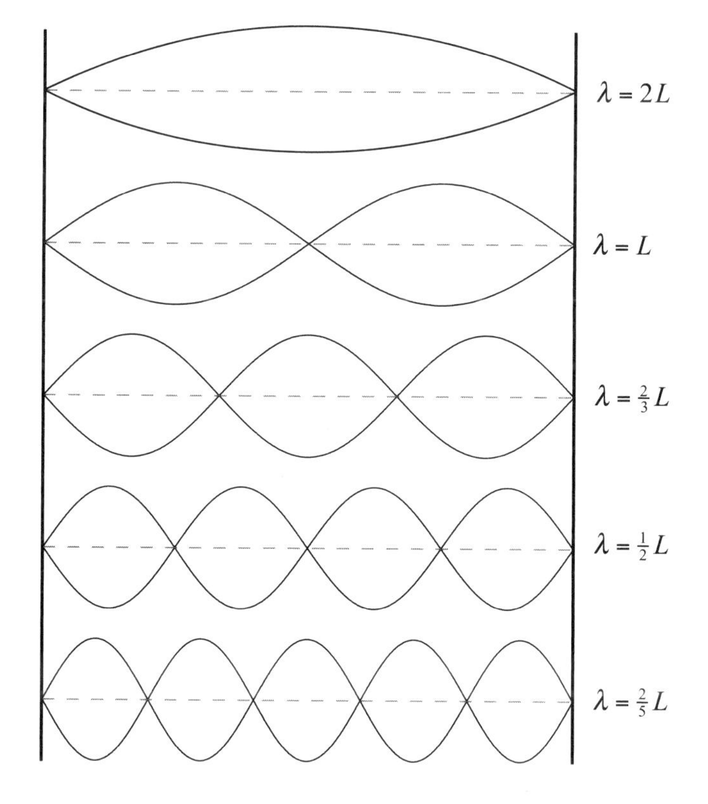

Figure 11.6

The first five possible standing wave patterns on a rope of length L *with fixed ends. In each case, we are only showing the limits of the motion.*

The top pattern in Figure 11.6 is the standing wave with the longest possible wavelength. In each case, λ is the wavelength of the component waves that are being reflected back and forth from each end.

Notice that halfway between every pair of nodes is a location where the amplitude of oscillation is at a maximum. Some books will call these locations "antinodes." It is convenient to label the standing waves by the number of antinodes they have, because then the number starts with one and increases by one for every successive pattern. The first standing wave, the one with the longest wavelength, is often called the "fundamental" frequency at which the string can vibrate. Because of the inverse relationship between frequency and wavelength, all of the other standing waves have higher frequencies than the fundamental. These other standing waves are also called "harmonics."

CONNECTIONS

On April 16, 1850, a 100-meter-long suspension bridge located in Angers, France, collapsed beneath a crossing French battalion, killing more than 200 soldiers. The source of this catastrophe was not a hostile enemy—it was the laws of physics. Just like our string fixed at both ends, a suspension bridge will have a set of resonant frequencies. If an external force were applied to the bridge at or near one of its resonant frequencies, it can induce significant standing waves along the structure. In the Angers case, that external force was caused in part by the soldiers' march. This created a large standing wave which had sufficient amplitude to bring the bridge crashing down.

TRY IT YOURSELF

24. A string that is 3.5 m long supports a standing wave with two nodes between the fixed ends (i.e., four nodes total). What is the wavelength of the waves which make up the standing wave pattern?

The Least You Need to Know

- When a disturbance propagates through a material medium, it is called mechanical wave motion.
- For transverse waves, the motion of the disturbance is perpendicular to the wave's motion, while for longitudinal waves it is parallel.
- The speed of mechanical waves is determined by inertia and the forces between adjacent parts of the medium.
- Displacements (positive or negative) of transverse waves get added when multiple waves occur on the same medium at the same time.
- Superposition of waves can lead to interesting interference phenomena, including standing waves.

CHAPTER
12

Sound

In This Chapter

- The true nature of sound
- Characteristics of sound waves
- How interference produces music

For the vast majority of us, the phenomenon called sound is a very familiar one. We hear a wide variety of sounds and noises every hour of every day. We even know how to make our own sounds, with and without the help of various tools. We have learned to use sound for our own purposes, from communication to entertainment. We have now reached the point in our study of physics where we can explore the scientific nature of sound.

In this chapter, we will explain how the sounds you hear are actually waves. Because these waves usually travel in the transparent air, you can't see them happening. We will try to bridge the gap between your experiences and the underlying wave nature of sound. We will be able to account for all of the familiar characteristics of sound, and maybe even explain some more puzzling observations.

What Is Sound?

It is quite remarkable how important sounds of all kinds have been, and continue to be, in the human experience. From ancient times, natural sounds warned us of danger—from falling branches to approaching storms to lurking predators—and helped us find food in the form of animals we hunted. We gradually learned to communicate using sound, first by simple noises, then with more complex and sophisticated sounds formed by our vocal cords, tongue, and mouth, which became spoken language. An experienced mechanic can tell what's wrong with a car or other machinery just by listening to the sounds it makes. Doctors and nurses listen to the various sounds in our bodies to make sure everything is working well. We even use sound to locate objects and reconstruct their shapes in cases where vision can't do the job. And besides all of the practical uses of sound, humanity has created an endless variety of musical sounds through the ages, for the pure pleasure they give us in hearing them. Along the way, we have devised numerous ways to record and reproduce sound, to the point where we can now digitize sound information, modify it, and even create sound from scratch with computers and electronics.

The phenomenon we know as sound occurs when longitudinal waves travel in any continuous distribution of matter. This very general, scientific definition actually goes a bit beyond what people commonly regard as sound. It includes sounds that can't possibly be heard by human beings. There are some sounds that are outside the range of frequencies humans can detect, and other sounds that are simply too quiet. Also, for us to hear a sound, it has to be carried by a fluid medium (usually air) that can make contact with our eardrums. Sound waves are able to travel in solids as well as fluids, but we can't hear them unless they subsequently get transferred to the air that is in our auditory canal.

CONNECTIONS

The high-pitched growl of a TIE Fighter from Star Wars is immediately recognizable to any science fiction fan. But what would it sound like in real life? The answer is nothing at all. The longitudinal waves that carry sound require a medium, such as air, in order to travel. Outer space lacks this medium, which means there is no sound at all. The universe, which is "filled" almost entirely by empty space, is thus a remarkably quiet place.

Let's take a closer look at these longitudinal waves that make up sound. As we discussed in the last chapter, such waves are a pattern of disturbance in a medium, a pattern that travels, carrying energy from one place to another. It takes energy to create sound in the first place. The molecules in the medium undergo small oscillations back and forth in the direction of the traveling wave. When we mentioned longitudinal waves in the previous chapter, we talked about air as the medium, and we confined the air to a tube so we could keep the waves moving in only one direction. The reality is that sound waves can travel in solid, liquid, or gaseous media, and that in general they can travel in all directions at once.

We have a choice of how to represent or think about these waves. One option is to visualize them as small variations in the density of the material, as we did in the previous chapter. The medium has an average density, but as a sound wave goes by, the local density increases and then decreases, swinging above and below the average density. Alternatively, we can think of sound waves as variations in pressure above and below the average pressure. This picture applies best to gases, and will make more sense when we talk about the characteristics of fluids in the next chapter. A third way is to describe the longitudinal waves of sound as the actual oscillations in the displacement of the molecules.

No matter which representation you choose, the actual disturbance can take almost any shape at all. Because of the principle of superposition, any actual shape can be constructed by adding together some number of sinusoidal waves with different frequencies and amplitudes. So we can explain most of the important characteristics of sound by discussing simple sinusoidal waves again.

Sinusoidal sound waves share the same characteristics as other waves, i.e., amplitude, frequency, wavelength, and wave speed. The speed of sound depends on the medium it is traveling in. Detailed study of the properties of a medium can tell us what the speed should be, but you can also just look up the speed of sound in reference tables, for just about any interesting case. Qualitatively, the speed is determined in a similar way to the speed of transverse waves on a rope or string under tension. In general, if the medium has more mass per volume (i.e., higher density), then you would expect the speed to be slower. And if the medium is stiffer, meaning more force acting between adjacent parts, that would lead to faster wave speeds. This latter factor tends to win out when talking about liquids and solids relative to gases. The speed of sound in air at room temperature and standard pressure is about 345 m/s. It is considerably faster than this in nearly all solids and liquids.

TRICKS AND HACKS

A good rule of thumb to remember is that the speed of sound under usual conditions (345 m/s) corresponds to about 770 MPH, such that it takes about five seconds for sound to travel one mile. This can help you gauge the distance from a source of lightning, as you will see the lightning before hearing the thunder, and the time difference is almost entirely the travel time of the sound.

As with other sinusoidal waves, the frequency and wavelength of sound are related to the speed at which the waves are traveling. As we saw in the previous chapter, the relationship is fairly simple: $v = \lambda f$. Thus, for waves traveling in a specific medium, the wave speed is well defined and constant. Any specific frequency will be associated with a definite wavelength. Longer wavelength sounds correspond to lower frequencies, and higher frequency sounds are associated with shorter wavelengths.

TRY IT YOURSELF

25. A certain musical note produced by a flute has a frequency of 1,100 Hz. What is the wavelength of these sound waves in air under normal conditions?

When it comes to our sense of hearing, we can't directly perceive the wavelength of sounds. Instead, our brains interpret frequency as pitch in the musical sense. High-pitched sounds (whistles, the tinkling of small bells, the high notes on a piano) correspond to the higher frequencies, while low bass notes and rumbles and growls are lower in frequency.

The range of frequencies of sound forms a continuous spectrum, theoretically limited only at the low end (at zero, because there is no such thing as a negative frequency). But human hearing only works for a limited range of frequencies. The exact limits are different for different individuals, but in general, people can't hear sounds with frequencies below 20 to 30 Hz, or greater than about 20,000 Hz. Not only does the frequency range of hearing vary from person to person, but many animals have significantly different frequency ranges over which they are able to hear sounds.

The last characteristic of sound waves we need to discuss is amplitude. As you recall, amplitude is the term we used for how far above or below average the pressure or density of the medium varies as a sinusoidal sound wave goes by. It can also be the distance that molecules actually travel away from their equilibrium positions. In any case, the greater the amplitude, the louder the sound. The way we perceive the loudness of sound is not so straightforward, however, and we can't say that the loudness or volume of a sound is directly proportional to any of the amplitudes. We'll explain why this is the case in the next section.

Sound Intensity

The most common way of quantifying the strength of sound waves is by considering the energy that they carry. It's that energy getting transferred to the eardrum which enables the sound to be detected by the nerves in your inner ear. In fact, it is the rate of the energy transfer that really matters, along with how concentrated that energy is in space.

Putting this all together, we characterize the strength of sound using a quantity called the *intensity* (I). Recall that the rate of any energy transfer is power. The intensity of sound is defined as the power per unit area in a sound wave. Imagine a small open window that is perpendicular to the direction in which some sound is traveling. The amount of sound energy per unit time that passes through the window, divided by the area of the opening, is the average intensity of the sound.

Sound intensity is the power per unit area carried by a sound wave where the area is perpendicular to the direction in which the sound wave is traveling. The standard units for intensity are W/m^2.

If sound gets too intense, you can now see why it can actually be destructive. Energy is neither created nor destroyed. A very intense sound wave packs a lot of energy into a small area in a short period of time. That energy can do work on whatever it hits, which can potentially cause damage. Sounds that are very loud are certainly capable of damaging the delicate tissues and structures in our ears that allow us to hear.

Energy conservation also explains how and why sounds get softer the farther they travel. Everyone knows that it is easier to hear something when you are closer to the source of the sound. Unless there are some special acoustical arrangements, a source of sound in air will send waves in all directions at once. You can think of each wave crest or each pulse of density as forming an expanding spherical shell, centered on the source of the sound. The radius of the shell increases at the speed of the sound in the air.

As the sound travels, the energy in the waves must spread out over the ever-growing surface of that sphere. For a sphere of radius r, the surface area is given by $4\pi r^2$. Imagine you are located a distance r from a constant source of sound. If the source is generating sound with a power equal to P (in watts) at the source, and if that sound is moving in all directions equally, then the intensity at your location will be $I = P/(4\pi r^2)$ in W/m^2. The important general result we get from this analysis is that the intensity of the sound from a given source tends to decrease as the inverse square of the distance from the source.

TRY IT YOURSELF

26. A very unhappy baby can produce sound with an intensity of 0.000018 W/m^2 when heard at a distance of 14 m. What would be the intensity of this same source at a distance of 42 m?

Because sound can be reflected from surfaces like walls and floors, clever arrangements of surfaces can be used to focus sound, which means forcing all of the energy to preferentially move in one direction instead of spreading out. When this happens, sounds will seem a lot louder than you'd expect after traveling a given distance. If you have ever experienced a so-called "whisper gallery," you know what we are talking about.

A further complication arises from the fact that the human ear is sensitive to such a wide range of intensities. The loudest sound we can tolerate has an intensity of about 1 W/m^2, but the softest sounds we can hear have intensities that are about a trillion times smaller. Not only that, but when we hear two sounds with different loudness, we don't perceive the difference in the absolute intensity, but more like a ratio of intensities.

So it makes more sense to use a logarithmic scale to measure how loud a sound is. Equal steps on a log scale are actually equal ratios. When intensity is measured this way, it is called the sound level (or intensity level) and it is measured in units called decibels. If we call the sound level β, then the level corresponding to an intensity of I is:

Equation 12.1

$$\beta = 10\log_{10}\left(\frac{I}{I_o}\right)$$

The intensity I of the sound we are measuring is compared to a reference intensity $I_0 = 1.0 \times 10^{-12}$ W/m^2. This reference intensity is chosen to be about equal to the quietest sound that the average person can hear, so that sound level is about 0 decibels (abbreviated dB). Any sounds louder than that will then have a decibel level greater than zero, while any sound that is quieter will have a negative decibel level.

The logarithmic scale used for sound intensity can be tricky until you've gotten the hang of it. Make sure you always remember the factors of ten when comparing one sound to another. For example, the relative difference between a normal speaking voice (60 dB) and a whisper (20 dB) is not simply a factor of 60 ÷ 20 = 3. Rather, the relative difference is a factor of $(1 \times 10^{-6}$ W/m$^2) \div (1 \times 10^{-10}$ W/m$^2) = 10{,}000$!

Now we are in a better position to describe the range of human hearing. Figure 12.1 illustrates the range of sounds that the average human can hear. Keeping in mind that there are individual differences, this gives you a rough idea of what sounds can be heard by most people. The horizontal scale is frequency, and the vertical scale is the intensity (loudness) of the sound in dB. The curved line at the bottom is the threshold for hearing; any sounds with an intensity below this line are too quiet to be detected by the unaided ear. The upper limit indicated on this diagram corresponds to the intensity where the sound would start to hurt your ears. Any more intensity, and permanent damage (hearing loss) is the likely result.

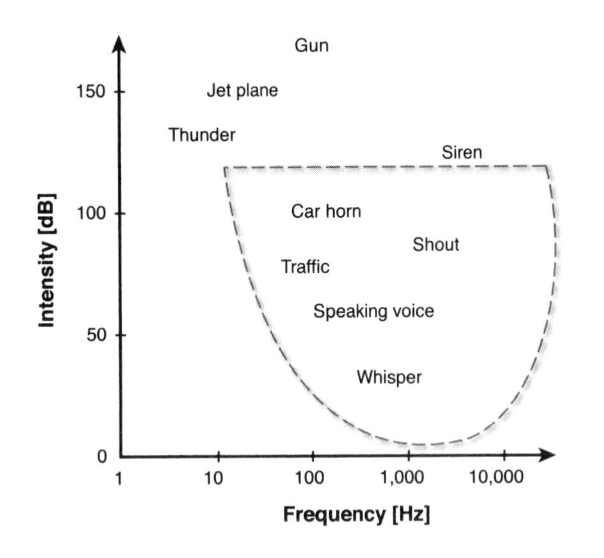

Figure 12.1

This figure shows the range of hearing for the average human being.

One important observation contained in this figure is that the quietest sound you can hear actually depends on the frequency. Humans are most sensitive to sounds in the range of frequencies between about 1,000 and 5,000 hertz (or 1 to 5 kHz). Sounds at lower or higher frequencies have to have more power/area in order to be heard.

Sounds on the left side of this figure, at frequencies too low to hear, are often referred to as *infrasound.* For longitudinal waves with very low frequencies, if they have enough amplitude, you actually don't need ears to detect them. You may be able to feel the vibrations on your skin, or see the effect of the vibrating air when it causes a leaf or piece of paper to oscillate. On the other end of the spectrum, frequencies too high for us to hear are called *ultrasound.* These high-frequency waves can actually be used to form images of objects that they hit. All you need is a way to generate the ultrasound and then a detector capable of measuring both the reflected intensity and the direction from which it came. This technology, sometimes referred to as a "sonogram," is frequently used for "seeing" babies before they are born.

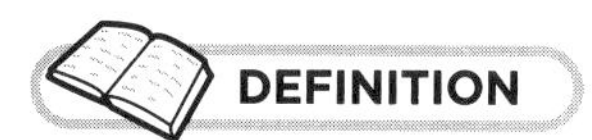

Infrasound is the range of sound with a frequency too low for normal human hearing.

Ultrasound is the range of sound with a frequency too high for normal human hearing.

Interference Applied to Sound

As with any other waves, sound waves obey the superposition principle. If multiple waves are traveling in the same medium at the same time, then the displacements of the waves from equilibrium simply add. Thus it is possible to have constructive or destructive interference with sound.

When sound has the freedom to travel in three dimensions, there are a lot of ways for interference to occur, and the situation can get pretty complicated. Sound from any source will naturally spread in all directions, until it encounters some obstacle. If the obstacle has a complex shape, the sound will tend to scatter in all directions when it hits. If sound waves hit a smooth, flat surface, they are more likely to be reflected. Reflected waves may travel back and interfere with waves traveling in the original direction. Today, the engineers who study the acoustic properties of enclosed spaces like auditoriums and concert halls use sophisticated software and powerful computers to figure out how sound energy will be distributed from various sources.

We can use a simple case, though, to illustrate the most important effects. Suppose two speakers are producing exactly the same sound at the same time. A single observer receives the sounds from the two sources simultaneously. If the observer is the same distance from the two speakers, then the two sound waves will reinforce each other. Peaks from the two sources will arrive at the same time, and we would call this constructive interference.

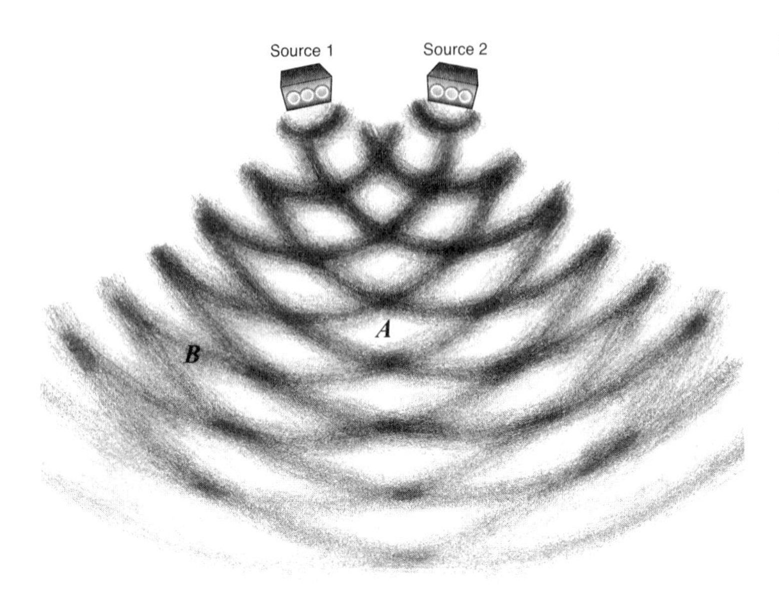

Figure 12.2

The two speakers in this figure are small sources of identical sound waves. Depending on where an observer is located, the intensity of the sound will vary greatly.

Figure 12.2 illustrates this concept. An observer located at location A is equally distant from the two sources. The sound heard at A is relatively loud due to constructive interference. Destructive interference at location B means almost no sound will be heard there. At other locations in the room, however, the distance from the two speakers to the observer will be different. If the difference in path length happens to be half a wavelength, or $3/2$ of a wavelength, or any odd multiple of half a wavelength, then the interference will be destructive. Peaks from one source will line up with valleys from the other source, and the result will be a wave with a much smaller amplitude. As you recall, we describe this situation as destructive interference.

Interference is put to practical use in many musical instruments. Recall it was interference between waves moving in opposite directions that led to the idea of standing wave patterns in the previous chapter. The exact situation we described then, standing waves on a stretched string with fixed ends, is used in pianos, guitars, and a host of other stringed instruments to create sounds with definite pitch, otherwise known as musical notes.

In such cases, a string is plucked or struck or bowed to get it vibrating. The strongest vibrations occur at the so-called fundamental frequency, corresponding to the longest wavelength standing wave that the string can support. This is why in the piano, for which the strings have different lengths, it is easier for the longer strings to support standing waves with lower frequencies. But the speed of the waves on the string is also a factor, which is why six guitar strings of the same length can sound different notes. You recall that both the tension in the string and its mass per unit length affect the speed. The tension of guitar strings can be adjusted (and the lower strings are heavier). With the same string length, but different wave speed on the strings, the fundamental standing wave will have a different frequency for each string.

After the string starts vibrating at its own frequency, there has to be a way to transfer those vibrations to the air in order to make sound. In a piano, the strings cause the whole body of the piano to vibrate, and that is enough to get the air oscillating. In a guitar or violin, the wooden body of the instrument plays that same crucial role. The vibrating strings are attached directly to the hollow wooden chamber, so when that starts vibrating it does a good job of transferring the energy to the air. In an electric guitar there is an intermediate step, where the energy from the vibrating strings is translated into electrical signals, which are amplified and used to create sound in a speaker.

RED ALERT!

If you ever get into more detailed calculations of sound generated by stringed instruments, don't make the mistake of assuming that the speed of the transverse wave on the string is the same as the speed of sound in air.

Wind instruments like trombones, flutes, or pipe organs also rely on standing waves, but instead of transverse waves in a string, the standing waves are sound waves in air right from the start. Longitudinal waves confined to a tube bounce back and forth from the ends just like waves on a string. They will also form patterns with nodes and antinodes, depending on how long the tube is and whether the ends of the tube are open or closed. If we think about the sound as a displacement wave, then a closed end must be a node, but an open end will correspond to an antinode of the standing wave.

Unlike stringed instruments, you can't tune a wind instrument by changing the speed of the waves, because that is just the speed of sound in the air they contain. Tuning can only be done by changing the length of the air column. In order to produce lower notes, larger instruments are required. That is why there are "families" of most of the wind instruments, with larger versions of the saxophone and tuba providing bass accompaniment. We see the same physics at work in the stringed instruments, progressing from violin to viola, cello, and bass violin. The study of sound as it relates to music and acoustics can be fascinating and very extensive. Too bad we have to move on!

The Least You Need to Know

- Sound is longitudinal waves that travel in any continuous distribution of matter.
- Sound waves carry energy, and the rate of energy transfer per unit area gives the intensity of the sound.
- Sound intensity from a localized source decreases as the distance from the source increases, by an amount proportional to the inverse of the distance squared.
- The frequency of sound is related to pitch, and high-pitched sounds correspond to higher frequencies.

CHAPTER

13

Fluids

We all know that most of the surface of our planet is covered by oceans of water. Our world is also covered by an "ocean" of air, which we call the atmosphere. Both of these oceans are extremely important to life on Earth. But they are also so common that we may not give them much thought. Instead of taking them for granted, it is worthwhile to try and understand something about how these oceans work.

In physics, we classify both liquids and gases as fluids. The fact that fluids are able to flow more or less freely gives them some interesting properties. While gases and liquids share many characteristics, they are also different in some important ways. Some of the unique behavior of gases will come up again later in Chapter 16, so in this chapter we'll focus mostly on liquids. So, without further delay, let's dive in!

In This Chapter

- Characteristics of fluids
- The variation of pressure with depth
- Why some things float and others don't
- The effects of fluid flow

Density and Pressure

Even though fluids don't have a definite shape, they do have some characteristics in common with large solid bodies. Like solids, all fluids (including air) have mass, which is distributed over some volume. Different kinds of solid or liquid substances manage to pack different amounts of mass into a standard unit of volume. We call the ratio of mass to the volume it occupies the *density* of a substance.

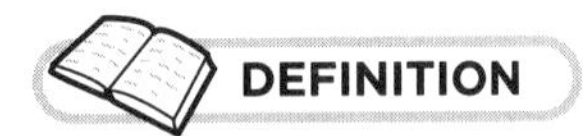

DEFINITION

Density is a property of anything that has mass. It is the ratio of the mass divided by its volume. The symbol ρ (the lowercase Greek letter *rho*) is usually used to denote density, and the standard units are kg/m^3.

Most of the phenomena we're going to talk about in this chapter are relevant to liquids such as water. In our somewhat simplified view, we will assume that the density of a liquid is constant and the same at every location in the liquid. In reality, a liquid's density can be changed slightly by changing the temperature or by compressing it. We will ignore these effects for now.

If you have a tub of water in front of you, and you try to apply a force to the water with your hand, you won't be able to apply much of a force, because the water will just flow out of the way when you try to push on it. Forces in fluids behave in a unique way. In order to get a handle on this behavior, we will often imagine the liquid completely contained in some vessel, perhaps with a piston or something with which to apply a force.

Because liquids are able to flow, and their density remains constant, the only forces that a fluid can exert are normal forces, directed perpendicular to some surface. This force only acts in the direction from the fluid toward the surface, like a push.

So, that gives us the direction of a force exerted by a fluid, but what about the strength of the force? It turns out that any time a fluid exerts a force on a surface, the magnitude of that force is proportional to the area of the surface it is pushing on. Over a relatively small area, the ratio of force to area is constant, so we give this ratio the name *pressure*.

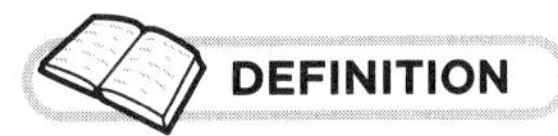

DEFINITION

Pressure is simply a force divided by the area of the surface over which it is applied, $P = F/A$. The standard unit for pressure is N/m^2, also called a pascal (Pa).

You may have noticed that we are reusing the capital P to stand for pressure, where previously we used it for power. These two physical quantities are not related to each other, but there are only so many letters in our alphabet. The context of the situation should make it clear which P we are talking about.

Although we used surfaces to construct our definition of pressure, it turns out that pressure exists everywhere within a fluid—even when there are no surfaces around. Every portion of the fluid exerts a pressure force on the adjacent parts of the fluid. In some sense, pressure is a way to talk about the force that a fluid exerts on itself. Actually, the concept of pressure can be used any time a force acts on a surface, not only in fluids. But pressure is a particularly useful concept when dealing with liquids and gases.

Pressure and Depth

Let's imagine an open container full of liquid, as shown in Figure 13.1. For now and for most of this chapter, we will assume that all of the liquid is at rest in the container, none of it is moving. We'll also assume that we're here on our home planet, where gravity acts downward and g is a constant 9.8 m/s^2.

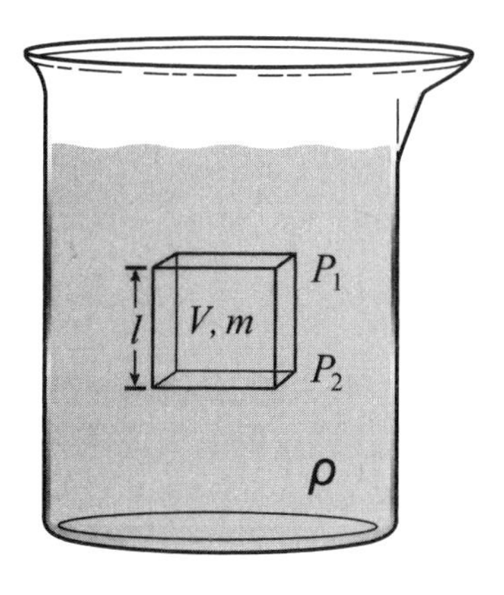

Figure 13.1

A portion of fluid is contained in the imaginary cube whose edges each have a length l. *At the top of the cube, the pressure in the fluid is* P_1, *and at the bottom it is* P_2. *The density of the fluid is* ρ *everywhere.*

Now think about a portion of the liquid somewhere below the surface. Let's say that the portion is bounded by an imaginary cube whose sides are all either vertical or horizontal. Let each edge of the cube have a length l. If the fluid is at rest, and is to remain at rest, then no fluid enters or leaves the cube. The net force on our cube of fluid has to be zero.

The four vertical sides of the imaginary cube each experience a pressure force from the fluid outside the cube that points horizontally toward the inside of the cube. Since the cube is not moving, any pair of horizontal forces on opposite sides of the cube must be equal and opposite in direction, and all four of these pressure forces clearly add up to zero.

But what about the top and the bottom? Pressure forces exerted by the fluid above and below the imaginary cube must be vertical. And let's not forget about another vertical force acting on this cube of fluid: gravity. Given the constant density of the fluid, this weight force must be downward with a magnitude equal to $mg = \rho Vg$ (where we've applied the definition of density). Here V represents the volume of the cube, which is equal to l^3. Now in order for the cube to remain stationary (with zero acceleration), the vertical forces on it must also add up to zero. There are three vertical forces: an upward pressure force on the bottom, a downward pressure force on the top, and the downward weight force. If we label the pressure at the top of the cube P_1, and the pressure at the bottom P_2, then for the sum of the three forces we can write:

$$P_2A - \rho Vg - P_1A = 0 \qquad \textbf{Equation 13.1a}$$

In this expression, A is the area of any cube face. Now since the cube's volume is $V = l^3$, and the area of a face is $A = l^2$, it is easy to see that this expression can also be written $P_2A - \rho Alg - P_1A = 0$. Now divide by the common factor A, solve for P_2, and you'll get:

$$P_2 = P_1 + \rho gl \qquad \textbf{Equation 13.1b}$$

Note that l is a positive number. So in order to support the weight of the fluid above, the pressure must get larger as you go deeper into a fluid! From this last expression you can see that the exact shape of a portion of fluid is not so important. All that really matters is the vertical distance l between the upper and lower locations that you are considering. In words, we have shown that pressure increases whenever you go deeper in a fluid, by a definite amount that depends on the density of the fluid and how much deeper you go.

RED ALERT!

Be careful when you take notes or write up a solution for problems like this. Make sure you can tell the difference between the symbol for pressure P and the symbol for density ρ. (Or even the symbol used for linear momentum p.) It is easy to confuse these in handwriting.

We can get a more generally useful form of this result if we move the imaginary cube upward until its top is at the surface, where the liquid meets the atmosphere. Now we'll call the pressure at the surface P_{atm}, and the pressure at the bottom of the cube just plain P. We'll also change the l to d because it's now equivalent to the depth below the surface. Our new expression then looks like this:

$$P = P_{\text{atm}} + \rho gd \qquad \textbf{Equation 13.2}$$

This equation says that whenever you have fluid in a location with gravity, the pressure at a depth d in the fluid is greater than the pressure at the surface by an amount equal to ρgd.

RED ALERT!

Students sometimes get confused because we use depth d like a typical coordinate axis, with its origin at the surface of the liquid. Since we usually orient our vertical coordinates with the positive direction upward, you might think that d should be negative for points below the surface. But this is not the case. Whenever you use an expression like Equation 13.2, you must consider the depth to be positive, and the deeper you go, the more positive d gets. The important physics to remember is that pressure increases as you go deeper in a fluid.

One of the interesting things about Equation 13.2 is that the variation of pressure with depth does not depend at all on the shape of the container holding the fluid. The only thing that matters is that there is a surface open to the atmosphere, and the vertical distance below that surface.

You may have noticed that we did not say that the pressure at the surface of the liquid is zero. So far, we have been assuming that fluids can only push against a surface, not pull, so all the pressures are positive numbers. This implies that pressure is an absolute quantity; there are no absolute pressures that are less than zero.

The air of our atmosphere is also a fluid, attracted to Earth by gravity. So there is some absolute pressure in the atmosphere we're walking around in right now. This is the pressure represented by P_{atm} in Equation 13.2, the pressure that exists at the surface of the liquid, the interface between the liquid and the air of our atmosphere.

The pressure in the air around us varies with several factors, particularly temperature and other weather conditions. As you might expect, the pressure decreases when you go to higher altitudes, for the same reason that it increases with depth in a liquid. For gases like air, however, we can't use the approximation of constant density. We'll leave those complications aside, and restrict ourselves to considering atmospheric pressure near sea level.

The average atmospheric pressure at sea level is taken to be $P_{atm} = 1.013 \times 10^5$ Pa. We are so accustomed to this pressure that we don't notice it. We live our entire lives with the atmosphere pressing on us with a force of about 10^5 newtons for every square meter of our surface area. Moreover, since atmospheric pressure is always around, we are often only interested in the difference between atmospheric pressure and the absolute pressure in a certain place. This quantity is called the *gauge pressure.*

Gauge pressure at a certain location is the difference between the absolute pressure at that location and atmospheric pressure: $P_g = P - P_{atm}$.

Look back at Equation 13.2. If we were to ask for only the pressure due to the liquid at a depth equal to *d*, the answer would be equal to $\rho g d$. This would also be the gauge pressure, according to our definition. Another way of looking at this is that the absolute pressure at depth *d* is the sum of the pressure due to the liquid ($\rho g d$) plus the atmospheric pressure pushing on the surface (P_{atm}).

Why do we use the term gauge pressure? Recall that pressure is force divided by area. Imagine a closed vessel of some sort. Let's say you want to measure the pressure that exists inside the vessel. Imagine also that you can attach a small tube with a moveable piston that doesn't allow any fluid to escape, as shown in Figure 13.2. You can make this into a pressure gauge just by measuring the net force on the piston. There is a pressure force pushing out from inside the vessel, but there is also a pressure force due to atmospheric pressure pushing in. The area over which the force is applied is the same for both of these.

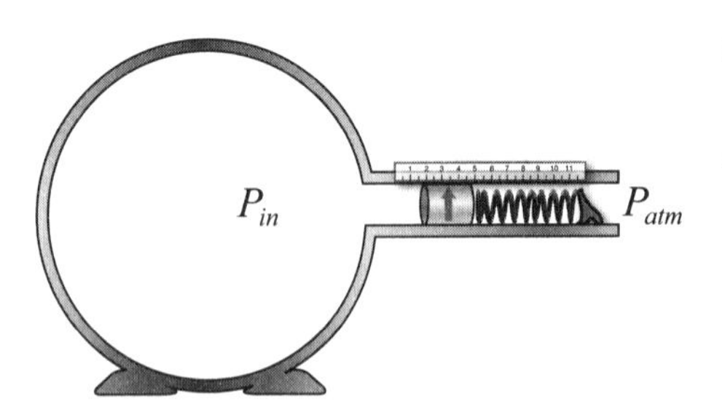

Figure 13.2

The net force on the piston, divided by its area, is a direct measure of the gauge pressure in the vessel.

If we assume the pressure inside the vessel is greater than atmospheric pressure, then the net force on the piston is outward: $F_{\text{net}} = P_{\text{in}}A - P_{\text{atm}}A = P_{\text{g}}A$. So $P_{\text{g}} = \frac{F_{\text{net}}}{A}$, and gauge pressure is just the net force on the piston divided by its effective area. It doesn't matter what the atmospheric pressure actually is, a gauge of this type is only sensitive to the difference between the absolute pressure inside and the atmospheric pressure outside. This is exactly the sort of gauge you use to measure the pressure in the tires of your car or bicycle.

Unlike absolute pressure, gauge pressure can be negative. Negative gauge pressure just means that an absolute pressure that you are measuring is less than the atmospheric pressure. In such a case, the net force on the piston would be directed toward the inside of the container in the figure above. There is a limit, however, since there is still no absolute pressure less than zero. The magnitude of the most negative achievable gauge pressure is just equal to atmospheric pressure.

TRY IT YOURSELF

27. The density of fresh water is 1,000 kg/m^3. Assume that the atmospheric pressure is 1.013×10^5 Pa. What is the absolute pressure at the bottom of a swimming pool that is 15 m deep?
28. What is the gauge pressure at the bottom of this same swimming pool?

Hydraulics

Because pressure exists throughout the volume of a fluid, we can put that pressure to some good practical uses. If you have a closed volume of fluid, and you increase the pressure at one location, that increase will automatically be spread to all other locations in the fluid. This seemingly trivial fact is now used in a great many industrial applications.

Imagine a fluid-tight piston that is free to move in a cylinder of a certain diameter. Now imagine you have two such piston/cylinder devices that are identical. One end of each cylinder is joined to the other with a strong but flexible hose, and the hose and cylinders are all filled with fluid.

Now, if you apply a force to one piston, it will increase the pressure in the fluid, and that same force will be exerted by the fluid on the other piston.

This turns out to be a great way to transmit forces from one place to another. It requires no complex mechanisms, and the force can be easily transmitted by a significant distance around obstacles or along moving pieces of machinery. If the working fluid is air (or another gas) we would describe a device like this as "pneumatic." On the other hand, if liquid is used in the tubes, then such a device would fall into the category of mechanisms called hydraulics. Hydraulic and pneumatic devices are widely used as controls, transmitting small amounts of force and displacement precisely.

Hydraulic lines are also used in large machinery like backhoes, bulldozers, etc., to not only transmit but also amplify forces. The next time you see a bulldozer or other piece of heavy equipment, look for the shiny cylinders or hoses that indicate hydraulics at work. Multiplication of a force with hydraulics is possible because it is pressure, and not force, that is transmitted. And since the pressure along the hydraulic line has a constant value, you can change the force at either end just by changing the size of the piston.

Imagine a small diameter cylinder on one end of a fluid-filled chamber and a much larger one on the other, as in Figure 13.3. A small force applied to the small piston results in a large force at the other end, where the ratio of the forces is exactly equal to the ratio of the areas of the two pistons (because the pressure is the same at either end). This effect is analogous to a mechanical lever that is a simple plank on a fulcrum, where the fulcrum is much closer to one end of the plank.

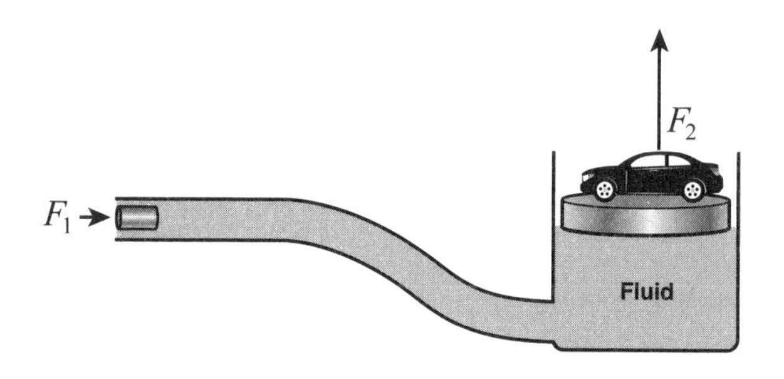

Figure 13.3

A small force F_1 *becomes a large force* F_2 *because the applied pressure spreads to the whole fluid and the piston at left has a much smaller area than the piston at right.*

Archimedes' Principle and Buoyancy

The imaginary submerged cube we constructed previously can also help us to understand why some things float and others don't. In the earlier case, the forces on the fluid cube were in balance because the cube had exactly the same density as the rest of the fluid. The weight force acting on the cube was exactly equal to the difference in the pressure forces from above and below, so there was no acceleration.

Now suppose our imaginary cube of fluid is replaced by a real, solid cube of exactly the same size. Nothing about the external fluid changes; the net force on the cube due to the pressure differences would have to be the same. But the weight force could be different, depending on what the new cube is made of. If the density of the solid cube is greater than that of the fluid, it will have more mass than the fluid cube did, and the net force will now be negative. In that case, you would expect the cube to accelerate downward. On the other hand, if the density of the solid cube is less than the density of the fluid, it will have less mass than the fluid cube did, and the net force will be up. It would then accelerate upward until it reaches the surface.

A long time ago, a clever Greek fellow named Archimedes realized that this idea applies very generally. The object doesn't have to be a cube; any shape at all will experience the same effect. For any object submerged in a fluid, you can imagine two forces acting. One is the old familiar weight force, and the other is the net effect of all the fluid pressure being exerted on all sides of the object. You don't have to know the direction and strengths of all these forces to realize that the net effect of the fluid pressure is an upward force on the object. That upward force would exactly balance the weight force, if and only if the object had the same mass as the fluid it displaces. Thus, it is relatively simple to calculate the strength of this force, which we call the *buoyant force.*

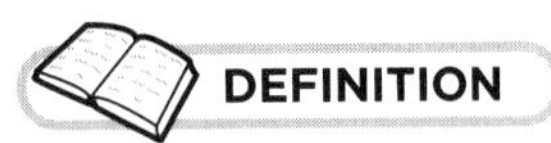

A **buoyant force** is exerted by a fluid on an object any time that object is partially or completely submerged, assuming there is gravity present. The buoyant force is directed vertically upward, and it has a magnitude which is exactly equal to the weight of the fluid displaced by the object.

The buoyant force is measured in newtons, of course. If we call this force $\boldsymbol{F}_B$, then the magnitude can be written $F_B = \rho_{fluid} V_{fluid} g$, where V_{fluid} is the volume of fluid that is displaced by the object. Only if the object is completely submerged will V_{fluid} also be equal to the volume of the object itself (V_{object}). Note that the buoyant force does not depend at all on the mass of the object. The density of the object determines whether it will sink or float in a given fluid, only because the weight of the object is a force of magnitude $F_g = \rho_{object} V_{object} g$ that is opposed to the buoyant force. Whichever vertical force is greater, that determines the direction of the total force and whether it accelerates up or down when submerged.

CONNECTIONS

If the density of an object is exactly the same as the density of the fluid it is surrounded by, that is called neutral buoyancy. If this is the case, then an object can be completely submerged in the fluid with zero net force: the buoyant force and the weight would be equal and opposite. Many fish can make small adjustments to their volume in order to achieve this desirable state. A submarine has a constant volume, but can adjust its weight by taking on or getting rid of extra water mass called ballast. By changing the mass contained in the submarine, it can either float, sink (dive), or cruise at a constant depth in the water.

An object with less density than a liquid can be in equilibrium on the surface, in which case we say the object is floating. Equilibrium is achieved by the fact that only part of the object's volume is displacing the liquid, and the rest is above the surface. The fraction of the volume below the surface is exactly what is required to make the upward buoyant force equal in magnitude to the downward weight of the object. Using the definition of the buoyant force, you can easily verify that the ratio of the submerged volume (V_{sub}) to the total volume of the object (V_{object}) is equal to the ratio of the object's average density to the density of the liquid:

$$\frac{V_{sub}}{V_{object}} = \frac{\rho_{object}}{\rho_{liquid}}$$ **Equation 13.3**

TRY IT YOURSELF

29. A piece of wood has a density of 600 kg/m^3. It floats on the surface of a barrel of fuel oil, which has a density of 890 kg/m^3. What fraction of the wood's volume is above the level of the oil?

It is also interesting to note that there is a buoyant force acting on you right now due to the air you are "swimming" in. We don't notice this upward force for two reasons: it is always with us, and it is much smaller than our weight. We spend our lives completely submerged in the fluid of the atmosphere, so the force measured by the bathroom scale is a little bit less than *mg*. How much less? Well, the density of air under typical atmospheric conditions is about 1.2 kg/m^3, and your density is just a little less than that of water, on average, say 950 kg/m^3. Therefore, the buoyant force due to the air is only about 0.13 percent of your weight.

Fluid Flow and Bernoulli's Equation

We've learned a lot of interesting things about fluids already in this chapter, and that was just for fluids at rest. When we start to talk about fluids in motion, things can get complicated pretty quickly. We will just introduce a few interesting tidbits about fluids in motion before we wrap up and move on.

Some of the complexity comes from the fact that different fluids can behave very differently. The fields of hydrodynamics (for liquids in motion) and aerodynamics (for gases) are related but distinct areas of study. Even the same fluid behaves differently depending on how fast it is moving, and on the shapes of things it moves against.

One distinction we should make early on is between turbulent flow and laminar flow. You may have heard of turbulence before, or experienced it when flying in an airplane. The best way to describe turbulent flow is that it's disordered and messy. In turbulent flow, different parts of the fluid are going at different speeds in different directions, even when they are relatively close to each other. Turbulent conditions occur most often when a fluid is moving very quickly, and/or when the boundaries are very rough or irregular. Turbulent flow is fundamentally complicated, and we won't deal with it any more in this book.

The opposite extreme is the very smooth kind of motion we call laminar flow. This kind of flow is much more predictable. It occurs at lower speeds and around or through smooth objects. This is the kind of fluid flow we will be able to talk about in the rest of this chapter. Along with our previous assumption that our ideal fluids have constant density, we will also assume smooth and steady flow for the fluid motion. This means that even though mass is moving, the velocity vector of the fluid at a particular location does not change in time. The same would be true for the pressure.

In steady flow of this type, one can imagine definite (possibly curved) paths taken by each piece of the fluid. Lines in space that correspond to such paths would not cross each other, because if they did, there would be a point where the velocity of the fluid was undefined. These nonintersecting lines are called streamlines and are characteristic of laminar flow. A bundle of adjacent streamlines are bounded by a surface through which no fluid passes. This surface can be real or imaginary.

Let's consider a smooth tube whose diameter varies, as in Figure 13.4. It is common sense that in any particular time interval Δt, however much mass enters the tube, the same amount must exit the other end. For a fluid entering the tube with a speed v_1, every bit of the fluid will travel a distance $\Delta x_1 = v_1\Delta t$. (We'll imagine keeping the time interval Δt pretty short, so that the diameter of the tube doesn't change much over the distance Δx_1.) Let's allow for the fact that the speed of the fluid exiting the tube may be different, so that at the exit end, fluid moves by $\Delta x_2 = v_2\Delta t$ in the same amount of time.

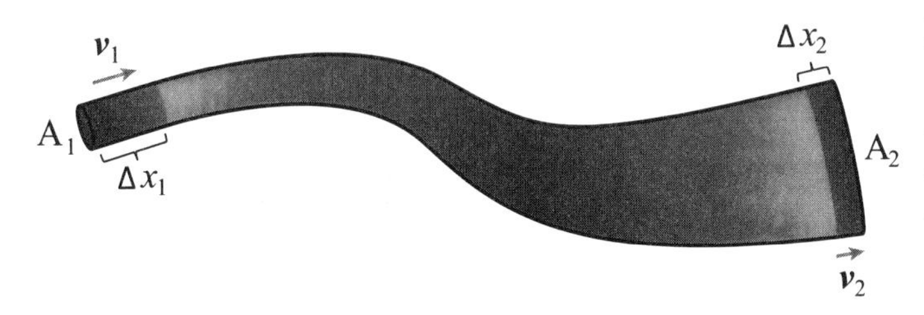

Figure 13.4

This figure illustrates the continuity condition for a fluid with constant density. In a given time interval, the same amount of fluid that enters one end must leave the other.

Remember that the same mass of fluid that enters must also exit. If we assume that the density of the fluid is always the same, then this means the same volume of fluid that enters must also exit. The volume of a cylinder is the area times the height, so equating the two volumes means:

$$A_1\Delta x_1 = A_2\Delta x_2$$
$$A_1 v_1\Delta t = A_2 v_2\Delta t$$
$$A_1 v_1 = A_2 v_2$$

Equation 13.4

This last simple expression is sometimes called a continuity equation. If a fluid flows smoothly in a pipe with constant diameter, then the speed of the fluid is the same all through the pipe. If, on the other hand, the pipe has a change in diameter, this equation shows that the speed of the fluid will also change to keep the product of speed and cross-sectional area constant.

If we make one more simplifying assumption, we can introduce another very useful equation for describing fluids in motion. We now want to assume that our fluids can flow without friction (in which case we call them "ideal"). This is quite accurate for gases flowing smoothly. For liquids, it depends on exactly what liquid you are talking about. Some liquids resist flowing because they have a lot of internal friction (think of honey). But others, like water flowing at low speed, have such sufficiently low friction that we can safely ignore it.

The equation we want to introduce is called Bernoulli's equation, which allows us to account for changes in pressure, elevation, and speed in fluid flow. Since streamlines can provide virtual "tubes" in which fluid flows, it can even be applied when there are no actual pipes or containers. But let's imagine a situation like Figure 13.4 above. Choosing two locations along the length of a tube, Bernoulli's equation states that the conditions at one location must match the conditions at some other, as follows:

$$P_1 + \frac{1}{2}\rho v_1^2 + \rho g y_1 = P_2 + \frac{1}{2}\rho v_2^2 + \rho g y_2$$

Equation 13.5

Here, P represents the pressure at a given point in the fluid, and y is the vertical height of the fluid at that point, relative to some arbitrary zero height. (The y-axis must now be positive upward for this to work.) Bernoulli's equation applies to the smooth flow of an ideal fluid, with all of the assumptions and approximations we have described.

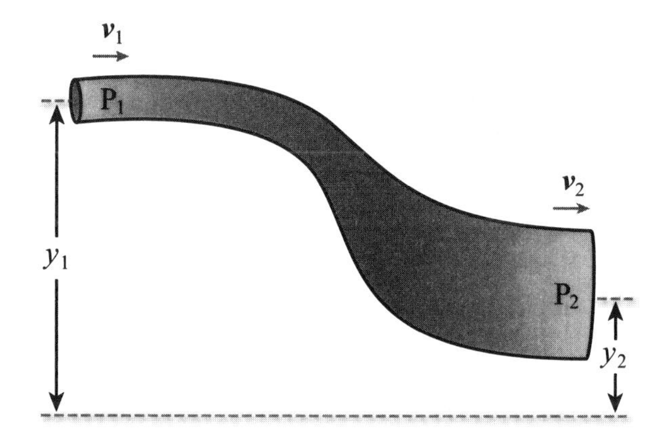

Figure 13.5

A fluid flowing smoothly in a closed pipe with varying diameter which can also rise or fall in vertical height. The pressure, speed, and height at any two locations are related by Bernoulli's equation.

We will not derive this equation here, but we can explain the physics on which it rests. It is essentially an expression of energy conservation in the moving fluid. We would expect mechanical energy to be conserved, since we are assuming no friction in the fluid. Since locations 1 and 2 are arbitrary, Bernoulli's equation essentially says that the quantity $P+\frac{1}{2}\rho v^2+\rho gy$ is constant everywhere in the fluid. You can easily see that the second two terms are the kinetic energy per unit volume and the gravitational potential energy per unit volume, respectively. So if you multiplied this whole expression by a certain volume, the second two terms would add up to the total mechanical energy in that volume.

But what about the first term? To explain this, consider what you get if you multiply a pressure times a volume. Dimensionally, it would be force/area × volume, which reduces to force × distance. You should recall that when a force acts over a distance, it does work, which is just a way to transfer mechanical energy. So in a moving fluid, pressure forces are doing work, and the Bernoulli equation is really just an expression of the concept of conservation of energy.

TRY IT YOURSELF

30. In your third-floor apartment, you want water to flow at a speed of 4.0 m/s from your kitchen faucet, which has a cross-sectional area of 3.6 cm^2. The water supply enters the apartment building through a large pipe 14 m below your kitchen faucet. The supply pipe has a diameter of 5.0 cm. What absolute pressure is needed in the water at the point where it enters your building? Note that the water will exit the faucet at atmospheric pressure and that ρ = 1,000 kg/m^3 for water.

The Least You Need to Know

- Pressure is force divided by area.
- Pressure in a fluid increases linearly with depth, as long as gravity is present.
- An object will float in a fluid if its overall average density is less than that of the fluid.
- A fluid exerts a buoyant force on any object that is equal to the weight of fluid that is displaced by the object.
- The Bernoulli equation, which is rooted in the conservation of energy, can tell you how pressure and speed vary in a moving fluid.

PART

3

Thermodynamics

For a long time, scientists had difficulty incorporating the phenomenon known as heat into the rest of mechanistic physics. They could measure temperature, and a lot of the properties that went along with it, long before they understood exactly what temperature was measuring and how it was related to heat.

Now we know that heat is actually another form of energy. It is associated with the motions of the tiny invisible particles that make up the matter that is all around us. Understanding this and the ways that heat moves around has led to many advances that have proven important for society, including the industrial revolution. In this part, we'll take a good look at what heat is, how it can be transported, and how conversions between heat and mechanical energy occur and can be controlled for our benefit.

CHAPTER

14

Temperature and Matter

Of all the concepts that we will cover in this book, there's a good chance that temperature is the one with which you are most familiar. For example, you may be accustomed to checking the outside temperature before leaving the house so that you can decide which jacket to wear. Most likely, you have a good feel for how "warm" 80°F is, or how "cold" 32°F will feel when you open the door. But, do you know what those numbers actually stand for? More fundamentally, do you know how physicists actually define the term "temperature?" By the end of this chapter, you should have a firm grasp on both of these questions.

Before moving on, we'll also talk about how temperature relates to the structure of matter, and in particular, how changes in temperature affect different types of matter. As you'll see, there is an intimate and nonseverable link between the properties of matter and temperature.

In This Chapter

- Temperature defined, and how it is measured
- The zeroth law of thermodynamics, and why it's useful
- The physical basis of temperature
- What happens to substances when heated and cooled

What Is Temperature?

The flippant answer to this question is "what the thermometer shows." But this does little to help our understanding of the physical basis for temperature, let alone give us any tools to do any physics. So, let's begin with a succinct definition of *temperature*, after which we'll explore what it actually means.

Temperature is a measure of how hot or cold an object is, relative to some standard value. It is closely related to the amount of internal energy within the object.

Firstly, the word "measure" should tip you off that temperature is something that we can measure with some sort of device, from which we can assign some quantitative value. Most of us know that the device used to do this is called a thermometer. We need to wait a few more sections before we can explain how exactly these devices work, but suffice it to say that by using a thermometer you can get a number that tells you something about an object's temperature or the temperature of the air around you. The higher the number you get, the higher the temperature and the hotter the object is. If two objects are measured, the colder object will exhibit the lower temperature.

This last example is actually important, because it hints that what really matters when speaking about temperature is relative differences. That's why we used the terms "relative to some standard value" in the definition above. This is what makes temperature a quantifiable property. It also allows for humans in one part of the world to communicate information about temperature to other humans elsewhere. For example, if you are baking cookies and the recipe says "bake at 350°F," you know how to set your oven such that you can expect the same results as the person who wrote the recipe (and who almost certainly used a different oven in the process).

The second sentence in our definition of temperature refers to the "internal energy" of an object. We put that in there so that you get a clue about the fundamental basis of temperature (i.e., it has to do with energy, and not so much with length or mass). In the next chapter, we will pull this string and explore in much more detail the relationship between temperature and energy.

Temperature Scales

You are probably aware that there is more than one type of thermometer out there. If you live in the United States, your thermometer probably gives you temperatures in °F, which is physicist shorthand for "degrees *Fahrenheit*." If you live in the Austrian Alps, your thermometer gives you temperatures in °C, which means degrees *Celsius*. What's more, if you ever snuck into a physics lab, you may have heard the guys in lab coats talking about temperatures in something called *kelvins*. These are the three most common temperature scales, and as we'll show, they are equivalent in the sense that they all measure the same thing, and you can reliably convert between any one and another.

Fahrenheit is a temperature scale set by the fact that water freezes at 32°F and that water boils at 212°F.

Celsius is the temperature scale used in the SI system of units. It is set by the fact that water freezes at 0°C and that water boils at 100°C. (For this reason, it is sometimes referred to as the "centigrade" scale.)

Kelvin is a temperature scale that provides information about absolute temperature. It says that water freezes at 273.15 K and boils at 373.15 K.

We have already explained that temperature is defined with reference to some standard, and in these definitions we have revealed what standard is actually used—the physical properties of water! Since water is the most prevalent chemical compound on Earth, it makes a lot of sense for it to serve as the basis for defining temperature scales.

Let's begin with the Fahrenheit and Celsius scales—the two most familiar in everyday life. You see from the definition that these are set empirically (i.e., based on observation and measurement) and that they are defined using two points—the point at which liquid water turns to ice when cooled (32°F and 0°C) and the point at which liquid water turns to steam when heated (212°F and 100°C).

The reason for using two values to fix a temperature scale is that this is the number of data points needed to determine exactly how big one unit is and also to anchor the scale in an absolute sense. For example, with the Celsius scale, we see that water that has just begun to boil is 100 units (called "degrees") warmer than water that has just begun to freeze. In other words, one Celsius degree is equivalent to one one-hundredth of the temperature difference between these two points. Also, by selecting the value of 0°C for the lower point, we have a well-defined scale in terms of unit size and starting point. This means that if we measure a glass of water and find that it is 50°C, then its temperature is exactly halfway between the point at which water freezes and the point at which it boils.

You may have noticed that the Celsius and Fahrenheit scales are not only shifted relative to one another, but that the size of the units is actually different. While a Celsius degree is one one-hundredth the temperature difference between water's freezing and boiling points, a Fahrenheit degree is $1 \div (212 - 32) = 1/180$th that same temperature difference. Since the size of a Fahrenheit degree has a larger number in the denominator, this means that a Fahrenheit degree is smaller (by a factor of 1.8) than a Celsius degree.

Sometimes you know the temperature in one scale and need to convert it to another. All that's needed to do so is a little bit of basic math. First, you need to account for the variance in unit size. Since one Fahrenheit degree is smaller than one Celsius degree by a factor of 1.8, any time you are converting a temperature reading in Celsius (T_C) to Fahrenheit (T_F), you first need to multiply T_C by 1.8. You also need to account for the absolute shift between the two. Let's pick a convenient point—the freezing point of water—to do so. We know that for this point, $T_C = 0°C$ and $T_F = 32°F$. This tells us that the shift that we need is therefore 32°F. Putting these two together, we see that:

$$T_F = 1.8\ T_C + 32°F \qquad \textbf{Equation 14.1}$$

TRICKS AND HACKS

You need to use Equation 14.1 to make an exact conversion between Celsius and Fahrenheit. If you're standing on a train platform in a foreign country one day, and you don't have a calculator handy, a good estimate can be made by multiplying the value in Celsius by two and then adding 30. Using this rule of thumb, we would say that 20°C is approximately 70°F, whereas the true value is 68°F.

The fact that both the Celsius and the Fahrenheit scales allow you to have temperatures below zero is an indication that these are relative scales only. However, physicists have determined that there is actually a lowest temperature physically possible. To reflect this fact, they call this temperature *absolute zero.* This turns out to be the basis for the third temperature scale we defined above—the Kelvin scale.

DEFINITION

Absolute zero is the lowest temperature that any physical system can theoretically attain. Its value is approximately -273.15°C.

Practicing physicists prefer to use the Kelvin scale because not only does it allow you to determine if one object is warmer or colder than another, it allows you to gauge how warm the object is relative to absolute zero. If you know that an object is sitting nicely at 300 kelvins, you know that it is 300 degrees (on the Kelvin scale) above the lowest achievable temperature. The most you could ever cool that object (in principle) is therefore 300 degrees kelvins.

Because the Kelvin scale has the added benefit of providing an absolute value, physicists usually reserve the symbol T (without any subscript) to represent temperatures on this scale. Using the same mathematical arguments we used to relate the Celsius and the Fahrenheit scales, we see that the conversion between Celsius and Kelvin is given by $T = T_C + 273.15$. Be aware that it is customary to skip the degrees sign (°) when writing temperatures in Kelvin. (Also note that we used T in a previous chapter to stand for the period of an oscillator, which is unrelated to temperature. This is just another consequence of having too few letters in our alphabet!)

TRY IT YOURSELF

31. The lowest temperature ever recorded on Earth was -89.2°C, as measured at Vostok Station in Antarctica in 1983. What is this temperature using the Fahrenheit scale and using the Kelvin scale?

To give you a sense of the differences between these three scales, the following table shows temperatures for various physical systems in Celsius, Fahrenheit, and Kelvin. You'll notice that on the Kelvin scale, temperature is always positive.

Temperatures of Various Physical Systems as Measured Using Different Scales

Physical System	Celsius Scale (T_C)	Fahrenheit Scale (T_F)	Kelvin Scale (T)
Surface of the sun	5,500°	9,930°	5,800
Candle flame	1,000°	1,800°	1,200
Nuclear reactor core	350°	660°	620
Highest temperature of typical cooking oven	250°	480°	520
Boiling point of liquid water	100.00°	212.00°	373.15
Highest temperature recorded on Earth	56.7°	134.1°	329.9
Human body	37.0°	98.6°	310.2
Freezing point of liquid water	0.00°	32.00°	273.15
Cold day in Moscow	-40°	-40°	233
Cold day on Mars	-150°	-238°	123
Boiling point of liquid nitrogen	-196°	-321°	77
Absolute zero	-273.15°	-459.67°	0

Thermal Equilibrium

Suppose you have just boiled an egg for 10 minutes and then you drop it in a pot of cold water. What will happen? Everyday experience tells us that the egg will cool down to a temperature that we can actually eat. But is the egg the only thing that changes temperature?

If you were to look carefully at the temperature of the water before and after the egg cooled, you will see that in addition to the egg cooling, the water will have warmed slightly. Moreover, while the temperature difference between the egg and water was probably large at first, after a few minutes you'll find that the egg and the water have reached the same temperature. The fancy term for this condition is *thermal equilibrium.*

Thermal equilibrium is the state achieved when two objects are brought into thermal contact and their temperatures approach one another until they reach exactly the same temperature.

Generally speaking, thermal equilibrium occurs any time two objects are placed in contact with one another (in the absence of any internal or external sources of heating). The first part is obvious from our daily lives—after an hour sitting on the table, a previously hot bowl of soup will cool down to room temperature. The part about "internal or external" heat sources may not be so clear, but this is the reason that we don't quickly freeze when we go stargazing on a cold winter night. Though our skin will be in contact with the frigid outdoor air, our bodies will stay at roughly the same temperature. One reason for this is the internal heating processes within our bodies.

We have now reached the first of four basic laws that physicists have come up with to govern the thermal properties of matter: the so-called "laws of thermodynamics." For some reason, these are numbered from zero to three, so we begin with the *zeroth law of thermodynamics.*

The **zeroth law of thermodynamics** states that if Object A is in thermal equilibrium with Object B, and Object B is in thermal equilibrium with Object C, the it follows that Object A is in thermal equilibrium with Object C (even if A and C are not in contact with each other).

Suppose that you have two large kettles of water sitting next to each other in your kitchen. You place a small thermometer in Kettle 1 and allow the thermometer-kettle system to come into thermal equilibrium. Then, you remove the thermometer from Kettle 1, insert it into Kettle 2, and then allow the second system to come into thermal equilibrium. If the thermometer does not heat up or cool down in the second kettle, the zeroth law implies that Kettle 1 and Kettle 2 must be in thermal equilibrium with one another. What's more, the reading on the thermometer will stay the same and gives us the temperature of the water in each kettle. This simple example therefore offers a rather profound conclusion—two objects that are in thermal equilibrium with each other must be at the same temperature.

The Physical Basis of Temperature

We now know that temperature is a measure of how "hot" or "cold" an object is relative to another object or to some reference value. Let's now look a little more deeply into the situation and try to understand the physical basis of temperature. We've already mentioned that temperature has something to do with the internal energy of an object. But what, exactly? Before we can answer that, we need to have a general understanding of how matter is constituted in the first place.

Let's start by considering a nice simple type of matter, such as a hunk of pure sodium metal. If you happened to have a very powerful microscope and looked deep into the metal, you would find that it is composed of little particles (called sodium atoms) that are arranged in a nice orderly pattern. If you watched the system for a little while, you would also observe that each of the atoms is jiggling around, almost as if each of the atoms were attached to its neighbors by a little tiny spring, as shown at left in Figure 14.1. What's more, if you heated the metal up, you would observe that the magnitude of the atomic jiggles increases, as shown in the right side of Figure 14.1.

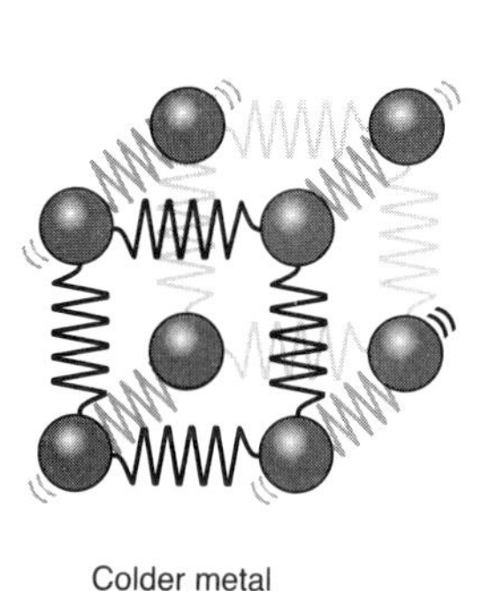

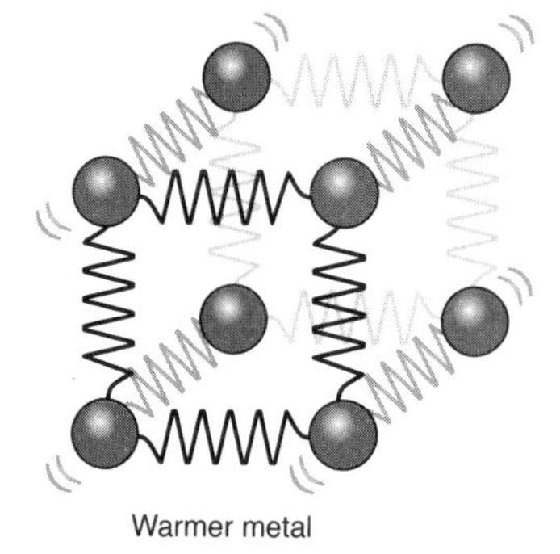

Figure 14.1

Many solid metals are composed of regularly spaced rows of atoms, and each atom moves as if bound by a spring to its immediate neighbors. The warmer the metal, the greater the amplitude of oscillation.

Conversely, if you then waved a magic wand such that the metal cooled down to absolute zero, you would see that the jiggling motion effectively stops. All of this indicates that the "internal energy" we referred to above is due to the jiggling about of the inner constituents of matter. And, as we learned in Chapter 10, the type of energy that we're dealing with is the kinetic energy of oscillation. The atoms are not really joined by little springs, mind you, but the forces that hold the atoms together can be nicely approximated as if this were truly the case.

But solid metals are not the only type of matter out there. Suppose we consider another form of matter with which we are intimately familiar—the air around us. If you peered through your powerful microscope to look at the fundamental structure of air, instead of a bunch of atoms bound together as if on springs, you would see a bunch of atoms whizzing about freely like balls thrown through the air (see Figure 14.2). In this case, it is the kinetic energy associated with these projectile-like motions that determines the temperature of the air. The molecules that compose air whiz around at faster speeds on warm summer days than on cold winter nights.

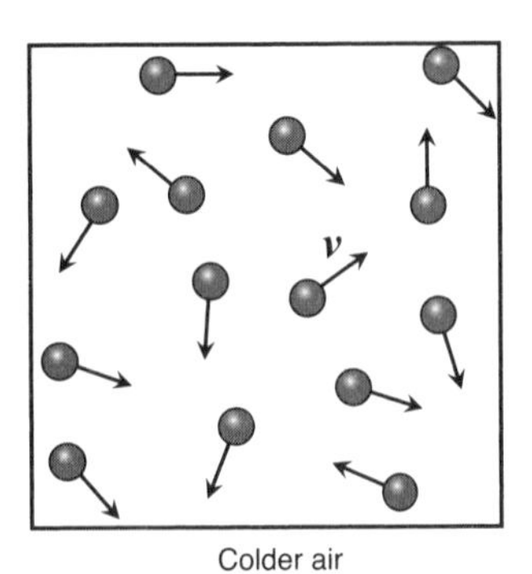

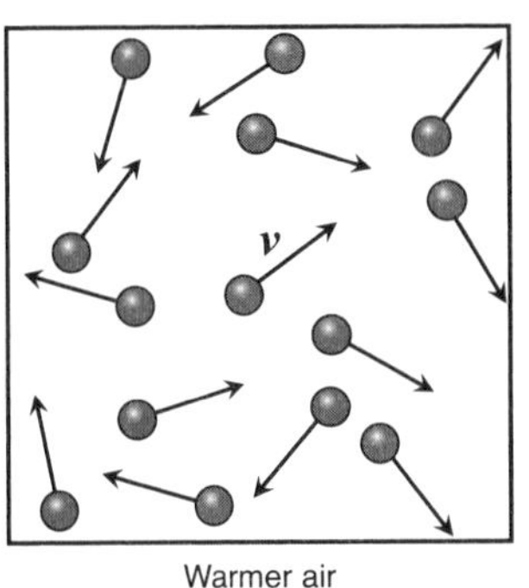

Figure 14.2

Air is composed of tiny molecules that whiz around nearly independently of one another. The warmer the air, the faster the molecules travel (indicated here by the velocity vector v*).*

Finally, let's take a quick look at water. Peering through the microscope, you would see that the molecules that form water behave somewhere in between that of the metallic atoms and the air molecules we just discussed. Liquids like water are characterized by molecules that are "loosely" connected with one another; loose enough, that is, that any given molecule can easily flow among and around all of the nearby molecules.

Thermal Expansion and Contraction

Now that we have a qualitative sense for how temperature affects the internal properties of matter, we have what we need to predict what happens to bulk materials (like chunks of metal or water in containers) when these are heated and cooled. Let's first consider our sodium metal. We've already seen that as the metal is warmed, the amplitude of oscillation of the interlinked atoms increases. This leads to a small increase in the average distance between the atoms. If you heat a reasonably sized hunk of metal, complete with trillions upon trillions of atoms, the cumulative effect of all those increases in the interatomic spacing will actually lead to a measureable increase in the size of the entire metallic hunk! Physicists refer to this as *thermal expansion.*

After studying this effect with lots and lots of solid and liquid materials, physicists have derived a nice, general formula that allows for quantitative predictions for the expansion due to increased temperature. Suppose your metallic chunk is shaped like a rod of length L_0 (where the "0" here simply means at the initial time before any heating). If you raise the temperature of the rod by an amount ΔT, you will find that the rod's length increases by an amount $\Delta L = \alpha L_0 \Delta T$. The symbol α, known as the coefficient of linear expansion, depends on the actual type of material that you are warming. This formula applies for objects of any shape (not just bars), so long as it is clear that L is measured along a single axis in one dimension.

To see how this works, let's consider a copper wire spanning from your house to the nearest telephone pole. We'd like to see how much its length increases from the dead of winter (where $T_F = 10°F$ and $L_0 = 20.000$ m) to the dog days of summer (where $T_F = 100°F$). To solve this, we also need to know that $\alpha = 1.7 \times 10^{-5}$ $(°C)^{-1}$ for copper. (The superscript "-1" applies only to the units here; when this constant is multiplied by a temperature in Celsius degrees the units cancel because °C ×(°C)$^{-1}$ = °C ×(1/°C) = 1.) For starters, since we have the linear expansion coefficient in °C, we need to convert our temperatures to °C. Using Equation 14.1, we see that the lower temperature is T_C = -12°C while the higher temperature is T_C = 38°C. Therefore, the temperature difference is ΔT = 50°C. Plugging this in to the formula for linear thermal expansion, we see that the change in length is $\Delta L = [1.7 \times 10^{-5}\ (°C)^{-1}] \times (20\ m) \times (50\ °C) = 0.017\ m = 1.7\ cm$. Since this is the amount of change, we also see that the wire's length at the hotter temperature is 20.017 meters.

So much for expansion due to heating. What happens if you instead cool a solid material by an amount ΔT? In this case, the atoms vibrate less vigorously and the average spacing between them actually decreases. This leads to a shrinkage of the solid known as *thermal contraction.* In this case, the exact same formula applies. Since cooling will give you a negative ΔT, the result simply will be a negative ΔL.

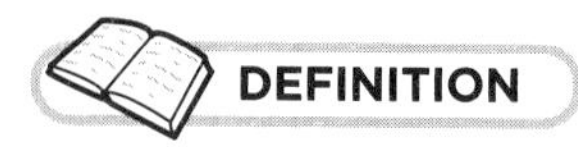

Thermal expansion is the increase in size of a solid or liquid due to heating.

Thermal contraction is the decrease in size of a solid or liquid due to cooling.

So far we have considered expansion in only one dimension. If you refer back to our figure of a sodium metal, however, you'll see that it looks the same in all three dimensions of space that it spans. This means that when heated an object will expand in each of these three dimensions which leads to an overall increase in the object's volume. The same basic formula applies, though it now takes the form $\Delta V = 3\alpha V_0 \Delta T = \beta V_0 \Delta T$, where V_0 is the starting volume, ΔT is the change in temperature, ΔV is the change in volume, and β is a material-dependent constant known as the coefficient of volume expansion. Note that due to the symmetry of the crystal structure of solid materials, $\beta = 3\alpha$.

It turns out that the aforementioned formulas for thermal expansion and contraction are applicable for both solid and liquid substances. They don't hold up so well when considering gases, though, and a different physical model will be required. We'll delve a little more deeply into this in Chapter 16.

TRY IT YOURSELF

32. If you have 80 mm^3 of liquid mercury at 0°C, what is its volume at 100°C? Note that for mercury, $\beta = 1.82 \times 10^{-4}\ (°C)^{-1}$.

We are now in a position to explain how a basic thermometer works. We'll consider a good old-fashioned mercury thermometer, which consists of a glass tube filled with mercury. The inner bore of the tube is very carefully prepared to be of uniform diameter throughout the length of the thermometer. As the temperature of the thermometer rises and falls, the volume of the mercury therein will expand and shrink as we previously described. With a uniform-bore diameter, the effective length of the column of mercury will expand in a very well-defined way.

To set the scale, you could first put the thermometer in an ice bath and draw a tick mark labeled 0°C at the height of the mercury. Then you put the thermometer in a boiling water bath and draw a tick mark labeled 100°C at the height of the thermally expanded mercury. Finally, you draw 100 even divisions between these two points, and you have a device that can give you the temperature for any substance in between. (You can also extend the scale below 0°C and above 100°C, provided the division spacing stays the same.)

Phase Changes

In the previous section, we discussed the microscopic characteristics of sodium metal, water, and air. These are just specific cases of the three more general states of matter: solids, liquids, and gases. Most materials can be found in all three states if they are subjected to a sufficiently wide range of temperatures. For example, below 0°C water will take its solid form (ice), between 0°C and 100°C it will take its liquid form, and above 100°C it will take its gaseous form (water vapor or steam).

This implies that there are a few "special" places along the way, at which a substance will make a sudden and dramatic shift from one state to another. The temperature at which a substance transforms from solid to liquid is called the *melting point* or *freezing point*, while the temperature at which a substance transforms from liquid to gaseous form is called the *boiling point* or *condensation point*. Moreover, the technical term for what happens at these special points is a *phase change*.

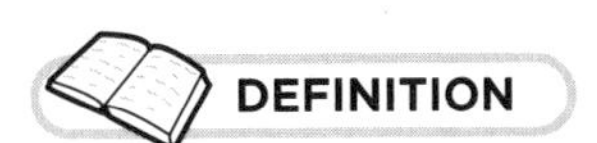

The **melting point** (or **freezing point**) of any substance is the temperature at which it transforms between the solid and liquid state when heated (or cooled).

The **boiling point** (or **condensation point**) of any substance is the temperature at which it transforms between the liquid and gaseous state when heated (or cooled).

A **phase change** is a transformation of a substance from one physical state to another, normally as a result of heating or cooling.

Thinking back to our earlier examples, we can now begin to see why phase changes occur. We saw that as you heat a solid, the atoms (or molecules) from which it is composed will oscillate more strongly. As you continue to heat the substance, there will be a particular temperature at which the oscillations are so strong that some of the particles can actually break free and begin flowing away from their neighbors. This is the beginning of the liquid state, which is characterized by a disorganized arrangement of loosely interacting constituent particles.

If you continue heating the liquid, you will eventually reach a specific temperature where the particles have sufficient kinetic energy that they can escape entirely from all the other particles in the substance. This is the beginning of the gaseous state, which is characterized by freely flying constituent particles that no longer interact with one another. Note that the process by which particles escape from the liquid into the gaseous state is also called evaporation.

RED ALERT!

In reality, the situation is not always this simple and some substances do not go simply from solid to liquid to gas when heated. The actual fate of a substance depends not just on temperature, but pressure as well. If the pressure is low enough, a substance can transform directly from the solid phase to the gaseous phase through a process called sublimation. Also, we have tacitly assumed that the substances under consideration are essentially pure. Real-world materials, however, are generally admixtures of various substances. Although the presence of additional substances will change some of the details, the general ideas we've discussed here will remain true for impure substances as well.

For a given mass of material, as an object warms, its volume expands. What does this mean in terms of its density? If you recall that density is given by $\rho = m/V$, you'll see that a slightly warmer (and therefore slightly more voluminous) object will have a slightly lower density. At an extreme case, when a certain mass of solid is warmed to its liquid state, the solid is denser than the liquid. For this reason, during the melting stage, solid materials normally sink in the surrounding liquid.

CONNECTIONS

On April 15, 1912, the RMS *Titanic* ended its infamous maiden voyage after colliding catastrophically with an iceberg. You may find this surprising, though, since we've just learned that solid substances generally sink when immersed in liquids of the same substance. Why was there an iceberg floating on the ocean surface, instead of being at the ocean bottom where it belonged? The reason has to do with a very unique property of water: the solid form is actually less dense than its liquid form. This ultimately boils down to the specific crystal structure of water ice, in which there is an unusually large amount of empty space. This property of water is rare among pure substances.

This concludes our basic introduction to temperature and how it relates to the microscopic properties of matter. So far, we have explained the ultimate fate of bodies at different temperatures that come into contact. But we have yet to discuss what actually happens as the energy of a warmer body transfers to a cooler body. To describe this, we must introduce a new concept known as heat. This topic is so rich that it deserves its own chapter, and it is to this that we now turn.

The Least You Need to Know

- Temperature is a measure of how hot or cold a substance is.
- Temperature can be measured using Celsius, Fahrenheit, or Kelvin scales, the latter being an absolute scale for which zero is the lowest temperature achievable ("absolute zero").
- When two objects at different temperatures come into thermal contact, they will eventually equilibrate at the same (intermediate) temperature.
- Solid and liquid substances expand when heated and contract when cooled.
- Changes in temperature can lead a substance to transform between solid, liquid, and gaseous forms, through a process known as phase change.

CHAPTER

15

Heat

Like temperature, heat is a word we frequently use in our daily lives, and you undoubtedly have some sort of intuition for both of these concepts. Unlike temperature, though, there is a big difference between what heat means in ordinary conversation and what it means to a physicist. In this chapter, we will define heat from a physical point of view. We will also learn about the important linkages between heat and energy.

This will enable us to deploy some new tools to model and predict the way that materials behave when heated and cooled. We'll get a sense for how much an object resists thermal change, which determines how much a hunk of stuff will warm when heated. We'll touch on how quickly (or slowly) various materials heat up, and explain why this is the case.

We will also return to a concept introduced in the previous chapter—the phase change—this time approaching it within the context of heat. This will improve our understanding of why things like melting points and boiling points are "points" in the first place rather than broader temperature ranges.

In This Chapter

- Thermal energy and heat
- The first law of thermodynamics
- The "hidden" heat of phase changes
- Three ways in which heat is transferred

Heat and Energy

In Chapter 7, we introduced the principle of conservation of energy. While everything we said at that point was true, you may have noticed that we used a good number of caveats and wiggle words. For example, we mentioned that things were much easier if the system you are studying was "isolated."

We also stated that, when solving mechanics-type problems using conservation of energy, it is important to separate "the objects you are interested in from the rest of the world." Well, dear reader, the time has now come to consider the rest of the world!

Let's start by considering a simple situation like a book lying on a level table. Let's give the book a little shove and then assume there is enough friction in the system that the book eventually comes to rest. From the energy standpoint, we know that initially (right after the shove) the book has some kinetic energy, but at the end it has none. Therefore, during the process of stopping, the kinetic energy must transform into something else. It could not have been gravitational potential energy, since all the movement happened on a level plane at the same height. So where did the kinetic energy go?

The answer is that it went into the rest of the world. To see where exactly, we need to enlarge our "system" to include not just the book, but also the surface of the table itself. (Some textbooks say at this point that we need to consider both the "system" and its "surroundings.") If you think back to our discussion on forces this makes sense, since to get correct answers in real-world situations like this we had to consider forces due to friction. These, of course, result from the surface properties of the table. So, the answer to our conundrum is that the book's kinetic energy was ultimately used to increase the temperature of the two objects affected by friction—the book and the table. If you had a very sensitive thermometer, you could actually measure this.

The type of energy we are considering here is called *thermal energy*, and it is basically the energy associated with the raising (or lowering) of temperature. Moreover, the actual amount of thermal energy transferred from one body to another during a physical process such as this is referred to as *heat*.

Thermal energy is the form of energy transferred from one body to another as a result of a difference in temperature. The amount of thermal energy transferred in the process is referred to as **heat**.

To explore a bit further the relationship between mechanical energy (that is, energy of motion) and thermal energy, let's consider an experiment originally conducted by British physicist James Joule. We'll begin with a well-insulated, cylindrical thermos bottle that is full of water. Inside is a paddlewheel that can rotate about the central axis of the thermos. A weight is attached via a pulley to the paddlewheel axis, such that when the weight moves downward, it turns the paddlewheel (see Figure 15.1).

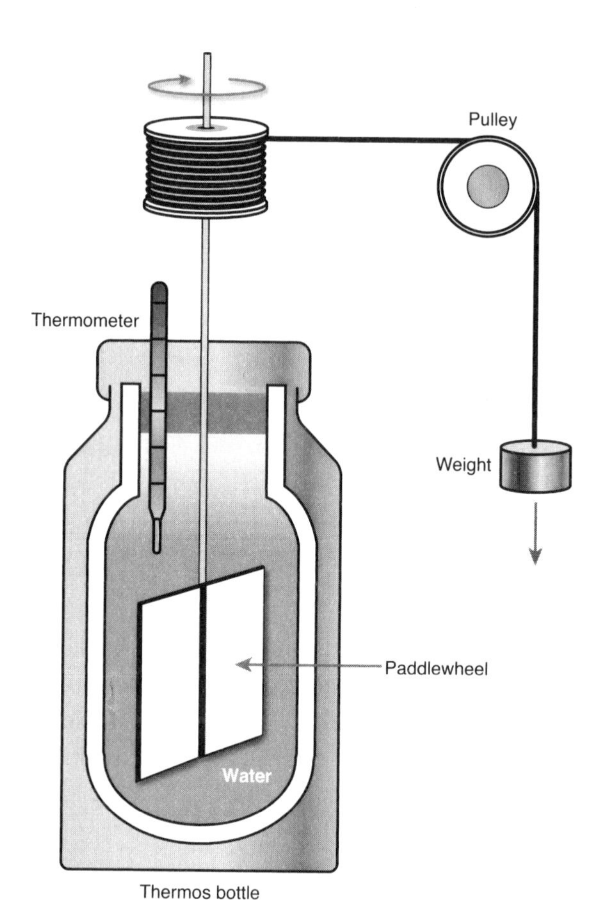

Figure 15.1

A schematic of the apparatus used by James Joule to solidify the relationship between mechanical and thermal energy.

If the weight falls due to gravity by a certain distance, we know that there is a loss in gravitational potential energy. Of course, it is not really lost, but rather transformed into another form. Basic mechanics tells us that some of the energy is transformed into kinetic energy of the paddlewheel. Eventually, though, the paddlewheel comes to a halt due to the frictional forces acting between the paddles and the water. We now know that the kinetic energy must have been transformed into thermal energy (or heat), which went on to raise the temperature of the water.

Sure enough, if you were to measure the temperature of the water over the course of this experiment, you'd find that it increased. You'd even observe that the increase in temperature is directly proportional to the amount of potential energy lost by the falling weight. You would find that about 4 joules of mechanical energy leads to a 1°C increase per gram of water in the thermos bottle. This special quantity of heat energy is known as the calorie, and it is abbreviated as "cal." Heat can be measured in both calories and joules, and the modern conversion factor between them is 1 cal = 4.186 joules. (Note that the term "Calorie" thrown around when speaking of nutrition is actually a kilocalorie—1,000 calories. That is why it is normally capitalized.)

The First Law of Thermodynamics

Recall our discussion from the last chapter about the structure of matter on the microscopic scale. Solid materials are composed of particles that oscillate relative to one another due to the internal forces holding them together. On the other hand, gases are composed of freely flying particles that interact negligibly with one another. Let's define the *internal energy* of a system as the sum total of the kinetic and potential energies associated with the particles that make up the material.

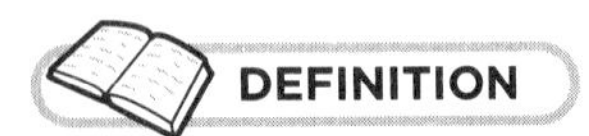

The **internal energy** of a system is the total kinetic and potential energy embodied by the particle constituents of the matter within the system. It excludes all kinetic and potential energy of the system as a whole (e.g., the energy of linear or rotational motion or gravitational potential energy).

Internal energy is very useful because it provides some intuition about what happens when heat is transferred from one body to another. Think back to Figure 14.1 from the last chapter that showed the microscopic structure of sodium metal at a colder and a warmer temperature. If you had two hunks of sodium metal at different temperatures, the atoms in the warmer hunk would have more kinetic energy and would oscillate with greater amplitude than those in the colder hunk.

If you were to put the two hunks in thermal contact with one another, the energy of the more energetic atoms in the warmer hunk would transfer by way of collisions to the less energetic atoms in the colder hunk. You can picture this beginning when the more energetic atoms on one side of the boundary collide with the less energetic ones on the other. The faster atoms will slow down and the slower atoms will speed up. Moreover, since the boundary atoms are "connected" to row upon row of other atoms as you move deeper into the metal, you can see how the thermal energy essentially travels through the entire material. Once thermal equilibrium is reached, the two objects would be at the same temperature, and all the atoms in both hunks would be oscillating with an equivalent amount of thermal energy.

We now have all the pieces we need to speak about the conservation of energy without any caveats and without the need to resort to unrealistic things like infinitely slippery surfaces. To do so, we'll formulate a precise mathematical relationship that embodies all the types of energy we've been considering. Here goes:

$$\Delta U = Q - W \qquad \textbf{Equation 15.1}$$

The symbol U in this equation refers to the internal energy of the system under consideration, so ΔU is just the change in internal energy between some initial state and some final state. The symbol W stands for work, which you'll recall from Chapter 7 is the (scalar) product of force and displacement. In other words, it has to do with mechanical motion of some sort. Finally, the symbol Q represents thermal energy, or heat. This simple formula is so important to physicists that it has been elevated to the status of physical law. In fact, it is the *first law of thermodynamics.*

The **first law of thermodynamics** says that the internal energy of a system can be increased either by adding heat to the system or by doing mechanical work on the system. It is the most general form of the principle of conservation of energy.

The first law tells us that if you add heat to a system or do mechanical work on the system, the result is an increase in the system's internal energy. For example, heating a block of sodium metal will lead to more vigorous jiggling of the constituent atoms. Conversely, if heat is removed from a system or the system does work on its surroundings, the result is a decrease in the system's internal energy. Cool down a block of sodium metal, and the atoms will jiggle around a bit less.

Internal energy, heat, and work are all forms of energy. For this reason, the SI unit for all three quantities is the joule. This is fitting, of course, since it was James Joule who recognized that you could equate thermal and mechanical energy as we have just done.

Heat Capacity

When you add heat to an object, you increase its temperature. How much you increase it has to do with the exact material from which the object is made. We've already seen that adding 4.186 joules of energy to a gram of water will raise its temperature by 1°C. This is but one special case of the more general mathematical relationship for this effect, which is given by:

$$Q = mc(\Delta T)$$ **Equation 15.2**

Here, Q is the heat applied, m is the mass of the object, and ΔT is the observed change in temperature. The symbol c is just a proportionality constant that takes into account the nature of the material under study. For water, of course, $c = 4.186$ J/g/°C. The name for this substance-dependent constant is *specific heat,* and it is a measure of how much heat is required to raise the temperature of a unit mass of a given substance by a given amount.

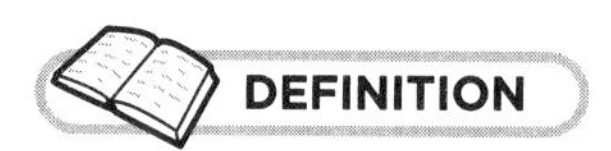

Specific heat (c) is the ratio of heat applied to an object to the subsequent rise in temperature, per unit mass.

Heat capacity (C) is the ratio of heat applied to an object to the subsequent rise in temperature.

You will often see a virtually equivalent form of Equation 15.2, where the mass is omitted: $Q = C\Delta T$. The constant in this case (C) is a quantity known as *heat capacity*. Again, this constant tells you how much heat is needed to raise the temperature of a particular object by a given amount. In this form, however, the heat capacity depends not only on the type of stuff the object is made of, but also how big it is. (In fact, "specific heat" is actually just contemporary shorthand for the term "specific heat capacity.")

Before delving further into an interpretation of heat capacity, let's first make sure we're comfortable with these two equations. Suppose you have sitting on your desk a 50-gram chunk of copper and a 50-gram chunk of aluminum. How much heat is required to raise the temperature of each by 7°C? To proceed any further, we need to know the specific heats of copper and aluminum. We can look these up and find that the values are 0.387 J/g/°C for copper and 0.900 J/g/°C for aluminum. Applying Equation 15.2, we see that 50 g × 0.387 J/g/°C × 7°C = 135 joules is required in the case of copper and 315 joules is needed for aluminum.

This example helps provide some intuition about what heat capacity means. We now see that it takes nearly three times as much thermal energy to raise the temperature of aluminum relative to copper. In a sense, then, heat capacity is like a sort of "thermal inertia" that acts to inhibit changes in temperature. The higher the heat capacity, the more effort (i.e., heat) is needed to increase its temperature. Along these lines, it's interesting to note that applying Equation 15.2 to water tells us that 1,470 joules would be required to raise the temperature of 50 grams of water by 7°C. That's more than 10 times the amount needed for an equivalent amount of copper! In fact, water has one of the highest heat capacities among commonly encountered substances, while the heat capacities of metals are generally pretty low.

We have so far considered only increases in temperature, for which ΔT is positive and for which heat flows into the system. But the equation works just as well for negative values, such that a decrease in temperature means that heat flows out of the system.

TRY IT YOURSELF

33. Suppose you have 3,785 g of milk sitting at 19°C on your kitchen table. If you put this into the refrigerator, how much heat must be taken out of the milk in order to reduce its temperature to 3°C? You can assume that the heat capacity of milk is the same as water.

Phase Changes Revisited

While Equation 15.2 is generally pretty handy, it turns out that there are a few special situations in which it does not apply. In the practice problem above, we considered cooling a gallon of milk from near room temperature (19°C) to just above its freezing point (3°C). Since you begin and end with milk in the liquid form, you can freely apply Equation 15.2 in this instance, and simply stick in $\Delta T = -16°C$. But what if we instead want to put the milk in the freezer and cool it down to -20°C? In this case, you begin with liquid milk and end up with solid milk, so the equation no longer applies.

Now two things are happening. In the process of cooling down over a temperature range that straddles its freezing point, the milk will not only cool but undergo a phase change (from liquid to solid) as well. Stated more generally, whenever a substance is warmed (or cooled) sufficiently to undergo a phase change, then the change in temperature is no longer simply proportional to the heat applied to (or removed from) the material.

The reason for this is that an additional amount of heat is required to carry a substance from the solid to the liquid state or from the liquid state to the gaseous state. The general term for the "extra" heat required is "latent heat," and it is given by the formula $Q = mL$. In this case, Q is the amount of heat required to complete the phase change for a given mass (m) of a substance. The latent heat (L) depends on both the identity of the substance and also the type of phase change exhibited.

In this section, we have pretended that heat capacity and latent heat are constant quantities that depend only on the type of substance considered. This is not quite exact, though, and these values also depend slightly on ambient conditions like temperature and pressure. These subtleties must be accounted for whenever you are seeking exact answers to specific problems, though they do not change the basic concepts we are discussing here.

Where does this "latent" (from the Latin term for "hidden") heat come from? To answer this, it helps to once again think of the microscopic structure of matter in the solid, liquid, and gaseous state. Solids, as we've seen, are characterized by nicely ordered rows upon rows of atoms or molecules. Liquids, on the other hand, flow and take the shape of their containers. For this reason, the order exhibited by solids must be broken during the melting process. It is this very "rearrangement" of the constituent particles that is behind latent heat. For a given mass of material, a certain amount of energy is needed to break all those interatomic bonds such that the substance can flow like a liquid, and this is called the "latent heat of fusion."

The same general argument applies during evaporation, in which case the loose binding that characterizes the liquid state is replaced by a bunch of free particles that whiz about more or less independently of one another. A certain amount of heat is needed to liberate all the atoms into the gaseous phase, and this is called the "latent heat of vaporization."

Imagine that you have a 100-gram ice cube that you've just taken from the freezer (temperature = -20°C) and you want to convert it to 100 grams of water at 70°C. Let's apply the above equations to see how much total energy is required. First of all, we need to warm the ice cube up from -20°C to the freezing point (0°C). For this we need to know that the specific heat of ice is 2.09 J/g/°C, but once we do, we can apply Equation 15.2 to see that 4.186 joules are needed to heat the ice. The next step is to account for the energy needed to melt the ice, for which we use the latent heat of fusion. Given that the latent heat of fusion for water is 333 J/g, we can apply the latent heat formula to see that 33,300 joules are required just to turn 100 grams of ice into 100 grams of water with no change in temperature. Finally, we can use Equation 15.2 again to calculate that 100 g × 4.186 J/g/°C × 70°C = 29,300 joules are needed to heat the now-liquid water from 0°C to 70°C.

We now have everything we need to solve the problem. In words, the total heat required is the heat needed to warm the ice + the heat needed to melt the ice + the heat needed to warm the water. In numbers, this is Q_{total} = 4.186 J + 33,300 J + 29,300 J = 66,780 J.

TRY IT YOURSELF

34. How much additional heat is required to transform the 100 g of water at 70°C into 100 g of steam at 150°C? For this, you'll need to know that the heat of vaporization for water is 2,260 J/g and that the specific heat for steam is 2.01 J/g/°C.

The situation can be nicely summed up with the following graph, which shows how the temperature increases as thermal energy is applied to 100 grams of water from the solid state to the gaseous state. A similar graph would apply to most other pure substances, just with different details like the slopes of the lines and the temperatures where phase changes occur.

From this figure, we see that there is a linear increase in temperature as the ice is warmed, followed by a brief period where the temperature is fixed at 0°C even though heat is still being applied. This heat is being "used" to rearrange the interatomic structure of the ice into liquid form, while the temperature remains the same. Once the ice has melted, the temperature once again increases linearly as the water is heated from 0°C to 100°C. There, the temperature levels off until a sufficient amount of latent heat is applied to liberate all the water molecules from the liquid to the gaseous state. Finally, the temperature increases linearly once more until the steam reaches its final temperature of 150°C.

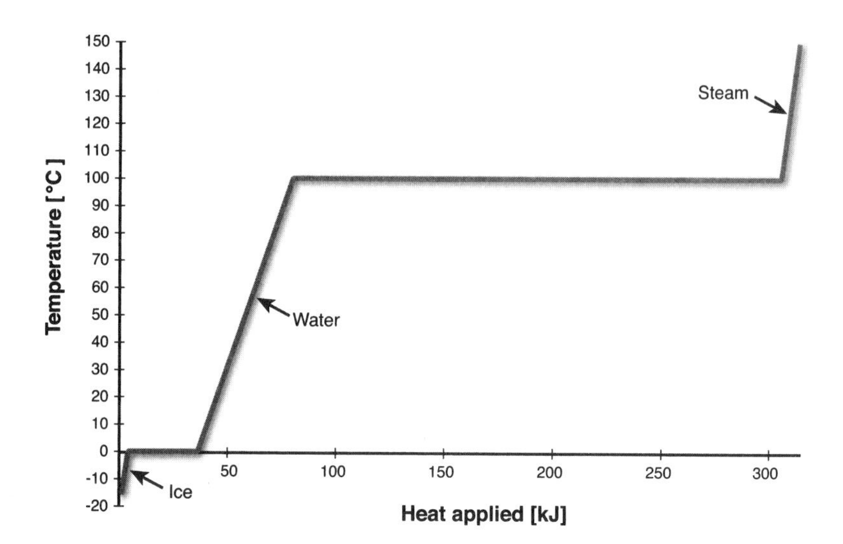

Figure 15.2

This graph shows how the temperature of 100 grams of frozen water (ice) increases as thermal energy is applied all the way through the boiling point.

Heat Transfer

Whenever two objects at different temperatures come into thermal contact with one another, heat will leave the warmer one and enter the cooler one until the two objects come into thermal equilibrium. While direct contact certainly helps the process, it turns out that heat can transfer from one body to another without ever even touching it. In the end, there are three primary ways that heat can transfer from a warmer to a colder body: like a branding iron, like a radiator, or like the sun. We'll explore each in turn in this section.

Conduction

The simplest of these to understand is the one we've already discussed—transfer of heat via direct contact. We discussed this when describing what happens to the jiggling sodium atoms when a warmer hunk of sodium metal is pressed against a colder hunk. The atoms on the hotter side of the boundary collide with those on the colder side, the process continues to additional atoms further from the boundary, and eventually both sides equilibrate at the same temperature. This direct process of heat transfer is called *conduction.*

Conduction is the transfer of heat through a substance from a region of higher temperature to a region of lower temperature. It relies on direct contact, and the energy is transferred mainly through collisions of the material's constituent particles. The matter itself does not need to move.

Let's now consider the slightly different case of a branding iron. If a cowboy wants to stamp his name on his cow, he may stick the end of the iron into a campfire. The far end will heat up quickly, but at first the handle end will remain cool. This difference in temperature between the cold and hot ends (called a "gradient") tells you that the iron is not yet in thermal equilibrium, so thermal energy will creep towards the handle until the whole iron is hot. How fast this occurs depends on the geometry of the iron as well as the type of material from which it is made. The compact expression that links this all together is as follows:

$$\frac{Q}{\Delta t} = -k\frac{A\Delta T}{\Delta x}$$ **Equation 15.3**

Here, Q is the amount of heat transferred over a length of time Δt, while ΔT is the difference in temperature between the two ends of the iron (separated by distance Δx). The remaining properties of the iron are described by A, the cross-sectional area of the iron, and k, a material-specific constant called thermal conductivity. Don't be confused by the minus sign; it simply says that heat transfers from the hotter to the colder end, which we already knew.

Suppose now that the branding iron is 0.2 meters long, has a cross-sectional area of 0.02 m × 0.02 m = 0.0004 m^2, and is made of pure iron (k = 79.5 J/s/m/°C). If the handle is at 20°C and the hot end is at 500°C, how much heat will be transferred through the bar in one second? Equation 15.3 gives us all we need to solve this, and we find that Q = 79.5 J/s/m/°C × 0.0004 m^2 × 480°C × 1 s ÷ 0.2 m = 76 joules.

TRY IT YOURSELF

35. Suppose that our cowboy repeats the experiment using instead a bar made of Styrofoam (k = 0.033 J/s/m/°C). Ignoring the fact that the Styrofoam will probably melt in the process, how much heat is transferred in 1 second?

If you solve the practice problem, you will find that nearly 2,500 times less heat will travel through the Styrofoam than through the iron during the same amount of time. This explains why beer coolers are made from Styrofoam and not iron—the rate at which cold beer warms up to room temperature will be much less with a cooler made from a poor heat-conducting material. Poor conductors of this type are generally referred to as insulators. Conversely, metals (like iron) turn out to be good heat conductors, which explains why cooking pots are made from metal.

Convection

Houses built in the early twentieth century were heated by large metal boxes called radiators. Hot water would pass through and thus warm up the metal box, which could then transfer energy to the room. But how? If you were to sit on the radiator, you'd warm up pretty quickly via conduction. But obviously conduction won't help warm you if you're instead seated across the room.

Radiators rely on the air in the room to circulate the heat from the radiator. The air molecules that collide with the radiator will pick up thermal energy from these collisions, and they will in turn transfer some energy (via collisions) to colder atoms a little further from the radiator. Eventually, currents of warmer air will begin to circulate away from the radiator. These currents result from gradients in air density caused by the radiator's heat, as well as the presence of gravity. As a result, warmer air will rise away from the radiator and replace cooler air from above. This will act to speed up the transfer of heat across the room. Thermal transfer of this sort is referred to as *convection.*

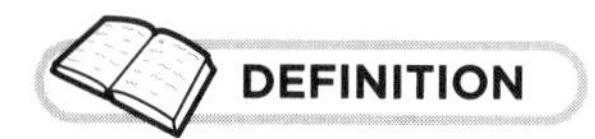

Convection is the transfer of heat through a fluid (liquid or gas), which is aided by currents established by temperature-induced density gradients in the fluid. This heat transfer requires actual motion of matter, as hotter material moves to replace colder material.

Air is not the only substance in which convection works. It will occur in any fluid—whether gaseous or liquid. Water is another typical fluid used for heating via convection (e.g., water in a tea kettle), as are antifreeze and various oils.

Radiation

You could reasonably argue that even when standing across the room from a radiator, you are in loose thermal contact due to the presence of the air in the room. But the third form of heat transfer that we will cover requires no such intervening medium at all. It can transfer heat across a perfect vacuum, and do so over millions of miles at that.

The process we are referring to is called *thermal radiation*, and in this case the thermal energy is actually carried by way of electromagnetic radiation. We'll get into electromagnetic radiation in Part 5 of this book, but suffice it to say that when the constituent particles of a hot object have an electric charge, the thermal-induced motion can generate electromagnetic waves. Electromagnetic waves at thermal energies are sometimes referred to as infrared radiation.

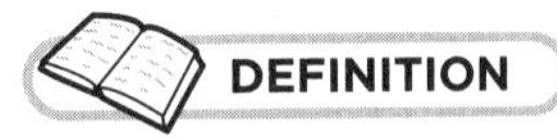

Thermal radiation is the transfer of heat via electromagnetic radiation, which stems from the thermal motion of charged particles within the hotter body. No matter at all needs to be present in the space over which heat is transferred by radiation.

Despite all the fires, power stations, and radiators on Earth, by far the greatest source of heat on this planet is thermal radiation from the sun. Every second, approximately 1,300 joules of thermal energy from the sun strikes (on average) every square-meter patch on Earth's surface—significantly more, you'll recall, than our cowboy transferred when he stuck his branding iron in the fire. If his iron got hot enough, it would have actually begun to glow, and the red-orange light emitted is another example of thermal radiation. Cows of the world beware—if the branding iron is glowing, it's going to be hot!

CONNECTIONS

In the fall of 1900, a German physicist named Max Planck was carefully studying the properties of thermal radiation emitted from hot objects like branding irons. Unfortunately for him, all of the known physics at the time was insufficient to predict the colors with which such objects glow. He therefore took a leap of faith and derived a new physical model for the radiation process. This new approach would serve as the basis for an entirely new branch of physics, quantum physics, which underpins a vast range of our modern technologies.

The Least You Need to Know

- To properly apply the law of conservation of energy, you must take into account thermal energy transferred to the surroundings as well as mechanical energy (i.e., kinetic or potential) of a system.
- Heat is the thermal energy that moves from a warmer body to a colder body.
- When you add heat to an object, you increase its temperature by an amount determined by the object's heat capacity.
- The rate at which heat transfers through an object depends on the object's thermal conductivity, which depends on the microscopic properties of the underlying substance.
- Heat can also be transferred through the processes of convection, which requires a fluid medium, or thermal radiation, which does not.

CHAPTER

16

Thermodynamic Processes

In This Chapter

- Ideal gases and their thermodynamic properties
- Heat engines and heat pumps
- The second and third laws of thermodynamics
- Entropy and the ultimate fate of the universe

In the previous two chapters, we learned about the basic workings of temperature and heat and why heat flows as it does when temperature differences are present. We also touched on the links between thermal energy and mechanical energy and explained how energy is only truly conserved when both are fully accounted for. In this chapter, we will learn how those basic ideas can be applied for our benefit through clever engineering.

An important theme throughout this chapter will be how heat flows can be used to generate motion, and how motion can be used to drive heat flows. To bring this all down to Earth, we're going to focus on three concrete examples—the hot air balloon, the steam engine, and the refrigerator.

The fundamental concepts that govern such cases are the laws of thermodynamics. We've already covered the first two of these; in this chapter, we will unveil the remaining two. In so doing we'll discover that not only do these important laws help us to predict and even apply thermal-mechanical processes, but they also help us determine what is and isn't physically possible.

The Ideal Gas Law

We've already discussed all the physics you need to describe the first example that we will consider—the hot air balloon. Chances are you've seen these graceful objects dot the sky on occasion, and you may have even heard the sound of their heaters roaring from time to time. Clearly, the heater is there to put the "hot" into "hot air balloon." But how do these things really work?

You'll actually find most of the answer in Chapter 13 of this book, in which the properties of fluids were introduced. Recall that we made a big deal at the time about how Earth is covered not by one "ocean" but two—the water kind as well as all the air in the atmosphere. This air is the very medium that is used by hot air balloons to reach their heights.

In this section we will treat the air as a so-called ideal gas, which means that any volume containing the molecules is very large compared to the size of any given molecule, the density and pressure are both relatively low, and the temperature is high enough that there is no chance of the gas condensing to a liquid. The air in Earth's atmosphere can be safely treated as an ideal gas, but use caution when applying the concepts of this section to other gases in more extreme conditions.

You may have already guessed that the upward force that lifts a hot air balloon is the buoyant force. Now, we're not thinking about logs or boats floating in water, but rather giant contraptions made of nylon and rope that float in air. Under the right conditions, the buoyant force can be increased enough that it exceeds the weight force pulling the balloon and its cargo and passengers back down to Earth. Once again, we can apply the basic rule of thumb that an object will float if its average density is less than the density of a surrounding fluid. Normal air has a pretty low density, though, so to reach the condition of floating, some component of the balloon must attain such a low density that the balloon's density as a whole (including the envelope, basket, heater, fuel, etc.) is less than that of air.

This is where the heater comes in. As you may have seen, the heater is pointed directly at a large opening in the bottom of the balloon. This allows the pilot to warm up the air inside the balloon directly. The idea is that, by heating the air, the overall density of the balloon will get low enough for it to fly. This is because the mass per unit volume of a gas decreases significantly as you increase its temperature.

Why might that be? Well, recall that on the molecular level a gas is composed of little molecules racing about. The higher the temperature, the faster their average speed as they move. Inside the balloon, the molecules will bump up against the inner surface of the balloon and exert an outward force, and therefore a pressure. The faster the molecules zoom, the larger this force and the larger the exerted pressure.

If you look carefully, though, the balloon isn't getting bigger as the heat is applied. This must mean that the pressure exerted on the inner surface of the balloon is in equilibrium with the pressure outside. (Recall that we already mentioned in Chapter 13 that the pressure of a fluid remains constant from one place in the fluid to the next.) Since there is a big hole in the bottom of the balloon (through which the heat is applied), air molecules can move into and out of the balloon as needed to ensure that the pressures inside and outside remain about the same.

If the pressure inside the balloon isn't changing, but the average speed of the air molecules is, then there must be fewer molecules crashing into the balloon's inner surface. The rest must exit the bottom of the hole and enter the vast "ocean" of molecules that make up the atmosphere. With fewer molecules inside the balloon, the total enclosed mass must decrease. The volume isn't changing. Hence, since density is the ratio of these two ($\rho = m/V$) it follows that a warm gas is less dense than a cold gas.

Physicists have come up with a nice compact relationship that summarizes how a gas responds when heat is applied. It is called the ideal gas law, and it looks like this:

$$PV = NkT$$ **Equation 16.1**

Three of the variables in the equation should be clear: P is the pressure of the gas, V is the volume in which it is contained, and T is its temperature on the absolute scale. The variable N simply stands for the number of molecules present. Finally, the symbol k is a constant value that must be present to make the dimension of both sides of the equation agree with one another; its value is known to be $k = 1.38 \times 10^{-23}$ J/K.

Note that this constant k is not related to the spring constant we talked about earlier, even though we are using the same symbol. This is not the only time we will reuse symbols, since even with the Greek alphabet, there are only so many letters to choose from. (For example, d is not always a lever arm.) Make sure you read carefully, and don't assume you know what a symbol stands for without checking for a definition or understanding the context of a particular situation.

As you can see, k is very, very small. The reason for this is that any reasonable volume of air, at a typical pressure, has a ridiculously large number of air molecules in it. The actual number depends on the exact type of matter and the ambient conditions. Scientists have adopted a useful benchmark, though, that helps us know how many molecules are present in everyday-sized objects. It is called *Avogadro's number* and it is equal to $N_A = 6.02 \times 10^{23}$. There are no units on this, since it really is just a number. Specifically, it is defined as the number of carbon-12 atoms in a sample with a mass of exactly 12 grams. What's more, any time you have 6.02×10^{23} molecules of any specific substance, you say that you have one *mole* of that substance.

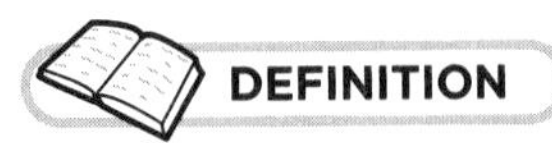

Avogadro's number, given by $N_A = 6.02 \times 10^{23}$, is a convenient quantity used when considering the large number of atoms present in everyday objects. It is equivalent to the number of atoms in a 12-gram portion of carbon-12.

A **mole** of any substance is the amount formed by 6.02×10^{23} atoms (or molecules) of that substance. It is sometimes abbreviated by "mol."

So we see that N_A is huge while k is tiny. To help get rid of those awkward exponents, let's see if we can't put them together somehow. Take the right side of Equation 16.1 and simultaneously multiply and divide by N_A. The result is as follows:

$$PV = \left(\frac{N}{N_A}\right) \times (kN_A) \times T = nRT$$ **Equation 16.2**

In the second step here, we have made two substitutions. First, we introduced a symbol that, by definition, represents the number of moles of gas present: $n = N/N_A$. Second, we introduced a new constant value $R = kN_A$. This is what physicists call the universal gas constant, since it is the same value no matter what ideal gas you are considering. That value is $R = 8.31$ J/mol/K. Our trick worked, and we got rid of those cumbersome exponents.

TRICKS AND HACKS

Students often get confused by the funny term "mole," not least because it conjures up images of visually challenged mammals. It's just a way of quantifying the amount of something that you have, based on Avogadro's number. For example, another way of saying you have 12 donuts is to say you have a dozen donuts. Similarly, if you have N_A atoms, then you have a mole of atoms. The number 12 is the analog to N_A while "a dozen" is analogous to "a mole."

To see how Equation 16.2 works, let's calculate just how many moles of helium gas, sitting at room temperature and atmospheric pressure, would fill a 1 m^3 container. First, let's make sure we have all the parameters we need. The pressure in this case is $P = P_{atm} = 101{,}325$ Pa. Room temperature on the Celsius scale is about 25°C. But the universal gas constant calls for temperature in kelvins, so we need to convert to the absolute scale. The result is $T = 25°C = 298$ K. All this, plus a little algebra gives us:

$$n = PV/RT = (101{,}325 \text{ Pa} \times 1 \text{ m}^3)/(8.31 \text{ J/mol/K} \times 298 \text{ K}) = 40.9 \text{ mol}$$ **Equation 16.3**

And, to see how many molecules this represents, simply multiply the number of moles by Avogadro's number: $N = n \times N_A = 40.9 \times (6.02 \times 10^{23}) = 2.46 \times 10^{25}$, or 24,600,000,000,000,000,000,000,000!

TRY IT YOURSELF

36. Suppose you hooked up a helium pump to the 1 m^3 container we just considered, and then pumped in another 40.9 moles of gas. Assuming the temperature and volume of the container remained unchanged, what would be the new pressure inside the container?

A hot air balloon is a simple device that can convert thermal energy (produced by the heater) into mechanical motion (the rising of the balloon in the sky). Let's face it, though. However beautiful it may be, a hot air balloon is not the handiest device on the planet. Let's now explore what is needed to construct a much more useful device—the steam engine.

Heat Engines

The basic function of a steam engine is to produce mechanical motion through the generation of steam. The steam is produced by heating water to a boil, and the motion is harnessed to do something useful, such as move a train forward or spin a turbine to generate electricity. The steam engine is just one specific example of the more general category of devices known as *heat engines.*

DEFINITION

A **heat engine** is a device designed to convert thermal energy into mechanical energy.

Heat, as you'll recall, flows whenever a hotter object is brought into thermal contact with a cooler object. Therefore two ingredients required for a heat engine are a hot object and a cold object (sometimes called the "hot reservoir" and "cold reservoir"). The third ingredient, which we will simply call the engine, is some device in between the hot and cold reservoirs that does the actual conversion of thermal energy to mechanical energy.

In the good old-fashioned steam engine, water is brought to a boil by a heater (the hot reservoir) and the expanding steam drives a piston and therefore does mechanical work (the engine). The steam is then brought into contact with a bath of cooling water (the cold reservoir) to condense the steam back to water so that it can repeat the process in cyclical fashion.

Since you are taking some of the thermal energy out of the system, the amount of heat that flows from the hot reservoir to the engine (Q_h) must exceed the amount of heat that flows from the engine to the cold reservoir (Q_c). If we let W represent the mechanical energy (or work) extracted by our system, then it should be clear that $W = Q_h - Q_c$. This is simply a consequence of energy conservation, and is illustrated in Figure 16.1.

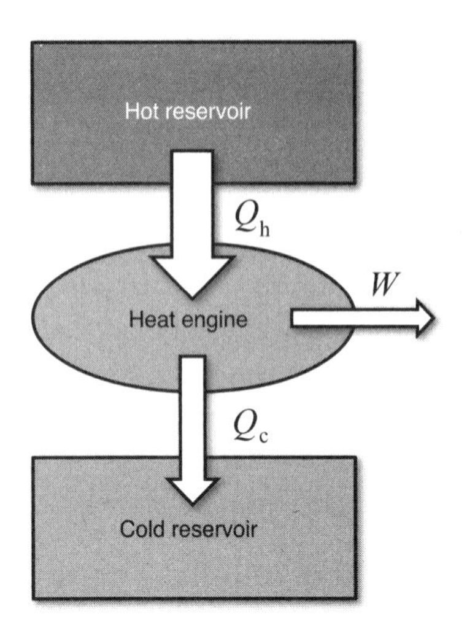

Figure 16.1

In a heat engine, heat flows from a hot reservoir to the engine (Q_h), and then from the engine to the cold reservoir (Q_c). Along the way mechanical energy is extracted (W).

To get the most out of your heat engine, you clearly want to maximize the amount of mechanical work relative to the amount of thermal heat you pour into the system. A good figure of merit to rate the quality of a heat engine is therefore the ratio of these two quantities. Physicists call this ratio the efficiency, which is abbreviated *e*:

$$e = \frac{W}{Q_h}$$ **Equation 16.4**

In practice, it turns out that most heat engines max out at efficiencies less than 50 percent. For example, automobile engines operate at about 25 percent efficiency. To some extent, this is simply a mundane limitation due to imperfections in our engineering. However, there is something more interesting at play here, a fundamental limitation due to the laws of physics. For reasons we'll explain later in the chapter, it is physically impossible to convert 100 percent of the thermal energy in a heat engine to mechanical energy.

What this means is that even the world's best engineers will never manage to develop the perfect heat engine, in which you extract a joule of mechanical energy for every joule of heat you put in. Some heat will always be lost in the process. But before we get into why this is, let's now examine a second thermodynamic system of great importance: a heat engine working in reverse.

TRY IT YOURSELF

37. What is the efficiency of a steam engine that produces 100 J of mechanical energy from an input of 150 J of heat? How much heat flows to the cold reservoir in the process?

Heat Pumps

Heat flows naturally from warmer objects to cooler objects. With a little bit of engineering, though, heat can actually be driven from a cooler object into a warmer object. The formal name for a device that can do this is the *heat pump.*

A **heat pump** is a device designed to use mechanical energy to transfer heat from a colder object to a warmer object.

What is required to do this seemingly impossible task? As you may have inferred from the preceding discussion, it is the application of mechanical energy. A generic schematic of a heat pump is shown in Figure 16.2. Once again we have the cold reservoir, the hot reservoir, and the pump that makes it all work. In this case you see that the energy flows are all opposite the case of the heat engine, although once again we see that $W = Q_h - Q_c$ (or, after a little algebra, $Q_c = Q_h - W$).

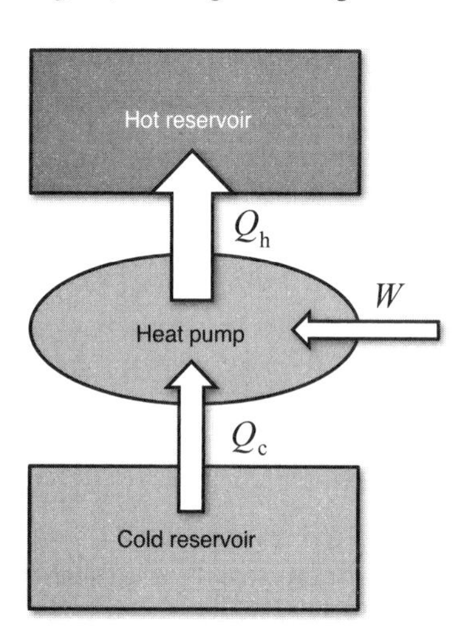

Figure 16.2

In a heat pump, mechanical work (W) is done on the system in order to drive heat from the cold reservoir to the pump (Q_c) and then from the pump to the hot reservoir (Q_h).

To see how this works, let's consider a concrete example: the refrigerator. You know that a refrigerator is a device that allows you to keep your milk and eggs at temperatures lower than room temperature. You're surely also aware that in order for the thing to work, it has to be plugged into the wall. The electricity drawn from the wall outlet is converted to mechanical energy and without it, your milk would quickly spoil.

Besides the big insulated box, there are three important elements to every refrigerator: an evaporator, a compressor, and a condenser. In addition, a refrigerator also uses a working fluid that runs in cycles through these three elements. Initially, the working fluid enters the evaporator—a long metal tube located in the region that you are trying to cool. Just before it enters, it flows through some sort of constriction designed to reduce its pressure to levels so low that it's on the verge of evaporating. As a result, the residual heat in the colder region (inside the box) is enough to cause the fluid to evaporate. This evaporation draws heat from the colder region (i.e., the latent heat of vaporization is absorbed by the fluid), which acts to make it even colder.

CONNECTIONS

A good working fluid for refrigerators (and air conditioners) is one that has relatively weak chemical bonds holding it together, so that it evaporates at low pressures and condenses at high pressures in ordinary temperature ranges. In the latter part of the twentieth century, the fluid of choice came from a family of chemicals known as chlorofluorocarbons (CFCs). These chemicals have the unfortunate side effect that they linger in the atmosphere when leaked, and the chlorine within them destroys ozone in the upper atmosphere. The ozone layer is important since it absorbs harmful ultraviolet rays from the sun. The famous ozone hole that caused so much consternation in the 1980s was due in large part to these chemicals, and a global response to the problem led to the phasing out of CFCs for refrigeration and other purposes.

The working fluid (in its gaseous phase) at low pressure exits the evaporator and flows to the compressor. The compressor can be viewed as a mechanical piston that reduces the volume containing the gas. As the volume is reduced, the pressure and temperature of the gas increases. The working fluid therefore exits the compressor as a hot, high-pressure gas.

Finally, the working fluid enters the condenser region, a long thin tube in good thermal contact with the room air. While passing through this condenser tube, the hot, high-pressure gas transfers heat to its surroundings and condenses back to the liquid phase. This time around the latent heat of vaporization is expelled by the fluid, further increasing the temperature of the surrounding air. Finally, the liquefied fluid is ready to enter the evaporator again and the cycle is complete.

Recall that $Q_c = Q_h - W$ for a heat pump. According to this, Q_c will be largest whenever $W = 0$. This seems to imply that the maximum cooling for our refrigerator should occur when no mechanical work is applied to the system at all. If that's the case, then we don't need to bother plugging it into the wall. If this sounds too good to be true, that's because it is. Once again there is something that prevents us from achieving 100 percent efficiency ("perfect" refrigeration in this case), no matter how good our engineering skills. The reason for this, and the reason that perfectly efficient heat engines are equally impossible, lies in a concept called entropy.

Entropy

We'll admit it up front—*entropy* as a concept is a little bit tricky to understand. It is easy enough to define, though, so let's start there. Whenever heat (Q) flows into or out of a system at temperature T, the entropy (represented by the symbol S) of the system changes. For very small changes, this can be expressed mathematically as follows:

$$\Delta S = \frac{Q}{T}$$ **Equation 16.5**

You may wonder how you can have a flow of heat if the temperature is not changing. This is why we emphasized that this relationship is really only true for small changes, in which case you can use the average temperature in the denominator when calculating ΔS.

Entropy is a quantitative measure of the amount of disorder in a system. Changes in entropy can be defined in terms of heat flows using the formula $\Delta S = Q/T$.

So much for calculating entropy. What is it, exactly, and why is it important? The best way to describe entropy is to say it is a quantity that embodies the amount of disorder in a system. To explain what we mean by that, let's consider two systems: the marching band at the Macy's Thanksgiving Day Parade and an elementary school playground at recess.

Next Thanksgiving, take a look at the marching band on television, and you'll see that every member is moving in the same direction at the same speed. Sometimes they even walk in lock step with one another. This is a pretty orderly way of moving a whole bunch of people around. Now, consider the elementary school playground. As soon as the bell rings and the pupils come outdoors, you'll immediately see children running in every which way at all different speeds. You'll even see collisions here and there, both of the elastic and inelastic variety! This movement of little humans is about as disorderly as can be. Physicists would say that the entropy of the school children is much larger than that of the marching band.

This analogy can be extended to explain an important difference between thermal energy and mechanical energy. Recall that thermal energy stems from the kinetic energy of the constituent particles in a system. In a gas, this means particles whizzing about every which way just like school children on a playground. In a solid, this means particles vibrating every which way relative to each other. Mechanical energy, on the other hand, is associated with unified and organized movement in one particular direction—just like the marching band—or even a piston being pushed into a cylinder or a weight falling smoothly to Earth. It follows that thermal energy is associated with more entropy than is mechanical energy.

Another important fact is that, as a rule, systems left on their own tend to become more disordered with time, and hence their entropy tends to increase. This all boils down to the laws of probability, as we'll explain. Take a large group of randomly selected people, and then turn them loose in a football field. After a few minutes, would you be more likely to find these people marching in lock step like the marching band or running around like a bunch of crazy kids?

The answer is the latter, because there are relatively few configurations with marching-band-level order, while there are countless ways for people to move around at random. For a given system, there will always be more possible configurations available with less order (more entropy) than configurations with more order (less entropy). As time moves forward, the odds are that the system will take on one of the configurations with more entropy. Hence, as time moves forward, entropy tends to increase.

We've been talking about band members and school children, but it turns out that this concept is also very relevant to the thermodynamic processes that take place in physical systems. It is so important, in fact, that it has been enshrined in the so-called *second law of thermodynamics*, which holds that the entropy of an isolated system increases with time or, at best, remains the same.

The **second law of thermodynamics** states that the entropy of an isolated system never decreases with time.

This fact is precisely what limits our ability to make 100 percent efficient heat engines and perfect heat pumps. Let's first revisit the heat engine. We saw previously that it is impossible to convert 100 percent of the thermal energy flowing from the hot reservoir into mechanical energy. Some residual heat will always escape to the cold reservoir. We now see that this is because if we were to convert all of the disorderly thermal energy emanating from the hot reservoir into orderly mechanical energy in the engine, the overall entropy of the system would decrease—in violation of the second law.

As for the heat pump, we see that thermal energy is being expelled by the cold reservoir to the engine and also from the engine to the hot reservoir. In both cases, since the sign of Q is negative, the corresponding entropy is decreasing. If that were the end of the story, we would again have a violation of the second law. To avoid this, something must be done to increase the overall entropy. That something, of course, is mechanical work done on the system. In the case of the refrigerator, remember the mechanical work is applied at the compressor stage, where a piston compresses the gas and raises its temperature. In other words, orderly mechanical energy is used to produce disorderly thermal energy. This equates to an increase in entropy, and this is what is required to ensure that the entropy of the overall process increases.

To determine whether or not a thermodynamic process will move forward, all you really need to know is whether or not the entropy of the system changes. Therefore, entropy derives its importance as a relative quantity. It's worth asking, though, if entropy has any significance on an absolute scale. In other words, does entropy ever reach a value of zero, and if so, under what conditions?

By the very definition of entropy, it should be clear that an equivalent question is whether or not there exists a physical system with perfect "order." For starters, this would have to be a solid material since the constituent particles in gases and liquids have no orderly configuration in space. But even solids can be disorderly, depending on how the constituent particles join together. The most orderly configuration possible is a crystalline solid with no impurities or imperfections whatsoever. In other words, a perfect crystal.

But by now you also know that the atoms in a perfect crystal are vibrating constantly in various directions, and that the amplitude of those vibrations is determined by the material's temperature. As temperature is lowered, the amplitude of the vibrations reduces more and more. If you could somehow cool a solid down to absolute zero, the overall kinetic energy of vibrations would go to zero and hence the particles would come to rest. Since no motion is the most "orderly" motion possible, it follows that a perfect crystal at absolute zero must be in a state of lowest possible entropy. This is the crux of the final law underlying thermodynamic processes: the *third law of thermodynamics.*

The **third law of thermodynamics** states that the lowest achievable entropy for any physical system occurs with a perfect crystal at absolute zero.

Before leaving the wonderful world of thermodynamics, it's worth reflecting on the broader meaning of entropy. The punch line is that mechanical energy can be transformed into thermal energy, but thermal energy cannot be transformed into mechanical energy (in the absence of an external influence). If you think about it, this has ramifications far beyond just heat engines and heat pumps.

Consider a swinging pendulum, for example. We have already seen that in the real world a pendulum will always lose some of its mechanical energy to thermal energy via frictional forces and/or air drag. Thus, the second law explains why an isolated pendulum will eventually come to rest. But it also explains why a static pendulum cannot spontaneously start swinging by, say, absorbing some heat from its surroundings and converting it to mechanical energy. This would lead to a decrease in entropy for the system, which we now know is forbidden. So, as you can see, entropy is in many ways the underlying factor that determines what is physically possible and what is not.

Another interesting aspect regarding entropy is the fact that for any isolated system, entropy can only increase with time. So far we've thought about this with respect to relatively small systems (like automobile engines and refrigerators). But what if we were to take the entire universe as our "isolated system"? From this point of view, we see that the entropy of the entire universe will only increase with time. This implies that, one day, the universe as a whole will reach a state of maximum entropy.

Another way of viewing this is that one day, far in the future, the universe will reach a state of thermodynamic equilibrium. When that happens, every part of the universe will be at the same temperature, so no thermodynamic processes (or anything else) will be able to happen. Scientists who think about this sort of thing have dubbed this the "heat death" of the universe. Fortunately for us, it appears that we are still a long way from this fateful condition, so we needn't worry that our cars will cease to start or that our refrigerators will cease to cool any time soon.

The Least You Need to Know

- Objects that are lighter than air will float in the atmosphere.
- Entropy is a measure of a system's disorder, and it governs which physical processes can occur and which cannot.
- The second law of thermodynamics tells us that the overall entropy of an isolated system can only increase with time.
- It is physically impossible to build a perfect heat engine or a perfect heat pump, since the entropy of such systems would decrease.

PART

4

Electricity and Magnetism

Electricity has turned out to be an extremely useful form of energy in the modern world, so in the next five chapters we'll take a good hard look at the nature of electricity and the closely related phenomenon of magnetism. As usual, we'll focus on the basics first: charged particles and the forces between them.

For many people, this is the most challenging part of introductory physics. Most students don't have much direct experience with the basic elements here to fall back on, and this part may seem more abstract and mathematical as a result. But forces and energy are still the ruling concepts, and with care and a little imagination, you'll be able to comprehend this important part of physics as well as any other.

A couple of chapters will be devoted to analyzing simple electric circuits and the components that make them. Then we'll get into the fascinating relationship between electricity and magnetism, from the basic construction of an electromagnet to the interplay that allows electromagnetic waves to travel through empty space.

CHAPTER

17

Electricity

In This Chapter

- Charge, the basis for electricity
- The force between charged objects
- Polarization of electric charges
- The concept of the electric field

In stark contrast to our discussion of forces, fluids, and heat flows, our study of electricity will bring us into a brand new realm, one that involves some very different concepts from those with which we are most familiar. We have a new fundamental force to reckon with, whose source is a new physical quantity called the electric charge.

This part of the book provides an introduction to the topics of electricity and magnetism. Physics has shown us that these two things are very closely related, which is why the word "electromagnetic" was invented. Electromagnetic effects can get pretty complicated, and are tough to comprehend at first. We will start simple, with charged objects that don't move very much, so the force that we will talk about is also known as the electrostatic force.

Like any other force, the electrostatic force affects masses according to Newton's laws of motion. In particular, that means that an electrostatic force can lead to motion, causing a charged object to accelerate unless some other force prevents it from doing so. Although we shall postpone any real consideration of charged particles in motion, you should not lose sight of this important connection between electricity and mechanics.

Electric Charge

Electromagnetism plays an absolutely crucial role in the way our world works. It is impossible to understand the fundamental structure of matter without getting deep into electrical forces and the particles that are their source. Unfortunately, the real action takes place on a size scale that is so small it boggles the mind. If we want to talk about where electric forces come from, we will have to talk about particles that are so small that the word "microscopic" doesn't do them justice.

Our strategy in this book has been to help you relate to physics concepts by providing examples from your everyday life. That's going to be a lot more difficult in this section. These days, of course, we have all kinds of electrical devices that use electricity and magnetism to do many amazing things, but they mostly rely on charged particles moving around in electrical circuits. We'll get to circuits in a couple of chapters, but first we have a good deal of basic ground to cover. And we'll have to do it without a lot of familiar examples.

The electric force is in the category of forces that act without requiring physical contact. In this way it is a lot like gravity. But whereas the gravitational force acts between all objects that have mass, the electric force only acts between objects that have another physical property called electric *charge*. The strength of the electric force has nothing to do with the mass of an object at all.

Charge is the property that allows objects to feel the electric force. Uncharged objects are called "neutral," and are not affected by the electric force. Electric charge is a scalar quantity, and the standard unit for measuring it is the coulomb (C).

Careful experimental observations reveal that charge comes in two forms, which we call positive and negative. It was originally an arbitrary decision as to which charge would be called positive and which negative. It was observed that two objects with the same kind of charge, either both positive or both negative, would repel each other with a small force. If two objects have charges of opposite signs, then they attract each other.

The standard SI units for charge are called coulombs (C). Electric charge is a new fundamental dimension, which is not derived from any of the basic dimensions we used in mechanics. In everyday life, we don't experience large electrical forces because it is difficult to put much charge of either type on real objects. Charge tends to "leak off" of objects, especially when there is significant humidity in the air. Under dry conditions, certain objects can be charged by rubbing them with certain other materials. You can actually charge yourself up by shuffling your feet on a carpet on a dry winter day. While you might feel a spark when your charge gets discharged by contact with something else, you won't feel any attractive or repulsive force from another object.

In part, this is because the objects we normally encounter tend to be electrically neutral. They possess equal amounts of positive and negative charges, and those charges effectively cancel each other out. That's why it makes sense to call them positive and negative. The total or net charge on an object is the absolute amount of negative charge subtracted from the amount of positive charge it possesses. There are lots of electrical charges all around you, but virtually every object you deal with has equal positive and negative charges, resulting in a net charge of zero.

Another reason that we have a hard time relating to electricity is that charge itself is invisible. It's not just because atoms and other fundamental particles are so small. An ordinary object like a basketball can have a significant amount of charge on it, but even if it does, it doesn't look any different from a neutral basketball. Nor can we taste, smell, or feel electric charges directly. So where does this mysterious thing called charge come from?

Electric charge turns out to be an intrinsic property of the most fundamental particles that are the basic building blocks of the universe. You may already know that all of the chemical elements are made of incredibly small atoms. Atoms of different elements bind together to form molecules and all the different compounds we are familiar with. They are so small that there are about a million billion (1,000,000,000,000,000) of them in the smallest speck of dust on your bookshelf. Incidentally, atoms and molecules are the same material constituents we referred to earlier in this book when we discussed the nature of heat and temperature.

Any picture or diagram of an atom that you see is bound to be a misrepresentation. Atoms are too small to "look" like anything at all. Roughly speaking, the diameter of an atom is about 1×10^{-10} m. In addition, their behavior is governed by quantum mechanics, a fairly bizarre kind of physics that defies common sense notions like the location of a particle. Nevertheless, we will use a fairly standard figure to indicate the constituent parts of a typical atom, and their relative locations, to a reasonable approximation.

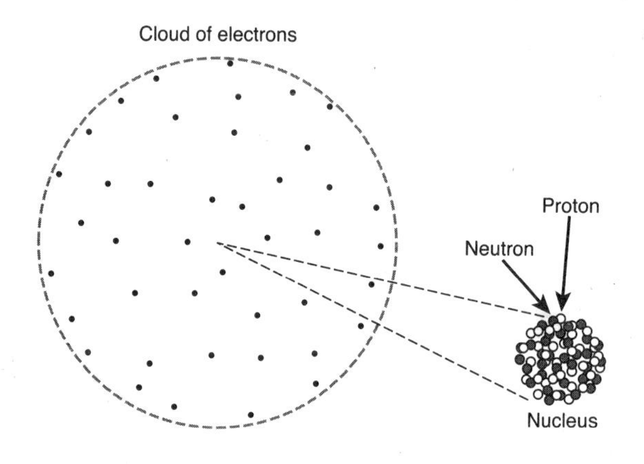

Figure 17.1

Every atom is composed of a positively charged nucleus surrounded by a cloud of negatively charged electrons. Most atoms are electrically neutral, with the positive charges balancing the negative charges exactly.

The atomic nucleus is where all of the positive charge in matter resides. Nuclei contain two kinds of particles: neutrons, which have no charge, and protons with positive charge. In diameter, the nucleus is about 100,000 times smaller than the whole atom, so practically all of the volume of the atom is just a bunch of empty space with some electrons lingering about. To put this into perspective, an atom the size of Madison Square Garden would have a nucleus about the size of a Concord grape located at center ice, while the electrons would occupy a space spanning to the rafters. But while the electrons hog the volume of the atom, the nucleus has almost all of the atom's mass. The individual electrons also have almost no physical size, but they do carry all of the atom's negative charge.

All of the electrons in the universe are identical to each other, and have the same amount of negative charge. All of the protons in the universe are also identical to each other, and each proton has the same amount of charge as an electron (only of the opposite sign). Your standard atom has exactly the same number of electrons as protons, so its net charge ends up being precisely zero.

This isn't always the case, however. It is possible for an atom to lose one or a few electrons relative to its neutral state, at least temporarily. When that happens, the net charge on the atom will be positive, because there are more protons than electrons. It is also possible sometimes for an extra electron or two to stick to an atom, giving it a net negative charge. Atoms that are not electrically neutral are called ions.

Electric charge is another conserved quantity in nature. Charge can't be created from nothing, nor can it be destroyed. Charges can only move around. If some object seems to change its charge, some kind of charge was added to it from someplace else, or charge was taken away—typically through the movement of electrons. And any change in the charge of an object can't be any smaller than the magnitude of the charge on a single electron or proton. This amount of charge is equal to about 1.602×10^{-19} C. Physicists call this kind of behavior quantization.

CONNECTIONS

The concept of quantization is extremely important to physics, and it lies at the heart of the important subfield of quantum physics. In addition to charge, quantum physicists have discovered a number of other physical properties that are quantized at very small scales, including energy and angular momentum. Quantization turns out to be critical in keeping an atom's negatively charged electrons from collapsing into the positively charged nucleus. In addition, the vast majority of today's high-tech technologies—from lasers to computer chips to the Global Positioning System—are applications of quantization.

Since there is a smallest possible unit of charge, the net charge on an object can only change in discrete steps. However, those steps are remarkably small, so small that we don't normally notice the quantization. When considering everyday-sized objects, it appears that the amount of charge can change continuously and that ordinary objects can have any value of positive or negative charge. Detectable amounts of charge will correspond to a very large number of electrons (either extra electrons for negative charge, or missing electrons for positive).

Conversely, one coulomb is a very large amount of charge, and you will never encounter any object with any significant fraction of a coulomb of charge on it. The charges we will deal with in our examples will be in the range of microcoulombs, which we abbreviate as μC. One μC is 10^{-6} coulombs.

Conductors vs. Insulators

Everything is made of atoms, and the vast majority of every atom's volume is dominated by its electrons. It is not surprising, then, that the behavior of charges in all ordinary objects depends mostly on the behavior of electrons. Although the electrical properties of all objects have this common origin, there are nevertheless important differences between substances with respect to the mobility of electrons.

Let's start with solids. Solid objects are defined by the fact that the atoms are stuck in fixed positions within the object. In some substances, each electron is very strongly attached to its own particular atom. So it is difficult for the electrons to move around within the substance. Electrically, such a material is classified as an *insulator.* Electrons can be removed from the surface of an insulator, or added to it, by rubbing it against other objects. But for the most part, electrons in an insulator do not move around very much.

In other substances, like most metals, the atoms share the electrons in common. There are still enough electrons to neutralize all of the positively charged nuclei, but the electrons are free to roam around without being anchored to one particular atom. Substances like these are classified as good *conductors* of electricity.

An **insulator** is a material in which electrons are unable to move around very much.

A **conductor** is a material in which electrons are able to move very freely.

Among solids, there is also an intermediate category, called semiconductors. The mobility of their electrons falls somewhere between insulators and conductors. The electrons are not as free to flow as in metals, but not as trapped as in the average insulator. These materials are very important in modern electronics, but we won't be discussing them any further in this book.

Liquids and gases are a different story, because the atoms themselves are able to move and flow relative to each other. Of course, if the atoms or molecules are electrically neutral, then this motion doesn't transport any charge. Gases in general are not good conductors of electricity.

Liquid metals (like mercury) are good conductors for the same reason that solid metals are: the electrons are free to move from atom to atom even faster than the atoms themselves can move.

Liquid solutions are a special case with regard to conducting charge. A solution refers to a liquid in which some other substance (or substances) is dissolved. This isn't a chemistry book, so we won't be discussing this in much detail, but you have probably experienced this plenty of times with ordinary water. For example, when table salt dissolves in water, it separates into positive and negative ions. These ions are then available to transport charge in the liquid. Depending on how many of these ions are present, the solution can be an excellent conductor. Pure water, on the other hand, is actually an insulator.

Coulomb's Law

Now let's get to the precise description of the force that acts between charged objects. This is the force we called the electrostatic force in the introduction, which is repulsive if the two charges have the same sign, and attractive if the two objects have opposite charge.

We start with the simplest case: two charged objects that are very small in size, like geometric points, located a distance r apart. Let q_1 be the charge on the first object, and q_2 be the charge on the second. Careful experiments have determined that such objects exert a mutual force on each other whose magnitude is equal to:

$$F_E = \frac{K|q_1 q_2|}{r^2}$$ **Equation 17.1**

This simple relationship is called Coulomb's law. We use the absolute value signs for the charges so that F_E is the magnitude of the electrostatic force, regardless of the sign of either charge. But we must not forget that force is fundamentally a vector, and this electric force must therefore have a direction as well. This force is directed along a straight line that connects the two small objects. As we mentioned previously, the force on one particle is toward the other if the two charges have opposite signs, and away from the other if the charges have the same sign. The force on q_1 due to q_2 is equal and opposite to the force on q_2 due to q_1, in compliance with Newton's third law.

The constant K in Equation 17.1 is required to give the force the proper units, when the charges are in coulombs and the separation distance is in meters. It is approximately equal to 8.99×10^9 N m^2/C^2.

Notice how Coulomb's law resembles Newton's law of universal gravitation. Instead of the product of two masses, now we have the product of two charges. Coulomb's constant K is just as universal and fundamental as the constant G in the case of gravity. Both forces are stronger when the objects are closer to each other and weaker when they are very far apart, with exactly the same dependence on distance. Because the constants are very different, the strengths of the two forces are very different. For fundamental particles with charge, like electrons and protons, the electric force is very much stronger than the gravitational force between them.

The possibility of a repulsive force makes for another big difference between gravity and the electric force. Gravity is always attractive. The electrostatic force can repel or attract. This makes it possible for electric forces to cancel themselves out, as described above. This is actually why we are more familiar with the gravitational force than the Coulomb force, even though gravity is fundamentally the much weaker of the two.

Coulomb's law also helps us to understand why it is difficult to put a large amount of charge on a real object. Imagine a basketball that already has some negative charge on it. If you want to make its charge even more negative, you would have to try to put some more electrons on the ball, but because it is already negatively charged, it repels those electrons. The more negatively charged the ball is, the stronger the repulsive force, and the more difficult it will be to add more negative charge.

It would be similar if you wanted to give the basketball a positive charge. You might do that by removing electrons, but the more positive the ball gets, the more difficult it becomes to pull electrons off of it since the positively charged basketball attracts the negative electrons. If you try to isolate the positively charged basketball, maybe by hanging it from the ceiling by a thread, it will actually grab electrons from nearby air molecules because of the strong Coulomb force.

TRICKS AND HACKS

Although we stated Coulomb's law only for a pair of pointlike charges, it applies just as well to multiple charges in any shape or distribution. We just have to remember that the Coulomb force is a vector. If there are multiple point charges located at different positions, we can get the net electric force just as we would expect, by adding the vector forces due to each charge in turn.

If some extended object has electric charge distributed continuously through it, a charge outside of that object will be attracted or repelled to every piece of charge in the body. The size of the attraction or repulsion will depend on how far away it is, as determined by Coulomb's law. You can mentally divide the body into small pieces, each with the appropriate charge, and separately calculate all the vector forces for each piece. Then you just add these vector forces to get the net force between the body and the other charge. As you might imagine, this can get pretty complicated, but with the appropriate math tools, it is straightforward to do.

There is another effect relevant to charge on extended objects that we should mention. Think of an isolated sphere made of some conductive material, like aluminum, and assume there is no net charge on it. Now bring a small positively charged rod close to the sphere. What happens?

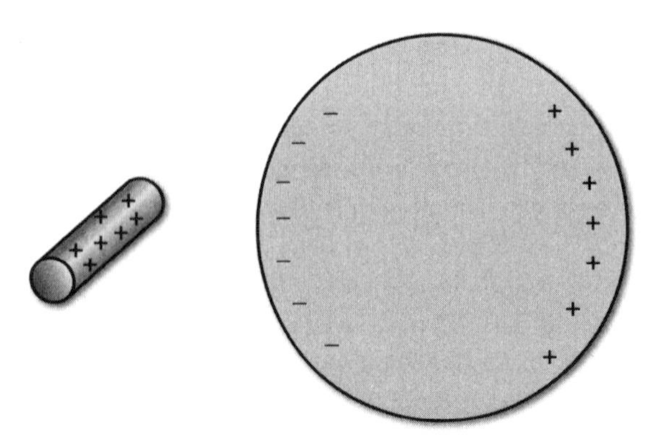

Figure 17.2

A small, positively charged rod is brought close to a larger conducting body that is neutral.

The positively charged rod will attract the electrons in the aluminum sphere, and they will move to get closer to it. As a result, the sphere will have a locally negative charge on the near side, and a locally positive charge on the far side (due to a shortage of electrons on that side), even though the charge as a whole will still be neutral. The interesting thing is that there will actually be a net force between the rod and that neutral sphere. This happens because the Coulomb force depends on distance. The attraction between the rod and the nearby negative charge will be stronger than the repulsion between the rod and the positive charge on the far side of the sphere.

The phenomenon of the spatial separation in the charge of a neutral object is called polarization. It even happens to some degree in insulators, because even though the electrons are stuck to their atoms, they can shift their positions slightly when an external charged object is brought near. That small shift, added up over trillions upon trillions of electrons, is enough to result in a net attractive force.

The Electric Field

As we have seen, the force between two charged objects depends on the amount of charge on each of the objects. For a lot of reasons, both conceptual and practical, we'd really like to define a property of just one of the charges (or a configuration of charge) that can tell us what force would be experienced by any other charge that was suddenly placed in its vicinity. We call this property the *electric field.*

DEFINITION

An **electric field** is a vector function of position in space that accompanies any distribution of electric charges. If a small "test charge" q' were placed at some location, it would feel a vector force $\boldsymbol{F}_E'$. Then the electric field is $E = \frac{F_E'}{q'}$. The magnitude of an electric field is measured in units of newtons/coulomb (N/C).

If a small "test charge" q' were placed in the vicinity of the charge(s) creating some electric field, it would feel a vector force due to the presence of this field. We want to think of this test charge as being very small, so that it does not itself influence the shape or magnitude of the electric field we are interested in. The electric field is a property of any specific configuration of charges. We'll call these latter charges the "source" charges, to distinguish them from the test charge we use to probe the field. Depending on how the source charges are arranged, the resulting electric field will have a magnitude and direction that varies with location in the surrounding space.

RED ALERT!

Occasionally, students want to include a test charge when trying to calculate the field due to another set of charges. Don't do this. The test charge is sort of an artificial, imaginary thing, necessary to determine the force that defines the electric field, but the electric field exists whether or not the test charge is present.

One of the main uses of the electric field is in calculating the force that would be felt by a charged particle introduced anywhere in the field. If you know the electric field at any particular location in space, then you automatically know the force that any charge introduced at that location would feel. If the electric field is $\boldsymbol{E}$ at a certain point in space, then a real charge q would feel a force $\boldsymbol{F} = q\boldsymbol{E}$ if it is placed at that point. If q is positive, the force is in the same direction as the electric field vector. If q happens to be negative, then the force it feels would be opposite to the direction of the field vector. Even though the charge can be positive or negative, q is a scalar, so we get a vector force when we multiply the charge by the electric field vector.

Let's use Coulomb's law to calculate the electric field due to a single point charge equal to q. Using Equation 17.1, we replace q_1 with q, which we consider to be the source of the field. Then we can think of q_2 as the test charge. Divide both sides of this equation by q_2, and we see that the magnitude of the electric field at any point that is a distance r from the source charge q is simply $E = \frac{K|q|}{r^2}$. The direction of the electric field is the same as the direction of the force that would be felt by a positive test charge. It is directed away from q if $q > 0$, and toward q if $q < 0$.

TRY IT YOURSELF

Refer to Figure 17.3 for problems 38 and 39.

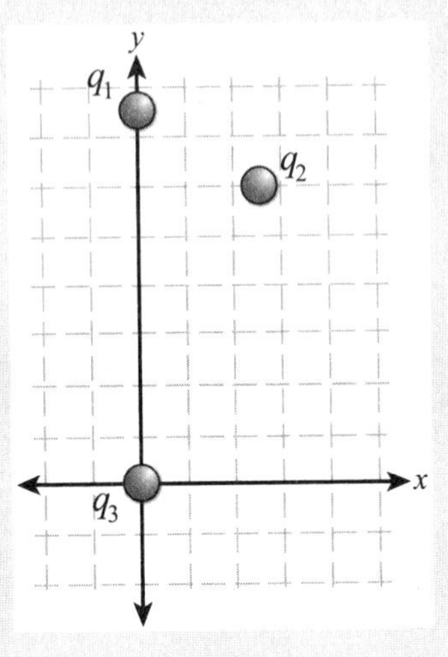

Figure 17.3

38. A charge $q_1 = 8\ \mu C$ is located at $x = 0$, $y = 15$ cm. Another charge $q_2 = 7\ \mu C$ is located at $x = 5$ cm, $y = 12$ cm. What is the net electrical force that is felt by a charge $q_3 = 3\ \mu C$ located at the origin (0,0)?

39. For the same charges q_1 and q_2 located at the same locations as in the exercise above, what is the electric field at the origin? You should assume q_3 is not present.

This same expression for the electric field works even if the charge q is spread out over some volume, as long as the distribution of the charge is spherically symmetric. (This means that the distribution is unchanged by rotation about any axis through the center of a sphere.) In this case, r would be the distance to the center of the spherical distribution. In other words, a spherically symmetric charge distribution behaves as if all of its charge were located at the center of the sphere. This is the same as the way spherical mass distributions are treated by the gravitational force.

TRICKS AND HACKS

In this book, we will use q for small charges, which can be assumed to be located at a single point, and uppercase Q for a charge that is distributed over some area or volume. However, this is not a universal convention and you should not assume it is the case in other books or references.

Since the electric field is a vector at every point in space, it is helpful to visualize the field as a set of lines that indicate the direction of the force a test charge would feel at various places. Such lines would be drawn in the direction of the field, and they can be curved or straight depending on the charges that establish the field. Since the range of the Coulomb force is infinite, field lines can't start or stop in open space, they can only terminate at the location of a charge.

We use arrows to indicate that the field points away from positive charges and toward negative ones. Because there is a definite direction of the field at any given location, electric field lines can never cross each other. (If two field lines did cross, that would indicate that the field had two different directions at one point in space, which makes no sense.) If you look at the left side of Figure 17.4, you will see that the field lines are close together near the point charge, and get less dense when you go farther from the charge. Recall also that the field is stronger the closer you are to the source charge. By insisting that field lines only terminate on charges, our diagrams will naturally show that the electric field is stronger in regions where field lines are close together, and weaker where the lines are farther apart.

As soon as they see electric field lines (especially curved ones), some students will assume that these lines represent the paths that charged particles must take in the field. That is not the case. These lines represent the direction of forces experienced by charged particles, and therefore their possible acceleration vectors. As we have seen before, an object's velocity vector does not have to be in the same direction as its acceleration, since acceleration is the change in velocity.

Figure 17.4 illustrates electric field lines for two cases. We can only show these lines in the two-dimensional plane of the page, but of course they extend into three-dimensional space as well. You can mentally rotate these illustrations about a line through the charges to visualize the 3D fields. Also, the field lines don't actually end except at charges. If it looks like a field line ends out in space somewhere, that's just because we ran out of room on the diagram; they would actually extend to infinity if no other charges were present. At every point, the electric field is tangent to a line in the direction indicated by the arrows. There are an infinite number of lines that could be drawn, only a selection of which are actually shown.

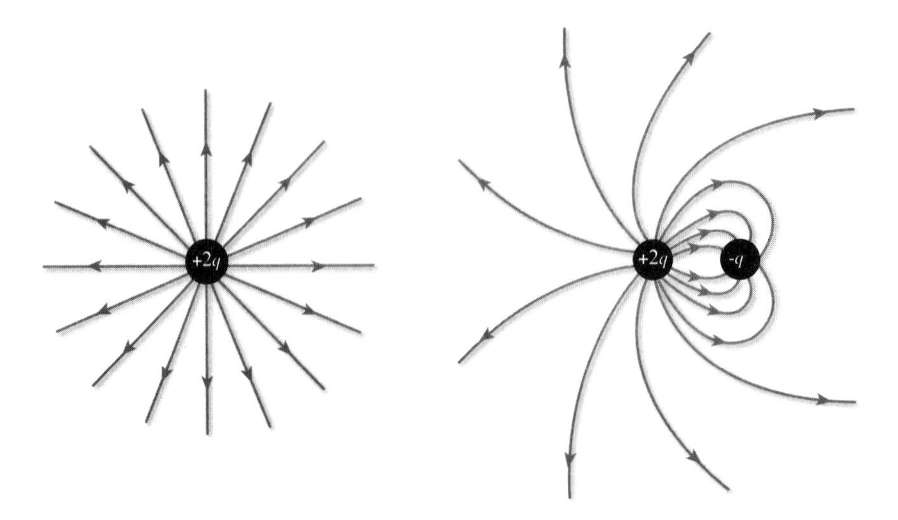

Figure 17.4

These lines illustrate the electric fields in two cases: a single isolated positive charge (left), and two unequal charges of different signs (right).

The electric field is another one of those concepts in physics with which beginners sometimes struggle, because we can't perceive it directly with any of our senses. We want to stress, however, that it is a real thing, as real as this book you are reading or the chair you are sitting on. It is produced by electric charges but exists all through space, even in regions far from any matter. It can exist within solids and liquids and gases, or in a pure vacuum.

An interesting thing happens, however, within solid objects that are good conductors of electricity. Recall that what makes a material a good conductor is the fact that electrons can move freely within the material. If you try to set up a constant electric field inside of a conducting object, these free electrons will all feel a force that is determined by the strength and direction of that field. They will then accelerate due to that force, and change their locations.

Those electrons don't keep moving forever, of course. They soon reach an edge of the object, and collect there (assuming no contact with another object). As in the previous discussion about polarization, the result is that certain parts of the object will be negatively charged, and other regions will be positive (due to the atomic nuclei left behind). These excess charges produce their own electric field.

The interesting part is that the field produced by the lopsided charge within the conductor will exactly cancel out the field you tried to impose, at every location inside the conducting body. There can be plenty of electric field outside of the conductor, but in general, it is not possible to maintain any electric field inside of a good conductor. The free electrons will move until they completely cancel out the field inside. This is why metal walls can be used to shield equipment from unwanted electric fields, and why you're generally safe inside your car during a lightning storm.

The Least You Need to Know

- Electric charges come in two varieties, positive and negative.
- Electric charges of the same variety (e.g., positive-positive) repel one another, while electric charges of opposite variety (positive-negative) attract one another.
- Electric charges exert a Coulomb force on each other proportional to the product of the charges, which also decreases as the square of the distance between them.
- The electric field due to any configuration of charges is the vector force that would be felt by a test charge, divided by the value of that test charge.

CHAPTER

18

Electrical Energy

In this chapter, we continue our study of the electric part of electromagnetism. We'll be focusing on our old friend energy, that conserved quantity that we know lies at the root of everything that happens in the universe. By its nature, electricity is a particularly versatile kind of energy, easily transported over long distances, and easy to convert to a variety of other kinds of energy on demand. This is why electricity has become such an important part of modern technology.

For static arrangements of charge, it turns out that the electrical force conserves energy, the same way gravity and spring forces did in previous chapters. As a result, we will again be able to define a potential energy, this time associated with the electric force. In this case, it will be most convenient to scale the potential energy to the amount of electric charge. The new quantity that results, electrical potential energy per unit of charge, will be crucial when it comes time to figure out how electric circuits work. It is actually the same as the voltage provided by the ordinary batteries that power your flashlights and cell phones.

In This Chapter

- Energy in electric fields
- Electric potential defined
- Storing charge and energy in capacitors
- The effect of dielectrics

We will also go beyond an oversimplification we made in the previous chapter. All of our discussions about electrical forces and fields presumed that there was nothing in between the charged objects we were talking about. In this chapter, we will see if anything changes if we consider the air that is all around us, or other materials filling the space between charges.

Potential Energy Revisited

We have discussed the importance of energy many times already, how energy changes forms but is neither created nor destroyed, and how it is behind most of the interesting things that happen in the natural world. On at least two occasions, we have used the concept of potential energy to account for mechanical work done against some force. Here we will do so again.

You should be able to recall two different forces to which we assigned specific potential energies: gravity and the spring force. The simplest case to deal with was the force of gravity near Earth's surface, which is nearly constant in magnitude and direction (for a given mass, that is). In that case, we made use of a potential energy function that simply increased with the vertical height above some arbitrary point, $U_g = mgy$. Notice that if you drop a mass from a height equal to y, such that it lands at $y = 0$, its gravitational potential energy will decrease (as the kinetic energy increases). The change in U_g would be negative, and exactly the opposite of the amount of work done by the gravity force during the fall. To raise the mass from zero to y, some external agent would have to add energy by doing work against gravity.

The spring or elastic potential energy was a little trickier to figure out, because in that case the force due to the spring was not constant, but changed as the spring was stretched or compressed. We gave a formula for the spring potential energy: $U_s = \frac{1}{2}kx^2$. In that case, the potential energy involved the spring constant k but not the mass of the object. That's because the spring force is the same no matter what mass it is pushing on. Again, if positive work was done against the spring force, the potential energy associated with the spring would increase by exactly the amount of that work.

Electric potential energy works the same way. It's just a lot more complicated, because the electric force can take all kinds of shapes. Instead of being constant, like the gravity we are familiar with, or proportional to some distance x, like for a spring, the electric force depends on the magnitudes and locations of all the nearby charges. Fortunately, the idea of the electric field can help us get a handle on this situation.

Let's begin with the simplest case we can find. Imagine two parallel flat plates, separated by some constant distance d. Let's say the plates are very large, with dimensions much larger than their separation d. Now imagine that electric charge is evenly distributed on both plates, one positive and one negative. We'll put the negative charge on the lower plate, as in Figure 18.1. Then we'll put an equivalent amount of positive charge on the upper plate.

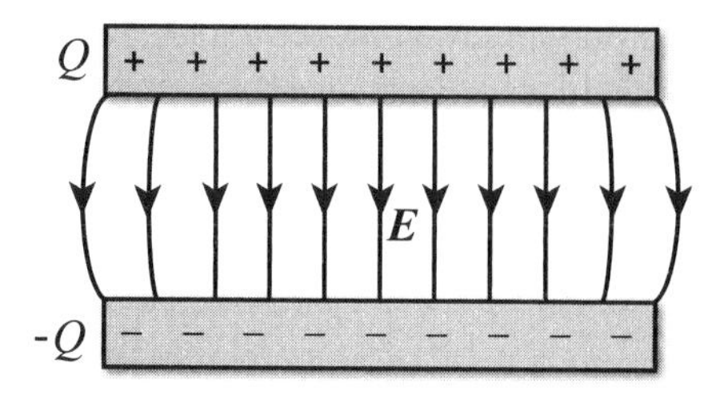

Figure 18.1

The upper plate in this figure has a total charge Q spread evenly across it. The lower plate has a total charge –Q distributed in the same way. In regions far from the edges, electric field lines are vertical and equally spaced.

Let's ignore any effects of gravity and just concentrate on the electric force that would be exerted on a test charge equal to q' located somewhere in the region between the plates (and far from the edges of the plates). If q' is positive, we know for sure that it will be repelled by the positive charge on the upper plate, and attracted to the negative charge on the lower plate. Not only that, but any horizontal components of these forces will cancel out, because there is about as much charge to the left as there is to the right. (This is hard to see in Figure 18.1 because we have expanded the vertical scale in order to see the field, but in reality, the separation between the two plates is much, much smaller than the size of the plates.) So for a substantial region between the plates, any force vector on a positive charge will point straight down, vertically.

Recall our rules for drawing electric field lines. These lines would be perfectly vertical, in the same direction as the force felt by a positive charge. Since the charge per unit area on each plate is constant, the density of the field lines would also be constant, and this arrangement would result in a uniform field between the plates. So for a given charge q', we have a constant field and a constant force in this region.

Consider the force first. A detailed analysis based on Coulomb's law would give us:

$$F_{\mathrm{E}} = 4\pi K q' \frac{Q}{A}$$ **Equation 18.1**

In this expression, K is the same constant that appeared in Coulomb's law, Q is the magnitude of the charge on either plate, and A is the area of either plate. It should not surprise you that the force on q' is proportional to the amount of charge per unit area on the plates, Q/A.

It is convenient at this point to define another fundamental physical constant, called the vacuum permittivity, or the permittivity of free space:

$$\varepsilon_o = \frac{1}{4\pi K} = 8.85\times10^{-12}\ \mathrm{C^2/(N\ m^2)}$$ **Equation 18.2**

Then the force experienced by q' takes on the simpler form:

$$F_{\mathrm{E}} = \frac{q'Q}{\varepsilon_o A}$$ **Equation 18.3**

Given the geometry of the setup, this situation has many parallels to uniform gravity. We can designate any location as having zero potential energy, so let's select the bottom plate. Let y be the vertical distance above the lower plate. Then the potential energy of a positive charge q' will have to increase if it is moved from the bottom plate toward the upper one. Any horizontal motion will not be opposed by the electric force, so it could not affect the potential energy.

Now, recall that for gravity $F_g = mg$ while $U_g = mgy = F_g y$. It would therefore be reasonable to expect that the potential energy of the charge q' would be equal to:

$$U_E = \frac{q'Q}{\varepsilon_o A} y$$ **Equation 18.4**

This is indeed the case, at least for any y that is less than or equal to d, the distance between the plates.

Electric Potential and Voltage

The potential energy of Equation 18.4 obviously depends on the magnitude of the test charge q' that is located between the charged plates, as well as its position. For the same charge density Q/A and the same location y, a 20-μC test charge would have twice the potential energy as one that is 10 μC. But this test charge is only needed to feel the force. Just as we defined the electric field to be a property of the charge configuration, independent of the test charge, it is useful to define some energy-related property that is just a characteristic of the charged plates, independent of whatever test charge might use. We can accomplish this by simply dividing the potential energy by the charge q'. The result of doing this is what we call *electric potential*, or often in this context, just the *potential.*

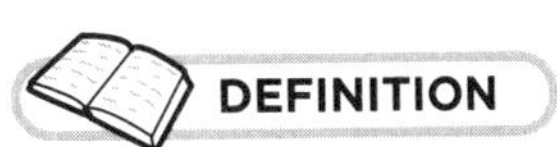

The **electric potential** is a scalar quantity that varies with location relative to electric charges. It is the potential energy that would be possessed by a small test charge located anywhere in the region of interest, divided by that charge. It is measured in J/C, a unit which is also called a volt (V).

The **potential difference** (or **voltage**) between two points is simply defined as the arithmetic difference between the electric potential at those same two points. It is usually written ΔV, and is also measured in volts.

As with potential energy, one has the freedom to decide where the location of zero potential will be. This is because, as before, only changes in potential energy (and hence electric potential) actually matter. This is why it is often more useful to think about the *potential difference* ΔV between two points in space. This potential difference is often referred to as the *voltage*, because it is measured in volts. We note that this is a slightly different use of the Δ notation than what we have been using. It is not a difference in voltage at different times, but at different places. This ΔV can be positive or negative, and there is no convention as to which location appears first in the subtraction.

There is potential for confusion here between the quantity called *potential* and the units it is measured in. We generally use the symbol V to stand for a potential, which is also measured in units abbreviated V. In this book, we will distinguish the quantity from the units by italicizing the quantity but not the units. For example, you should understand what is meant by $V = 50$ V. Also, be careful not to mix up the quantities *potential* and *potential energy.* The former is a property of the source charges alone while the latter also depends on the size of the test charge in the field.

The key insight is that in the static situations we are talking about, there is a definite value of the potential at every point in space. The potential does not have direction because it is not a vector. How the potential varies with position is determined by whatever configuration of source charges creates it (which may be very different from our example of the parallel plates). If you start at some point and move to a different point, the electric potential may change. But if you return to the original point, no matter what path you take, you will also find yourself back at the original value of the electric potential.

If you know the potential at two different locations, you can easily determine how the potential energy of a charged particle will change when going from one location to the other. Let's say for example, that $V_A = 50$ V is the potential at point A, and $V_B = 20$ V is the potential at point B. A particle with positive charge $q = 70$ μC will experience a reduction in potential energy if it moves from A to B. That change in U_E will simply be the potential difference times the charge of the particle:

$$\Delta U_E = (V_B - V_A)q = (-30 \text{ V})(7\times10^{-5} \text{ C}) = -0.0021 \text{ J} \qquad \textbf{Equation 18.5}$$

If the positive particle were released from rest at point A, it would arrive at point B with less electric potential energy. If no other forces were doing work, then it would arrive with 0.0021 joules of kinetic energy. On the other hand, a negatively charged particle would gain potential energy in this same situation, and that additional energy would have to be supplied by some external agent.

So if you know how the voltage changes from place to place, it is easy to use that knowledge to see how particles with any charge will behave in that region. The tricky part is actually figuring out the electric potential established by a given configuration of source charges. Rather than get into how to do this, we'll show you the potential for one more specific case and then see how we might build from that to more complex situations.

For a single, pointlike source charge equal to q, the electric potential turns out to be:

$$V = K\frac{q}{r} = \frac{q}{4\pi\varepsilon_o r}$$ **Equation 18.6**

Note that we are now using q to stand for a source charge, the source of this potential V. We have given two alternative forms for V, using either Coulomb's constant or the permittivity of free space. Notice that this simple expression implicitly sets $V = 0$ at an infinite distance from the source charge. At all other locations in space, V is positive if q is a positive charge, and negative if q is negative.

Once you have Equation 18.6, you can pretty easily determine the potential at any location due to a small number of source charges, wherever they are. The total potential is just the algebraic sum of potentials due to each source charge. All you have to do is keep track of how far away each charge is, and get the signs and magnitudes right. Since the potential is a scalar, you don't have to mess with any vector addition.

TRY IT YOURSELF

40. Calculate the potential at a distance of 2 cm from a point charge equal to -0.005 C.

Capacitors

The parallel plate arrangement we described previously turns out to be a very useful component in electrical devices. If we assume that the two plates are made of conducting material, and the plates are not connected to each other, then we call that component a *capacitor*.

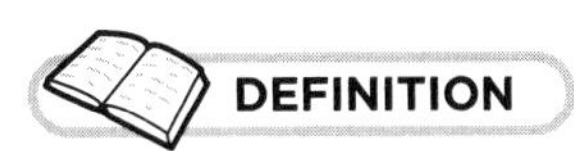

DEFINITION

A **capacitor** is an electronic device that is capable of storing electrical energy. It is typically constructed from a pair of conducting surfaces placed near to one another, across which a voltage may be applied.

A capacitor has the ability to store electrical energy. Imagine starting with no charge on either plate, and let's say our goal is to make the top plate positive and the bottom negative, as in Figure 18.1. In order to do that, we remove an electron from the top plate, and put it on the bottom one. That requires us to do some work, because the plate from which we took the electron immediately gained a net positive charge, which pulled on the electron as we removed it. If we were to continue shifting negative charge from the top plate to the bottom, we would have to do more work each time, because the increasing charge means ever more force is opposing our actions.

At a given time, let's call the total amount of positive charge on the top plate Q, as we did above. Let the area of each plate be equal to A. Using the uniform Coulomb force between the plates, we derived Equation 18.4 for the potential energy of a test charge q' in terms of the distance from the bottom plate, which you'll recall looked like this:

$$U_E = \frac{q'Q}{\varepsilon_o A} y$$

Now let's say the total distance between the plates is d. If we want to know the potential difference between the two plates, we just let $y = d$, and divide by the test charge q':

$$\Delta V = \frac{Q}{\varepsilon_o A} d \qquad \textbf{Equation 18.7}$$

Notice that during the time that we are increasing Q, the potential difference is increasing right along with it. It is conventional to define a constant of proportionality between Q and ΔV, which we will call C, such that $Q = C(\Delta V)$. From Equation 18.7 we see that for our parallel plate capacitor, $C = \varepsilon_o A/d$. This constant thus depends on the actual dimensions of the capacitor. We call this constant the *capacitance.*

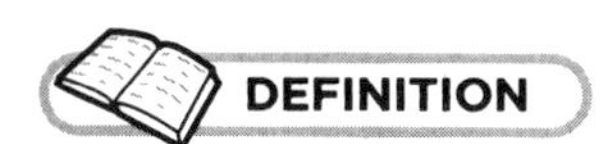

Capacitance is a fixed characteristic of any particular capacitor. It is defined as the ratio $Q/\Delta V$, where Q is the amount of positive or negative charge on either plate of the capacitor, and ΔV is the resulting potential difference between the two plates. The standard units of capacitance are coulombs/volt, where 1 C/V is also known as a farad (F).

Now that you know what a capacitor is, let's consider just how much electrical energy one of these can store. To get a handle on this, we will think through what is needed to establish a potential across a parallel plate capacitor in the first place. Initially, assume that the two plates are both electrically neutral. Now, start moving positive charge little by little from the bottom plate to the top plate. If you stop moving charge when the upper plate has a total charge of Q, then the potential difference ΔV is simply given by Equation 18.7.

Now recall that if you know the potential difference, you also know the potential energy that would be gained by any charge moving across that potential difference. But when we started moving charge, the potential difference was zero. So to get the total energy stored, we don't multiply Q times the final ΔV, but by the average potential difference, which is $\Delta V/2$. Therefore, whenever the charge on the two plates is Q and $-Q$, the energy stored in a capacitor is exactly equal to:

$$U_E = \frac{1}{2}Q(\Delta V) = \frac{1}{2}\frac{Q^2 d}{\varepsilon_o A}$$

Equation 18.8

$$U_E = \frac{1}{2}\frac{Q^2}{C}$$

Here we have made use of the equations for ΔV and C from above. Finally, note that if you just know the voltage across the capacitor, you can use the fact that $Q = C(\Delta V)$ to turn this into the following form:

$$U_E = \frac{1}{2}C(\Delta V)^2$$

Equation 18.9

In this form, all references to the charge inside and the geometric specifics of the capacitor have dropped out. So long as you know the capacitance and the potential difference, you can always use Equation 18.9 to calculate the energy stored, independent of the shape of your capacitor.

TRY IT YOURSELF

41. Say you want to store 35 joules of energy in a capacitor whose capacitance $C = 0.003$ farads. How many volts do you need to put across it?

Dielectrics

"Dielectric" is another term for an insulating material, which can be solid, liquid, or gas. Our prior calculations assumed that there was no material at all between the conducting plates of our capacitor. We saw that the capacitance was inversely proportional to the distance of separation between the two plates *d*. Therefore, one way to increase the capacitance, and thus the energy stored for a given voltage, would be to make *d* as small as possible.

Don't forget, however, that the plates of the capacitor may have opposite charges on them. They can't be allowed to touch, or they will immediately be discharged. For that reason, it is natural to try to put a thin layer of insulator between the plates, then press them together as close as possible. But if you do this, you will discover that the capacitance is actually larger than you expect! Somehow, the presence of the dielectric gives the capacitance a boost.

Why is this so? The explanation goes back to our discussion in Chapter 17 of the phenomenon called polarization. When the capacitor is charged, there is an electric field between the two conductors. If you place an insulating material in that region, there will be a slight separation of its charges, even though it remains neutral on the whole. Some excess positive charge will end up on the surface of the insulator near the negative plate of the capacitor, and some extra negative charge will end up on the other surface. The insulating layer will look like a parallel plate capacitor in its own right, making its own electric field within its volume. This electric field will be in the opposite direction from the one due to the charges on the capacitor. It will not be quite as strong as the capacitor's field, so its effect will be to reduce the net electric field in the space between the capacitor's plates.

If we go all the way back to the force relation in Equation 18.3, and then recall the relationship between electric force and the electric field, we see that the magnitude of the electric field in the parallel plate capacitor with no dielectric in it would be $E = Q/\varepsilon_0 A$. Plugging this into Equation 18.7, we see that the potential difference in our empty capacitor can be written $\Delta V = Ed$. The effect of a dielectric between the plates will be to reduce the electric field (and thus ΔV) for the same amount of charge. Therefore, since capacitance is defined as $C = Q/\Delta V$, we can finally say with confidence that the presence of a dielectric will increase the capacitance of a capacitor (compared to the case of nothing in between).

The polarizing effect of an electric field is different for different materials. We define a dimensionless parameter called the *dielectric constant* (κ) to be a measure of how "polarizable" any given material is, and therefore how much any given material can boost capacitance. Utilizing this, we see that the actual capacitance of two parallel plates of area A, separated by a distance d, is actually given by $C = \frac{\kappa \varepsilon_o A}{d}$, when the space between the plates is filled by a material with dielectric constant κ.

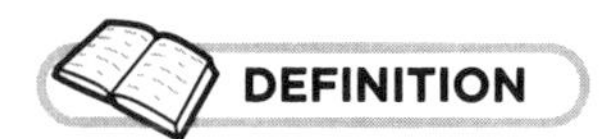

The **dielectric constant** is a dimensionless scalar that quantifies the ease with which a given insulating material can be polarized. It is equal to one for a pure vacuum, and is larger for other substances, depending on how easily they are polarized.

This new expression for capacitance will work fine without a dielectric because we say that for a perfect vacuum, $\kappa = 1$ exactly. Air is not much different, with a typical $\kappa = 1.00054$, so for electric fields in air, we don't make much of an error if we ignore κ. Most other materials have larger dielectric constants, ranging from about two up to a few hundred. Dielectric constants can be measured by various means. In practice, all we do if we need to know a dielectric constant is look up the material in the appropriate data table.

Energy in an Electric Field

By now it should be no surprise to learn that any electric field configuration embodies energy. We want to reinforce this important connection before leaving this chapter.

First recall the parallel between the concepts of the electric field and the electric potential. These are both entities that are established in the presence of some source charge configuration. They both affect test charges in their vicinity, but they are independent of the test charges themselves. The electric field is the vector force felt by a test charge, divided by that charge. The electric potential is the scalar potential energy of a test charge, divided by that charge.

Whenever there is an electric field in some region of space, it means that a positively charged object would feel a force in the direction of that field. That same object would necessarily experience a reduction in potential energy if it should happen to move in the direction of the field/force. Thus it must be true that the electric potential decreases in the direction of any electric field. The reverse is true as well.

For visualization purposes, it is useful to imagine a set of points in space where the potential has a constant value. For every different value of the potential, there would be a different set of points with that same potential, of course. In reality, such points always join up, so we call any such set of points an equipotential surface. In three-dimensional space, such a set of points defines a surface, but in two-dimensional drawings, these surfaces appear as lines.

Equipotential surfaces for a capacitor are just planes parallel to the plates (since the potential is simply a function of the distance from one of the plates). As an exercise, you might try to draw equipotential lines for the field illustrated in the second half of Figure 17.4, the case of two unequal charges.

It turns out these equipotential surfaces provide valuable information about the electric field: electric field lines are always exactly perpendicular to equipotential surfaces. Moreover, if you think about how the potential varies with position, this implies that the electric field always points in whatever direction corresponds to the quickest decrease in the potential. The is based upon the relationship between force and potential energy, which says that force always points in the direction of steepest reduction of potential energy.

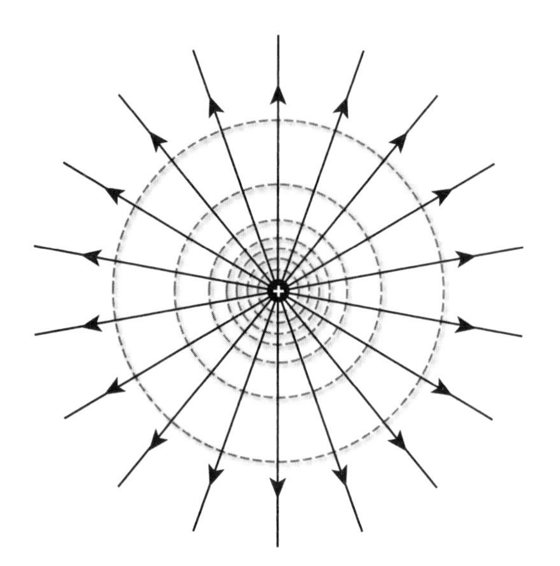

Figure 18.2

Solid lines represent the electric field for a positive source charge that is pointlike, while dashed lines indicate where the equipotential surfaces would be.

Finally, we can use our analysis of the energy stored in a capacitor to learn a general fact about the energy associated with an electric field. As we have seen in Equation 18.9, the energy stored in a capacitor is $\frac{1}{2}C(\Delta V)^2$. From Equation 18.7 we know that $\Delta V = Ed$. Now, we also know that the capacitance $C = \frac{\kappa \varepsilon_o A}{d}$. Substituting both of these into the expression for energy gives:

$$U_E = \frac{1}{2}\frac{\kappa \varepsilon_o A}{d}E^2 d^2$$

$$U_E = \frac{1}{2}\kappa \varepsilon_o (Ad)E^2$$

Equation 18.10

Now notice that the volume of the space between the capacitor plates is *Ad*. If we divide both sides of the equation by this, we see that the amount of energy per unit volume is $\frac{1}{2}\kappa \varepsilon_o E^2$, a quantity we call the energy density. In the case of the capacitor, where the electric field is constant, the energy density is constant. But it turns out that this result is a general property that is always true. Even if the electric field is not uniform, this expression gives the energy density at any point where the electric field has a magnitude equal to *E*.

The Least You Need to Know

- Electric potential is a measure of potential energy per unit charge.
- Parallel plates isolated from each other create a uniform electric field between them if the plates carry opposite charge; this arrangement is also called a capacitor.
- The potential difference, or voltage, between the plates of a capacitor is proportional to the amount of charge stored.
- The capacitance in a capacitor can be increased by inserting dielectric material between the plates.
- An electric field has an energy density equal to $\frac{1}{2}\kappa\varepsilon_o E^2$ at any point in space.

CHAPTER

19

Current and Resistance

In this chapter, we begin our discussion of electric charges in motion. Electric current, flowing in circuits made of conducting materials, is the key to a huge variety of practical uses for electricity. Current is the best way that electrical energy can be transported over large distances in order to provide light, heat, cooling, or even mechanical energy to do all kinds of work. Electric circuits are also used for communication, transmitting information and data while using relatively little energy. All of this depends on understanding and controlling the flow of charged particles in conductors.

There are entire college courses and major programs devoted to the study of electric circuits, and all of the ways that electrical energy is generated, transmitted, controlled, and utilized. We will only scratch the surface here with an introduction to direct current (DC) and alternating current (AC) circuits. We nevertheless hope to provide a good account of the basic building blocks that support this important application of physics.

In This Chapter

- The relationship between current and voltage
- Electric power
- Analyzing simple direct current circuits
- Alternating current and you

Electric Current

Electric current is simply defined as the flow of electric charge through conducting materials. A typical example of such a conductor is a wire made of a metal, such as copper. If you recall the distinction we made previously between insulators and conductors, you know that most metals are classified as good conductors. Because of their structure at the atomic level, the tiny electrons in the metal are not anchored to specific atoms, but are free to move all through the bulk of a solid piece of the metal.

Electric current is the rate at which electric charge passes through a particular point or area. If an amount of charge Δq crosses in a time Δt, then the average current is $I = \frac{\Delta q}{\Delta t}$. The standard units for current are coulombs/second. One C/s is also called an ampere, or amp for short (A).

Electric current doesn't flow in an isolated piece of wire or metal that is just sitting there. Something needs to cause the charges to move, and that something is the potential difference we introduced in the last chapter. Whenever there is a potential difference between one end of a conductor and the other, charged particles will move and there will be current. The fact that there is a potential difference between one end of a conductor and the other means that there is also an electric field within the conductor. It is this field that exerts force on the charged particles and gets them to move. If there is some source of energy that maintains the potential difference as the charges move, such a source is called an electromotive force (emf).

But didn't we say a couple of chapters ago that an electric field could not exist in a conductor? That is true, but only in the static case we were talking about back then. Static means nothing moving, and without any moving charges, it is true that there will be no electric field within a conductor. But with charges in motion, that is no longer the case.

Since current is the flow of charge, we need a convention for defining which direction to consider positive. Long ago, the decision was made to take as positive the direction that positive charge would flow. If we had a conductor full of positive charged particles that were free to move, then those particles would naturally move from the end with the higher potential toward the end with the lower potential. We call that the positive direction for current flow.

Unfortunately for generations of students, the direction of current flow that is considered to be positive is the reverse of what actually happens within a conductor. As you know, the positive charges in solid materials are locked in place. The particles that actually carry charge in metal wires are electrons. Since these have been designated as having negative charge, the direction that electrons travel is opposite the direction assigned for current flow. The negative electrons actually flow from the lower potential end to the higher potential, but mathematically that is exactly the same as if an equal amount of positive charge were flowing in the opposite direction. Whenever we try to quantify current, it really makes no difference whether the mobile charge carriers are positive or negative. Just remember that the direction of current that we call positive will always be from higher potential to lower.

The flow of heat that we talked about in Chapters 15 and 16 is one good analogy for understanding electric current. The laws of thermodynamics dictate that heat will naturally flow from a warmer region to a cooler one. We also saw that a larger temperature difference will cause heat to flow faster. Likewise, the laws of electrodynamics say that current will naturally flow from a region of higher potential to a region of lower potential, and the greater the potential difference, the larger the current.

An even better analogy for an electric wire is a hose full of water. If the ends of the hose are closed or turned upward at the same height, it can be full of water that does not move. The water plays essentially the same role as the electrons in the wire. If the ends of the water hose are open, then water will flow if and only if there is a higher pressure at one end than at the other. The difference in pressure plays the same role as the potential difference (or voltage) between the ends of the wire. The higher pressure forces the water toward the lower pressure end, and if the water is able to move within the hose, it will.

Ohm's Law and Resistance

In most materials, electric current is not completely free to flow. In fact, the ease with which electrons can move through real materials varies widely, and it varies with environmental conditions as well. Even in very good conductors, the flow of electrons is impeded by collisions with the atoms that are fixed in their positions.

The parameter we use to measure this is called *resistance.* Resistance is a characteristic of a physical object, the material it is made of as well as its dimensions. In general, resistance is also affected by temperature and possibly other external conditions, but we will ignore these small effects for now. When we talk about a resistance value in the rest of this book, we will mean the resistance under normal conditions and around room temperature, assumed to be constant.

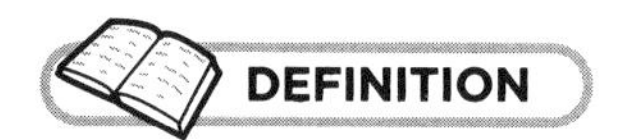

Resistance is a property of an object that describes precisely how difficult it is for charges to move from one end to the other. The standard unit of resistance is the ohm (Ω).

The higher the resistance of an object, the more difficult it is for current to flow. Resistance is never a negative number, and with rare exceptions, it is always greater than zero.

Resistance increases with the length of a wire, and is also greater for wires with smaller diameter. For a real wire, made of a specific material with a certain cross-sectional area, a relevant specification is therefore the resistance per length. Thin copper wires have a resistance of approximately 0.02 Ω/m or less, for example. Even insulating materials have resistance, with a value that is much larger than it is for conductors. For a typical insulator like glass or plastic, a comparable figure for resistance per length would be around 10^{18} Ω/m. For all practical purposes, this is equivalent to infinite resistance. Air has even greater resistance, but the only way to truly achieve infinite resistance is with the complete absence of matter, which we call a vacuum.

CONNECTIONS

In 1911, the Dutch physicist Heike Kamerlingh Onnes discovered that when certain materials are cooled to extremely low temperatures, their electrical resistance could drop suddenly to zero. For obvious reasons, this type of conductor is called a "superconductor." It was later discovered that this phenomenon is rooted in quantum physics and has to do with correlations between all electrons within the material. Unfortunately, physicists have yet to develop a superconductor that can operate at normal, everyday temperatures, and even so-called "high temperature" superconductors must be cooled below about -100°C to enter the superconducting state.

Over the many years since electricity was discovered, scientists and engineers have made a huge number of observations of how current is affected by potential differences across a vast variety of materials. These observations have revealed a very simple relationship, called Ohm's law:

$$\Delta V = IR \qquad \textbf{Equation 19.1}$$

Truth be told, this is not a universal law of physics, but it is an extremely good approximation for a wide variety of real materials, and we will consider it to be true for all examples covered in this book.

Ohm's law reveals the true utility of the concept of resistance. We have defined voltage in terms of more fundamental quantities, i.e., energy per charge. We did the same for current, as in charge per time. We haven't done the same for resistance. For practical purposes, Ohm's law allows us to define the ohm unit as one volt per ampere. Measuring how much current flows through an object for a given potential difference is the most common way to measure the resistance of the object. A constant resistance means that the current through the object is directly proportional to the voltage across the object.

Electric Power

Before we take a closer look at circuits, we want to consider the fate of electrical energy as charges move. Recall that current flow is just electric charge moving from a higher potential to a lower one. Since electric potential is really potential energy divided by charge, consider what happens when a certain amount of charge Δq falls through a potential difference ΔV. The total energy will be reduced by a number of joules equal to the product of that charge and the potential difference: $\Delta V \times \Delta q$. Now consider the rate of that energy reduction. We can find this rate by simply dividing the energy reduction by whatever time it takes for that amount of charge Δq to move. Call this time interval Δt. If we make use of the definition of current I, then the rate of change in electrical energy can be written as follows:

$$\Delta V \frac{\Delta q}{\Delta t} = (\Delta V) I$$ **Equation 19.2**

What happens to this energy? If the electric current is flowing through a conductor with resistance, then this corresponds to electrical energy being converted to heat. Any rate of energy conversion is called power, as we have seen previously, and should be measured in watts. Whenever you have a current I flowing in a conductor with resistance R, electrical energy is being converted to heat at the rate given in Equation 19.2. We often make use of Ohm's law to replace the ΔV with IR, and use the following expression for the power associated with the heating of a resistance:

$$P = I^2 R$$ **Equation 19.3**

This phenomenon is also known as "joule heating" or "ohmic heating." It is consistent with energy conservation as stated earlier, because it is just the rate at which electrical energy is being converted to thermal energy. From the point of view of transmitting electrical energy from one place to another, this effect is considered as an energy loss, just as friction effects cause a loss of mechanical energy, even though the energy has really only been converted to another form.

TRICKS AND HACKS

Here is another chance to practice your skills at dimensional analysis. Does I^2R really have proper power units? First look at the resistance $R = \Delta V/I$. Since voltage is energy/charge and current is charge/time, the unit for R (ohm) can be expressed as J s/C^2. Multiply this by the units for current squared, (C/s)^2, and sure enough, I^2R is measured in J/s, or watts.

The concept of electric power goes beyond losing energy to heat, however. Electrical energy can also be converted to other useful forms of energy, including light or mechanical energy (by a light bulb or an electric motor). Alternatively, electrical energy can be generated by various means, being converted from other forms. In these cases, $P = (\Delta V)I$ is often the more useful form of

Equation 19.3 for figuring electric power. For example, a generator delivering 50 amps of electric current at 100 volts would be producing electric power equal to 5,000 watts, or 5 kilowatts.

TRY IT YOURSELF

42. An electric heater has a resistance of 6 Ω. If it is connected to a voltage source of 120 V, how much current will flow through it?
43. For this same heater, how much heat energy is produced during one minute of operation?

DC Circuits

As we have seen, two things are required for current to flow: a potential difference and a conductive path through which charges can move. The conductive path is what we call a circuit. In order to have a circuit, there has to be at least one path between two points at different potentials. Circuits can be simple or complex, and may involve several possible paths in a network of connections.

There are many possible sources for the driving potential difference. It could be a battery, a generator, or another electrical device simply called a power supply. We'll start by considering the simplest case, a potential difference that is constant in time. Chemical batteries are the classic example here, and are designated by the voltage they provide when fully charged (1.5V, 9V, etc.). When the voltage source is constant, so is any current that flows in the circuit (as long as the circuit itself doesn't change). We use the term "direct current" or "DC" to describe such situations.

Diagrams are very useful for representing and analyzing DC circuits. Before going any further, let's define some of the symbols that are commonly used to illustrate DC circuit elements.

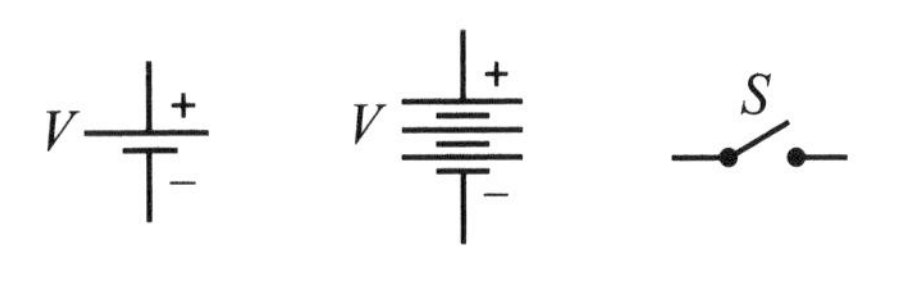

Figure 19.1

Some common circuit elements. Top row from left: a DC voltage source or battery, a multicell battery, and a switch. Bottom row: a resistor, a capacitor, and the reference for 0 V, also called "ground."

For the relatively short wires and conductors that connect components in circuits, the resistance is extremely small. When we analyze circuits, we will assume that a wire has zero resistance, and the error we make by doing so will be completely insignificant. Therefore, any line that is drawn to connect elements on a circuit diagram is assumed to have zero resistance. Since Ohm's law tells us that potential difference is proportional to R, this means that all points connected by a simple line are at the same potential.

The first rule for analyzing DC circuits is that there is a definite electrical potential at every point in the circuit. Only potential differences matter, but we can use absolute voltages if we arbitrarily determine one location in the circuit where the potential is equal to zero volts. That is the purpose for the last symbol in Figure 19.1. This potential difference is called "ground" because in the real world, it is made to be the same potential as Earth's surface, which means the floor and walls of the room you are standing in are all likely to be at this same potential. Therefore touching a circuit at any location that is connected to ground cannot give you an electric shock.

The simple statement that there is a definite voltage at every point in the circuit is useful when there are multiple loops, or many different paths between different points. Often, the solution to a circuit problem begins by following the change in voltage around any closed loop; the sum of all voltage drops must be zero for a complete loop. You can also learn a lot by comparing the voltage difference along two paths that join the same two points.

The conservation of charge leads to a second useful rule for analyzing DC circuits. Consider a point in a circuit where three conductors come together. Any way you look at it, the amount of current flowing into that point must be equal to the amount flowing out. This gives a simple, arithmetic relationship between the currents in the three branches. The specific relationship depends on the directions of the various currents. We generally use an arrow to indicate the direction of positive current in any particular conductor. This same rule applies no matter how many lines are joined at a given point.

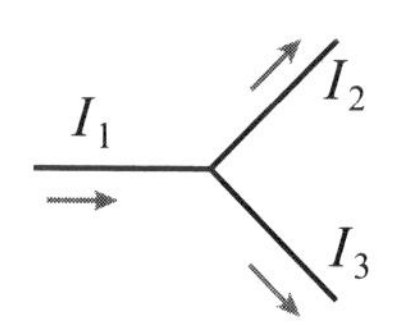

Figure 19.2

Three wires are joined at a single point, somewhere in a larger circuit. If currents flow in the indicated directions, then it must be true that $I_1 = I_2 + I_3$

The two rules we introduced previously—that the sum of potential drops must be zero for a complete loop, and the sum of currents into any point must equal the sum of currents out of it—are the primary tools with which to attack circuit problems. These are often referred to as "Kirchhoff's rules" in honor of the physicist who pioneered them.

Combining Resistors and Capacitors

When multiple resistors and capacitors are connected in the same circuit, it is often helpful to simplify certain combinations of like elements. In the cases that follow, there is a simpler arrangement of components that has exactly the same electrical behavior as the more complicated arrangement. In all of these cases, we are looking for combinations that are only connected to the rest of the circuit at two points.

TRICKS AND HACKS

If you are not sure whether or not a particular combination of resistors (or capacitors) qualifies for replacement, draw a box around it as it appears in the circuit. There must be two and only two wires coming out of the box connecting it to the rest of the circuit. It may also be helpful to then redraw the contents of your box off to the side, straightening out any crooked lines to make it look as simple as possible.

There are basically two ways to connect two components electrically. These are called series and parallel, and are illustrated in Figure 19.3. For example, if both ends of more than one resistor are connected together, we say they are connected in parallel, even if they are not drawn next to each other in the circuit. If only one end of a resistor is connected to only one end of another (and nothing else is connected at the junction between the two resistors), then we call it a series connection.

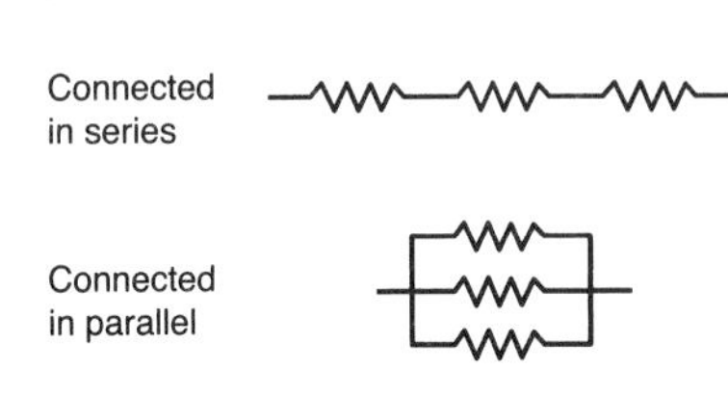

Figure 19.3

Examples of series and parallel connections. Top: three resistors connected in series. Bottom: three resistors connected in parallel.

For any number of resistors connected in series, the effect when viewed from end to end is the same as one single resistor whose resistance is equal to the sum of all the connected resistors. For three resistors connected in series, the equivalent resistance is given by $R_{eq} = R_1 + R_2 + R_3$. For any number of resistors connected in parallel, the equivalence is a little more complicated. Here is the formula for the equivalent resistance of three resistors in parallel:

$$\frac{1}{R_{eq}} = \frac{1}{R_1} + \frac{1}{R_2} + \frac{1}{R_3}$$ **Equation 19.4**

It should be clear how this formula is applied when there are only two resistors, or more than three.

The prescription for series resistors should make sense if you remember that long wires have a certain resistance per unit length. Stringing identical resistors in series would be just like increasing the length of a wire. The formula for parallel resistors is a little more difficult to predict. It actually follows from Ohm's law and the fact that the potential difference across every resistor is the same. For three resistors in parallel, the total current through the whole parallel bunch has to be the sum of the currents through each resistor:

$$I_{\text{tot}} = I_1 + I_2 + I_3$$ **Equation 19.5**

Ohm's law says $I = \Delta V/R$, which is true for the total current through an equivalent resistance, as well as the currents in the individual resistors. Thus we have:

$$\frac{\Delta V}{R_{\text{eq}}} = \frac{\Delta V}{R_1} + \frac{\Delta V}{R_2} + \frac{\Delta V}{R_3}$$ **Equation 19.6**

The same potential difference ΔV exists across each resistor in this arrangement, as well as across the whole combination. Divide both sides of this expression by ΔV, and you get Equation 19.4.

Note that for resistors in parallel, the equivalent resistance is always less than the smallest individual resistance. This is because the current has multiple ways to go instead of being forced to go through any one resistor. To really make this clear, let's take a look at the situation illustrated in the circuit diagram shown in Figure 19.4.

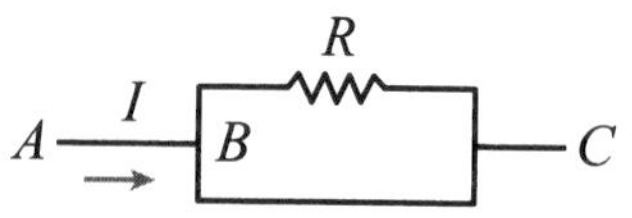

Figure 19.4

Viewed between points A and C, this circuit appears to be a resistor in parallel with a perfect conductor.

We can try to use Equation 19.4 to analyze this case. But if we assign a resistance equal to zero for the conducting branch, we seem to have a problem. You may have been taught to never divide a number by zero. Theoretically, you actually can do that, but the result is an infinitely large number. So in this case, we are saying that $1/R_{\text{eq}}$ is infinite. This just means that the equivalent resistance of this combination is zero! The value of the resistor doesn't matter at all. Although the current could split at the junction labeled B, in reality none of the current will bother to go through the resistor. It will all go through the lower branch, so that this combination is equivalent to a resistor with zero resistance. To reference a common figure of speech, current always "takes the path of least resistance." Bypassing a resistance (or any other component) with a conductor is also called "short circuiting."

For combinations of capacitors, the procedure is complementary in a way. You may recall from the previous chapter that the capacitance of a given capacitor was proportional to the area of the plates which get charged up. If you connected two identical capacitors in parallel, it's not hard to imagine that they would function the same as one capacitor with twice the area of either one, and thus with twice the capacitance. It doesn't have to be just two capacitors, and they don't have to be identical. For any number of capacitors connected in parallel, the total plate area would be the sum of the areas of each capacitor. Therefore, any number of capacitors in parallel can be replaced by a single capacitor with capacitance equal to the sum of the individual capacitances. For three capacitors, the equivalent capacitance is given by $C_{eq} = C_1 + C_2 + C_3$.

For capacitors in series, the equivalent capacitance is related to the individual capacitances in this way:

$$\frac{1}{C_{eq}} = \frac{1}{C_1} + \frac{1}{C_2} + \frac{1}{C_3}$$ **Equation 19.7**

The derivation of this mirrors the derivation of resistors in parallel. For capacitors in series, it is the voltage differences that add up to the total voltage drop, instead of the currents that were added for resistors in parallel. Throw in the fact that charge is conserved, and you end up with Equation 19.7.

Alternating Current

The electric power supplied to your home is not provided at a constant potential, so is not the simple case of direct current we've described so far. The electrical grid that supplies our homes and businesses operates with what we call "alternating current," or "AC." This is current that reverses direction in a periodic way. In subsequent chapters, we will see why AC is the natural way for generators to produce electric power from mechanical energy, and why AC is easier to transmit and adapt to various uses.

We can use the same tools we used for other oscillations to describe this situation. The electromotive force at your electrical outlets is time dependent:

$$v = V_0 \cos(2\pi f t)$$ **Equation 19.8**

In the rest of this chapter, we will use lowercase v and i to stand for voltages and currents that vary in time, to distinguish them from the DC voltages and currents of the previous sections. The constant V_0 in this expression is the amplitude of the oscillating voltage. The varying voltage v swings between the extremes of $+V_0$ and $-V_0$. In this equation, t is time, and f is the frequency of the oscillations, just as in Chapter 10. In the United States, $f = 60$ hertz for distributed electric power, while in many other countries $f = 50$ hertz.

If you plot this voltage v as a function of time, you may notice that the average potential over a long period of time will be zero. How can such an electromotive force actually accomplish anything useful? The key to understanding power in AC circuits is to go back to Equation 19.3 for electric power: $P = I^2R$. If the AC voltage above is connected to any resistive circuit (with resistance R), the current will be proportional to the voltage at all times, with the result that:

$$i = I_o\cos(2\pi ft)$$ **Equation 19.9**

Here, I_o is the amplitude of the oscillating current and it is given by $I_o = V_o/R$. With the dissipated power equal to i^2R, we can see that the power will not average to zero, because even when i is negative, i^2 is not.

Let's look more closely at AC power for some equivalent resistance that we'll call R. At any instant in time, this can be written as follows:

$$P = I_o^2 \cos^2(2\pi ft)R$$ **Equation 19.10**

What we're really interested in, though, is the average power delivered to the circuit. The average of the function $\cos^2$ over any number of complete cycles is $\frac{1}{2}$. Therefore, the average power can be written as follows:

$$P_{avg} = \frac{1}{2} I_o^2 R$$ **Equation 19.11**

If we define $I_{rms} = \frac{I_o}{\sqrt{2}}$, then $P = (I_{rms})^2R$, just like for the DC case. This rms_{value} stands for the "root mean square." An oscillating current (or voltage) can be completely specified by its frequency and either its amplitude or rms value. The rms value is handy because a DC current equal to I_{rms} would give the same average power as an AC current with amplitude I_o.

We can also define a root mean square value for the voltage:

$$V_{rms} = \frac{V_o}{\sqrt{2}}$$ **Equation 19.12**

We generally use this V_{rms} to specify the voltage of the electricity in your home. That means the 120 VAC voltage at an average American electrical outlet actually oscillates between positive and negative 170 V.

The Least You Need to Know

- Electric current is the motion of charged particles in a conducting medium, from a higher to a lower potential.
- In a given conducting object, Ohm's law states that the potential difference is proportional to the current, and the constant of proportionality is called the resistance of the object.
- Power in an electric circuit is equal to the product of current and potential difference $(\Delta V)I$, or equivalently I^2R; this is the rate at which electrical energy is turned into heat for a simple resistance.
- The current into any point of a circuit must be exactly equal to the current out of that point.
- If you follow the potential differences around any closed loop in a DC circuit, the total of all changes must be zero.

CHAPTER

20

Magnetism

In This Chapter

- The nature of magnets
- Magnetic fields
- The relationship between magnetism and moving charges
- Electric motors and electromagnets

In this chapter, we begin studying the magnetic half of electromagnetism. Magnetism is closely related to the electrical phenomena we've been discussing, but it is a bit more complicated. For one thing, it demands that we have electrical charges that are in motion. Never fear, though, we'll take it one step at a time.

Magnetism has plenty of applications in the real world. Magnetic forces are essential for electric motors, for example, as well as for generators. Electromagnets are used to pick up metal in the junk yard, or to separate metal pieces mixed in with other materials. Electromagnets are especially useful for physicists who study subatomic particles. They allow us to focus and direct beams of charged particles. These beams act as probes for learning about forces on length scales that are way too small to see or measure by other means.

Magnetic Forces and Fields

The magnetic forces you are most familiar with probably come from permanent magnets, so we'll start our study by discussing these. Various materials can be magnetized, iron being one common example. Magnets come in different shapes, but to keep things as simple as possible, we will begin by talking about permanent magnets that are shaped like straight bars. The horseshoe-shaped magnets you see in cartoons are just long bar magnets that have been bent into that shape.

It will help us talk about bar magnets if we realize that the needle of an old-fashioned compass is also a small bar magnet. It is a permanent magnet that is supported in a way that allows it to turn freely about a vertical axis. People observed hundreds of years ago that such a magnetized needle would always point in the same direction on Earth. One end would always point toward the North Pole, and the other would always point toward the South Pole. This was obviously a nice discovery that was very useful for navigating unfamiliar territory. If properly supported, any simple bar magnet will behave this way. This allows us to label the ends of a bar magnet, and in general, those ends are also called "poles." By convention, the end that tries to point north is called the north pole of the magnet (N), and the other end is called the south pole (S).

Once we have established this, we can play around with magnets and learn some interesting things. One of the first things we learn is that two bar magnets exert forces on each other, and these forces don't require any contact between the magnets. We also soon learn that this force can be attractive or repulsive, like the electric force. The N pole of one magnet will attract the S pole of another, but two N poles (or two S poles) will repel each other. Again, in a way that is similar to the force between electric charges, these forces are stronger when the poles are closer together.

RED ALERT!

The interaction between a permanent magnetic in a compass and planet Earth indicates that Earth itself is somehow like a bar magnet. Oddly enough, though, in order for the compass magnet's N pole to point to the north, we see that Earth's north geographic pole is like the S pole of a magnet, and Earth's south geographic pole must be a magnetic N pole. But that's fine; no law of physics prevents us from naming things that way.

So, unlike poles attract, and like poles repel. That would seem to indicate that the poles of a magnet are like electric charges. But magnetic poles are different in one very fundamental way. It is apparently impossible to isolate either pole of a magnet. If you were to cut a bar magnet in half, you would not get a separate N pole and S pole, but rather two magnets, each with their own N and S poles. The two smaller magnets would each be weaker than the original magnet. You could also stick them back together easily. The new N pole would strongly attract the new S pole, and the combination would behave just like the original bar magnet.

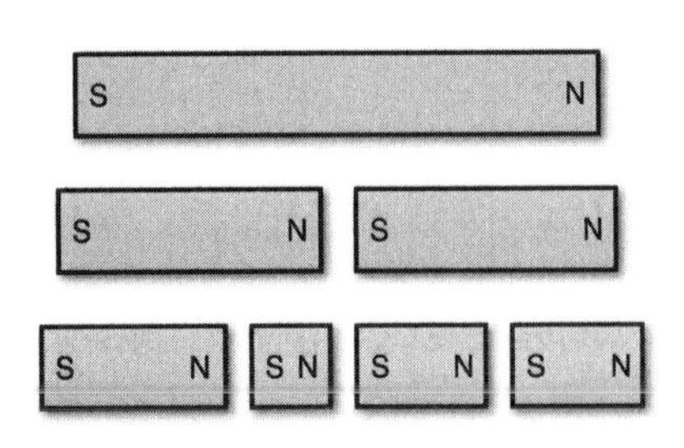

Figure 20.1

Dividing a bar magnet into pieces does not separate the poles, but only creates more magnets, each with two poles.

Cut the magnet into as many pieces as you like, which are as small as you can make them, and every piece will also be a magnet with its own two poles.

Not every magnet is as simple as a bar magnet. It is possible to construct magnetic objects that are composites with lots of poles, arranged in different ways. They will always have the same number of N poles as S poles. Most refrigerator magnets are made like this.

Let's get back to simpler magnets. The technical term for a magnet with the usual two poles is a *magnetic dipole.* And, the quantity that reflects the strength of such a dipole is called the *magnetic moment.* As we have seen, magnetic dipoles can exert forces on each other over a distance. We want to handle this the same way we dealt with other noncontact forces. We suggest that a dipole alters the space around it similar to the way an electric charge does, in the form of a force field. We say that the dipole is surrounded by a magnetic field. It is hard to define this field the same way that we did for electric fields, because magnetic poles can't be isolated the way charges can. That makes it difficult to define the strength of the field. Instead of a single charge, the source of a magnetic field is a dipole, and only a magnetic dipole can feel the force of that field.

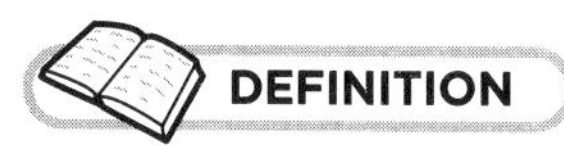

Magnetic dipole is a source of magnetic field, any magnetized element with one north pole (N) and one south pole (S).

Magnetic moment is a vector parameter which indicates the orientation and strength of a magnetic dipole.

We can, however, illustrate the shape of a magnetic field in a way that is analogous to what we did for electric fields. The magnetic field is a vector, and we will use the symbol $\boldsymbol{B}$ to represent it in general. The magnetic field can be visualized as directed lines in space, and a magnetic field line can only terminate on a magnetic pole. By convention, field lines point from N poles to S poles. The field vector $\boldsymbol{B}$ has a direction at every point in space, indicated by the direction of the field line. That direction is the direction of the force that would be felt by a north magnetic pole, if there was one at that location. As before, field lines can never cross each other, because then there would be an ambiguity in the direction of the force. And again, the magnetic field is stronger where field lines are closer together.

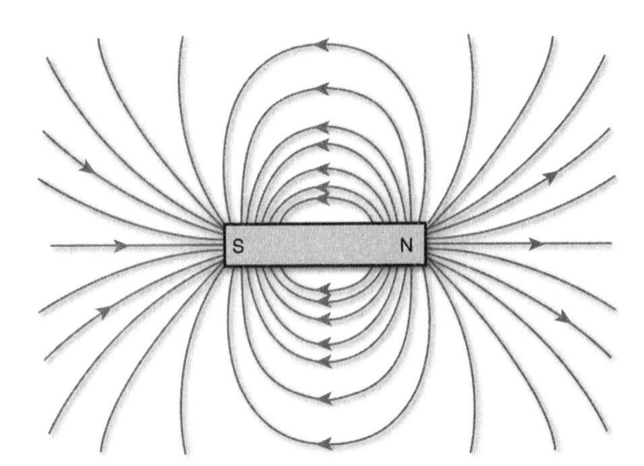

Figure 20.2

Magnetic field lines in the space around a dipole bar magnet. An infinite number of such lines could be drawn; this is only a representative sample.

A magnetic force will simultaneously act on both the N and S poles of any small "test" dipole that we put in a field. As a result, it will tend to rotate the dipole as the dipole tries to align itself with the magnetic field at its location. That means that the best way to quantify this effect is by using a torque instead of a simple force. That's also why small compass needles are useful for visualizing and plotting magnetic field lines in real situations. A compass needle will come to rest in the direction of the magnetic field at its location.

In the SI system that we are using in this book, the strength (or magnitude) of the magnetic field should be measured in units of newtons/(amp meters). We abbreviate one N/(A m) as a tesla (T). At this point, you may be surprised that dimensions of electric current (amps) are somehow related to magnetic field strength, but we'll see the reasons for this pretty soon.

In order to get any farther, we must explore the microscopic origin of magnetic fields. This is also how we begin to understand the fundamental connections between electricity and magnetism. The bottom line is that magnetic fields are always created by electric charges that are in motion. In permanent magnets, the source is electrons' rotational motion within their atoms. As a result of this motion, electrons essentially act like tiny magnetic dipoles.

CONNECTIONS

Within an atom, electrons exhibit two different types of motion—one associated with their orbits about the nucleus and one that acts as though electrons spin about their own axes. In reality, though, the electron doesn't really spin about its axis, since a careful look shows that this would violate the theory of relativity. The so-called electron "spin" is a ramification of quantum physics and it turns out to be critical for the formation of the different elements, their chemical properties, and even the order of the periodic table.

As we have seen, all atoms contain electrons, and even small objects contain a very large number of atoms. If the electron dipoles in an ordinary piece of matter were all aligned with their N poles in the same direction, that piece would be an incredibly strong magnet. But that never happens. It is much more typical for electrons to pair up so that their magnetic dipoles point in opposite directions. This is largely why most objects do not produce strong magnetic fields. Even in matter with unpaired electrons, small regions where electrons do align end up pointing in random directions with respect to each other. At the macroscopic level, the magnetic fields from all of these regions will tend to cancel each other out.

If such a material is brought near a strong magnet, however, it may be possible for the alignment of these different regions to change. In certain materials, a net magnetic dipole can be induced by the external field, much as electric charges get separated in the polarization phenomenon we discussed in Chapter 17. When this happens, the material will be attracted to the magnet. This is how magnets attract nails and paper clips, which themselves are not normally magnetic.

Moreover, if the alignment persists after the strong external magnetic field is removed, then the object has effectively become a permanent magnet. The degree of persistence varies among different metals, and can be improved by forming certain alloys.

In most materials (glass, plastics, wood, liquids, gases, and even many metals that are not in the iron group) there are not enough magnetic dipoles present to get aligned in the first place, so there is nothing available to be attracted by the magnetic force. This is why magnets only attract certain metals, and most things are immune to the magnetic force.

Charged Particles and Magnetic Fields

Since magnetic fields are caused by the motion of charged particles, we should be able to use that fact to create magnetic fields for our own purposes. But before we get into that, we first consider how charged particles are influenced by magnetic fields.

First of all, magnetic fields have no effect on charged particles that are stationary. Only moving charges feel the magnetic force. The force that a moving charge feels depends in a unique way on the magnetic field, the particle's velocity, and the charge of the particle. But motion alone is not enough, and in some cases even a charge that is moving in a magnetic field doesn't feel any force at all. If the velocity of the particle is exactly aligned with a field line (or in exactly the opposite direction), then the magnetic field has no effect on the charged particle at all. But if the velocity vector makes some other angle (call it ϕ) with the field, then there will be a force exerted on the particle.

This force is peculiar because of its direction. The force felt by a charged particle moving in a magnetic field is not in the direction of the field itself as you might expect. Rather, the force always makes a right angle with both the magnetic field and the velocity vector of the particle.

This is hard to illustrate on the two-dimensional page, but we give it a try in Figure 20.3, for a positively charged particle in motion.

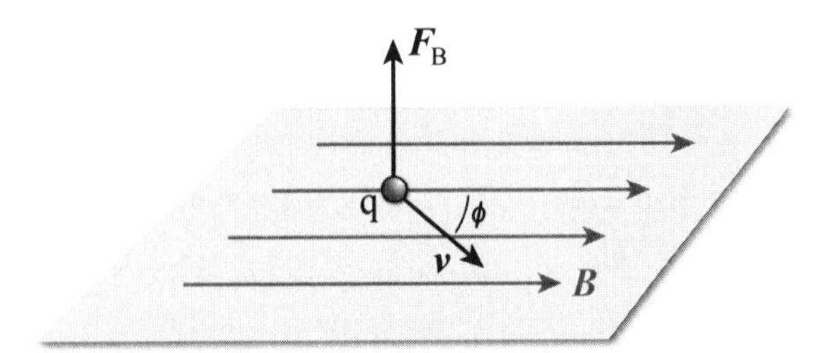

Figure 20.3

A positively charged particle moves with velocity **v** *in a region with a magnetic field. The particle feels a force* $\mathbf{F}_B$ *at right angles to the velocity vector and the local field lines as shown. The force is zero if the angle* ϕ *is either 0° or 180°.*

The magnitude of this force depends on the strength of the field (B), the amount of charge (q), and how fast and in what direction the particle is moving relative to the field ($\boldsymbol{v}$). When all of this is taken into account, the magnitude of the magnetic force turns out to be $F_B = qvB\sin(\phi)$. Note that for a given charge, speed, and field strength, the maximum magnetic force occurs when the particle moves at a right angle to the magnetic field.

Saying that the direction of the force is perpendicular to both the velocity vector and magnetic field vector is not quite enough to pin that direction down completely. You can think of the two vectors $\boldsymbol{v}$ and $\boldsymbol{B}$ as defining a plane that contains them both. There are actually two directions in space that are perpendicular to that plane. For a positively charged particle, physicists use something called the "right-hand rule" to choose which of those two directions is the correct one.

There are various ways to apply a right-hand rule, but here is one of them. First of all, there must be some angle between the $\boldsymbol{v}$ and $\boldsymbol{B}$ vectors. If that angle is 0° or 180°, then the force is zero and we don't worry about its direction. Otherwise, it is always possible to assign ϕ so that it is less than 180°. With that in mind, orient your hand in a loose fist with your thumb sticking out, in such a way that your fingers curl in the direction that would rotate $\boldsymbol{v}$ into the direction of $\boldsymbol{B}$. Then your thumb tells you which normal direction to choose for the force vector. This will be the correct choice if the moving particle has a positive charge. If it is negatively charged, the magnetic force will be in the opposite direction.

In the previous chapter, when we talked about moving charges, we thought of them as moving in conductors, like the wires in a circuit. It is also possible for charged particles like protons and electrons to simply move through space, without the benefit of conducting materials to move through. This is best done in a vacuum, and these days electric and magnetic fields are routinely used to create, steer, and focus such beams of charged particles. Large dipole magnets make it possible to build the big particle accelerators that are used to investigate the tiniest and most fundamental building blocks of matter.

Magnetic Force and Current

Let's try to apply what we have just learned to the force on a wire that is carrying some current. We know that current is defined as electric charges in motion, so we expect that magnetic fields will exert some force in this case as well.

It's pretty easy to derive this force for a segment of a straight wire that makes a 90° angle with a magnetic field surrounding it. If the wire carries a current I, we know that charges are moving in it at some average speed. Call the speed v, and concentrate on a finite segment of the wire that has a length equal to l.

From our previous discussion, we know that each individually charged particle feels a magnetic force with a magnitude equal to qvB. We can add up all of these little forces to get the total force on the wire. If we just label the total moving charge in that segment of wire Q, the result would simply be $F_B = QvB$.

Now all we have to do is substitute for the average speed of the charges v. If the length of the segment of wire is l, then we can also define a time interval Δt, the time it takes for a charge to move through the whole segment of wire. Then, by definition, $v = l/\Delta t$. Substituting this into our expression for the force above yields:

$$F_B = Q\left(\frac{l}{\Delta t}\right)B = IlB$$ **Equation 20.1**

The last step just required us to recognize that $Q/\Delta t$ is exactly equal to the current I flowing in the wire. This expression is fine if the wire happens to be perpendicular to the magnetic field. If instead there is an arbitrary angle ϕ between the wire and the field, then the force would be this instead:

$$F_B = IlB\sin(\phi)$$ **Equation 20.2**

As long as we consider the direction of the current I to be the direction in which the equivalent positive charges move, we can again use the right-hand rule to find the direction of this force. Just imagine $\boldsymbol{I}$ to be a vector pointing in the direction of positive current, with a magnitude equal to the current in amps. Then we can define the angle ϕ in a way that is consistent with our prior discussion. Curl the fingers of your right hand from the $\boldsymbol{I}$ direction into the $\boldsymbol{B}$ direction, and your thumb again points in the direction of the magnetic force on the wire.

This force is the basis for transforming electrical energy into mechanical energy, in order to do useful work. It is what allows us to build electric motors of all shapes and sizes. All we have to do is wind wires on an assembly that is allowed to rotate (usually called a rotor). Surround the rotor with permanent magnets to create a magnetic field. Then by controlling when current flows in the wires of the rotor, we can assure that the resulting magnetic force will cause the rotor to rotate.

TRY IT YOURSELF

44. A 1 km length of power line runs east-west, carrying a current to the west that is equal to 600 A. Let's say there is a magnetic field pointing due north that has a magnitude of 0.02 T all along the line. What force does the field exert on the power line? (Be sure to include the direction.)

Magnetic Field Due to a Current

Now we return to the question of how to create magnetic fields that we can control more easily than those in permanent magnets. By carefully studying electric current, we discover an interesting symmetry in nature. Magnetic fields are able to exert forces on moving charges. But moving charges also create magnetic fields. If you have been following along carefully, this shouldn't be a big surprise. We said early on that magnetic forces can only be felt by magnetic dipoles, and magnetic dipoles are also the source of magnetic fields.

Again we will start with the simplest case we can think of. We imagine a long, straight, conducting wire, isolated from any other electric or magnetic fields. Part of such a wire is shown in Figure 20.4. Experiments show that at every point outside the wire, there are circular magnetic field lines, centered on the wire. The strength of this magnetic field is greater the closer you get to the wire. To find the direction of this field, you would use your right hand in a slightly different way. Now if you point your thumb in the direction of the current, your fingers naturally curl in the direction of the curved magnetic field.

The magnitude of this field will be proportional to the current, as you would expect, and will also depend on how far from the wire you are. If we call r the distance from the center of the wire to the location where we want to know the strength of the field B, then it turns out that the strength of the field can be written as follows:

$$B = \frac{\mu_o I}{2\pi r}$$ **Equation 20.3**

This expression has the expected behavior with distance, such that the magnetic field gets weaker as you get farther from the wire. To arrive at this formula, we had to introduce a new fundamental constant μ_0, which is called the permeability of the vacuum. In our system of units, this constant is given by $\mu_0 = 4\pi \times 10^{-7}$ T m/A.

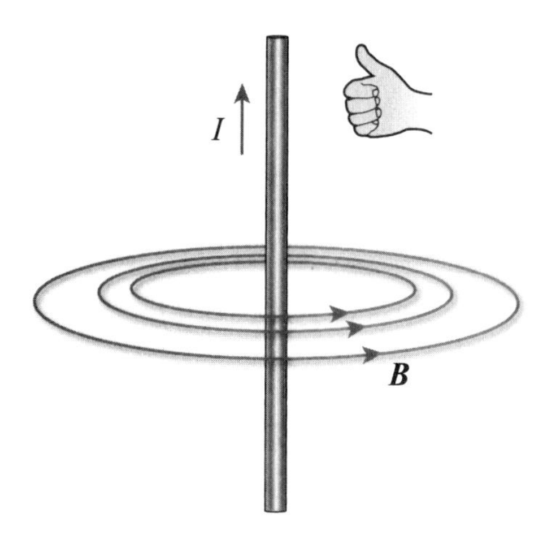

Figure 20.4

A current I *is flowing upward in the single straight wire shown here. A magnetic field is created by this current all around the wire, whose field lines are circles centered on the wire, perpendicular to it.*

Because current creates magnetic field, and magnetic fields exert forces on currents, it is possible that two wires can exert forces on each other if they are in the same vicinity, and both are carrying current. You should be able to use the right-hand rule now to determine the directions of these forces for parallel wires separated by some small distance. The conclusion is that two such wires will attract each other if their currents are in the same direction, and they will repel each other if their currents are in opposite directions.

Solenoids

Instead of simple straight wires, let's consider now what would happen if we have some current in a circular loop of wire. We know that each piece of the wire has circular magnetic field lines all around it. Qualitatively speaking, if we bend the wire itself into a circle, these field lines will bunch up on the inside of that circle, and spread out on the outside. Near the center of the circle, all of those field lines from all the parts of the wire will be in the same direction. Thus we expect a pretty strong magnetic field near the center of the current loop, pointing in a direction that is perpendicular to the plane formed by the loop's geometry (that is, pointing through the loop). We put this phenomenon to practical use by stacking a large number of such loops as a coil, a configuration that is also called a *solenoid.*

A **solenoid** is a device formed by coiling many loops into a wire, though which a current is run. It generates a large magnetic field within the coils and can be used to generate a relatively strong magnetic field with modest electric current.

If we look first at a single circular loop of current, with a radius equal to R, the magnitude of the magnetic field at the center of the loop turns out to be as follows:

$$B = \frac{\mu_o I}{2R}$$ **Equation 20.4**

Now consider a tightly wound solenoid made of N turns of wire, so that all of the loops carry the same current and have approximately the same radius. The magnetic fields due to all of these loops add up, so along the central axis of the solenoid, the magnetic field can be written as:

$$B = \frac{\mu_o NI}{L}$$ **Equation 20.5**

Figure 20.5 shows the shape of the full magnetic field where L is the length of the solenoid. Outside of the coils, the shape of the field resembles that of the bar magnets we considered at the beginning of this chapter. It is effectively a dipole with N and S poles. Once again, a right-hand rule helps relate the direction of the central magnetic field to the direction of the current spiraling around the solenoid.

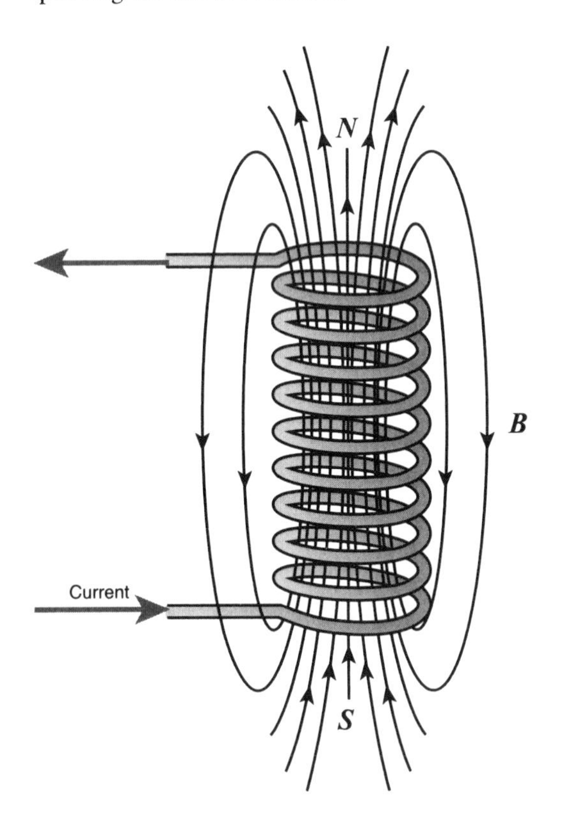

Figure 20.5

The solenoid generates a strong dipole field with the shape shown. The strongest field is located at the center of the coil and directed along the solenoid's axis.

Wrapping a solenoid around an iron rod is a practical way to create an even stronger magnet. You can try this for yourself with a large nail and some ordinary wire, and you won't need very large voltages to do it. This kind of magnet is not a permanent magnet, because the magnetic field disappears if current is not flowing. Therefore this arrangement is called an electromagnet, given that electric current is needed to create the magnetic field.

The Least You Need to Know

- Magnetic fields are produced by and act on magnetic dipoles. There are no isolated magnetic poles.
- A magnetic field also exerts a force on a moving electric charge, at right angles to the charge's velocity and the direction of the field.
- Currents, which are streams of moving charges, generate magnetic fields around them.
- When shaped appropriately, wires can be used to produce magnetic fields whenever electric current is passed through them; these are called electromagnets.

CHAPTER

21

Electromagnetic Induction

In This Chapter

- Creating electric current without batteries
- The magnitude and direction of induced currents
- Converting mechanical energy to electricity
- How a changing magnetic field creates an electric field, and vice versa

We've come a long way in our survey of electricity and magnetism. We've learned about electric and magnetic fields, and how they are created. For electric fields, it was pretty simple: electric charges produce electric fields around them, and electric fields exert forces on electric charges. Making magnetic fields was trickier, since we needed charges to be in motion, as when current flows in a wire or electrons move in their atoms. In a nice bit of natural symmetry, a magnetic field can also exert a force on electric charges, but only if those charges are moving.

In this last chapter of this part on electromagnetism, we'll close the loop, so to speak. The magnetic force we have seen so far only exerts a force on charges that are already moving, and the direction of that force is perpendicular to their direction of motion. Thus a steady magnetic field can't get charges to start moving if they are stationary. It would be very useful if there were also some way that a magnetic field could be used to make current flow. This turns out to be possible, and it is the key to generating electric power on a large scale for a wide range of applications.

An Example of Induction

Once you know the secret, there are a lot of different ways to produce electric current using a magnetic field. But what is the secret? As we have seen, an electromotive force (emf) is required to cause current to flow. Any process by which a magnetic field produces an emf is called *electromagnetic induction.*

Electromagnetic induction is a process by which an electromotive force is induced by a magnetic field that is changing with respect to some charge. If the charges are in a circuit, there will also be induced electric current as a result.

If we are interested in a flow of current, we need to have a circuit. At a minimum, that means a closed loop of wire. But even if you have a stationary loop of wire in the presence of a static magnetic field, you still won't induce any current, no matter how strong the field is. The secret to inducing a current is to have something that changes with time.

There are a lot of ways to actualize a changing relationship between some electric charges and a magnetic field configuration. All the different options can get pretty confusing pretty fast. So we'll start with a very simple setup that illustrates the essential features and builds directly on the concepts introduced in the previous chapters.

Start with a simple metal rod, which is a good conductor, located in a large region in which the magnetic field is uniform and steady, constant in magnitude and direction. This situation is illustrated in Figure 21.1. If the rod is stationary in the field, then nothing happens. Electrons are more or less stationary and evenly distributed through the rod, exactly balancing the positive nuclei so that every chunk of the rod is electrically neutral.

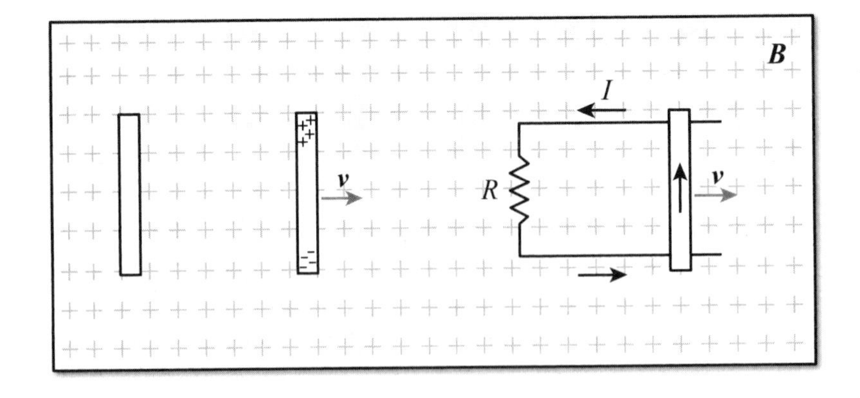

Figure 21.1

A constant magnetic field points directly into the page, indicated by the crosses. At left, is a stationary conducting rod. At center, the same rod is moving to the right, creating a charge separation and thus a potential difference between the ends. At right, the rod is sliding along a stationary U-shaped conductor, and current flows in the counterclockwise direction.

Now we quickly bring the rod to a constant speed, in a direction that is at right angles to the length of the rod and to the magnetic field lines. What happens to the electrons then? They are initially forced to move to the right, along with the rod. But as they move, they feel the magnetic force that we talked about in the last chapter. You recall that the magnitude of that force is equal to $qvB\sin(\phi)$, and in this case $\phi = 90°$. The negatively charged electrons are free to move, but only within the confines of the conductor. Using our rule for the direction of the force, we conclude that the mobile electrons will gather at the bottom of the rod (as shown in Figure 21.1).

The result of this motion is a rod that has excess negative charge at one end, and excess positive charge (because of the atoms that lost electrons) at the other. Therefore, a potential difference now exists between the two ends of the rod, which wasn't there when the rod was stationary. That potential constitutes an induced emf. Moreover, if the rod stops moving, the electrons would return to their original positions, and the emf would disappear.

If we now provide another conducting path, with some resistance, between the ends of the rod, we can turn the induced emf into an induced flow of current. We could do this by introducing a long, stationary, U-shaped conductor that our rod can slide along. Electrical contact between the rod and the other conductor would need to be maintained as it slides to the right, of course. The motion of the rod relative to the magnetic field causes electrons to move down as we noted, which is equivalent to positive charge moving up. But now those charges can flow out of the rod and around the rest of the conducting loop.

As long as the rod is moving at a constant speed, there will be a steady current in the loop that is counterclockwise, as illustrated in the figure. This is an example of one type of electromagnetic induction. The magnetic field is constant, but current flows because of some motion. Therefore, we call this "motion-induced" current.

Magnetic Flux

In order to figure out how much current flows, we would need to know both the resistance in the circuit and the voltage of the induced emf. Figuring out the resistance is straightforward, but to determine the magnitude of this induced emf, we must first define a new physical quantity, which we will call *magnetic flux*.

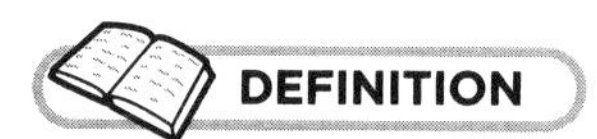

Magnetic flux is the total strength of magnetic field passing through any given area, multiplied by that area. Only the component of the magnetic field that is perpendicular to the area contributes to the flux. The symbol for flux is Φ and its standard units are T m^2, also called webers (Wb).

Magnetic flux is difficult to describe, so we'll use a figure to illustrate it. Take for example a nice, flat square with an area A. Put that square area into a region of space with a uniform magnetic field of magnitude B, this time pointing to the right. We set the square perpendicular to the page, so that we are viewing it nearly edge on. We also define an angle (θ) between the direction of the magnetic field and a line perpendicular (normal) to the square's flat area. Finally, we'll use a capital Greek phi (Φ) to represent flux.

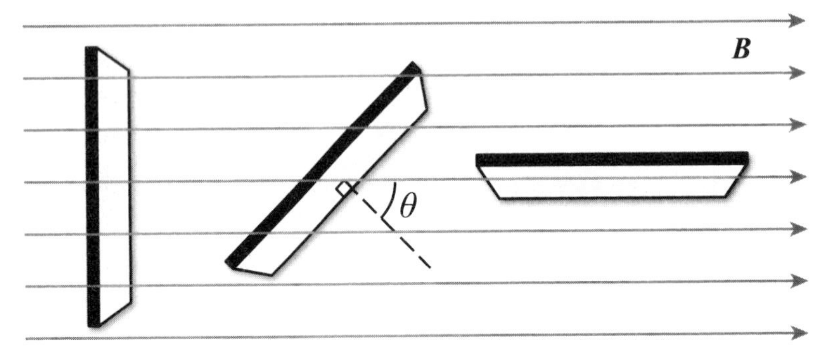

Figure 21.2

A square area in a uniform magnetic field, with three different orientations.

Let's assume that the magnetic field is constant in magnitude and direction across the area we are looking at. Then the magnetic flux is given by the following expression:

$$\Phi_B = BA\cos(\theta)$$ **Equation 21.1**

(We are using the subscript B on this flux to remind us that it is associated with a magnetic field. It is also possible to define a flux for electric fields in the same way.)

The flux in general depends on the strength of the magnetic field and the size of the area we are talking about. The cosine factor is there to make sure that only the component of ***B*** that is perpendicular to the area contributes. If the magnetic field is directed along the plane of the square's surface, then $\theta = 90°$ and there is zero flux through the area. For a given area and field strength, the maximum flux is achieved when the area is perpendicular to the field ($\theta = 0°$). Qualitatively, you can think about the flux as being proportional to the number of field lines that pass through the area in question. Note that, by definition, magnetic flux is never negative.

Faraday's Law

In the real world, we have to deal with magnetic fields whose strength varies with position, and whose direction changes, like the dipole field illustrated in a previous chapter. We may also have occasion to calculate the flux through curved surfaces. In such cases, we use more sophisticated mathematics (calculus) that essentially allows us to divide such surfaces into small, basically flat areas. Then we simply add up the flux through all of those little areas to get the total flux. We won't get into all of that here. In this book, we're more interested in what the flux is used for.

Now that we have a definition for magnetic flux, it becomes very easy to express the magnitude of an induced emf. Consider a conducting loop located in a plane. The loop defines a flat area, which can be any shape at all. The simple expression we need is called Faraday's law:

$$\mathcal{E} = \left|\frac{\Delta\Phi_B}{\Delta t}\right|$$ **Equation 21.2**

In this expression, $\mathcal{E}$ stands for the magnitude of the induced electromotive force. This emf is still measured in volts, but unlike a potential difference, it is always positive. All we are saying is that the magnitude of the induced emf is proportional to how fast the magnetic flux through the loop is changing.

Equation 21.2 really gives an average emf induced during the time period Δt, but the concept works for instantaneous emf in the usual way (by letting the time interval shrink towards zero). It doesn't matter whether the flux is increasing or decreasing, as long as it is changing in time. If the flux through a loop is not changing, then there is no induced emf and therefore no current. Also, this very general definition of flux hints at all the many ways that you can induce an emf (e.g., changing the area of the loop, the strength of the magnetic field B, or even the angle between $\boldsymbol{B}$ and the loop).

In practical applications, we often pile up multiple loops of conductor around the same area, as in a tightly wound solenoid. We're imagining a single wire wrapped in multiple loops, so it is as if each loop is connected in series with the next. In that case, any magnetic flux through one loop of the wire will be pretty much the same as the flux through all the rest. If that flux changes, the emfs will add, and a more useful form of Faraday's law is the following:

$$\mathcal{E}_{\text{total}} = N\left|\frac{\Delta\Phi_B}{\Delta t}\right|$$ **Equation 21.3**

Here, N is the number of turns in the coil. This obviously reduces to Equation 21.2 if there is only one loop. But while Faraday's law nicely gives the magnitude of the emf, it does not tell us which direction a resulting current will flow. For that we need another rule.

TRY IT YOURSELF

45. A solenoid with 150 turns of wire has an inner diameter of 4.0 cm. Over a period of 5 seconds, the magnetic field inside the solenoid is raised at a constant rate from zero to a strength of 12 T. What is the emf induced in this solenoid during that time?

Lenz's Law

Fortunately, there is one single rule that tells us the direction of the induced current in every situation of changing magnetic flux. It is called Lenz's law, and it simply states that any induced current goes in a direction that would oppose whatever action is causing it to flow.

To see what we mean, we apply this first to the example of motion-induced current illustrated in the rightmost panel of Figure 21.1. Recall that a steady magnetic field was pointing into the page. Even though the strength and direction of that field were constant, the flux through the current loop began changing as soon as we started moving the rod to the right. The flux was increasing because it is a product of the field strength and the area enclosed by the loop. By moving the rod to the right, we were increasing the area, at a rate determined by how fast we moved the rod.

TRICKS AND HACKS

A handy way of remembering Lenz's law is to imagine that loops of wire are opposed to any change of magnetic flux through them. If an external agent causes that flux to change, then the loop will generate its own current in order to produce a magnetic field opposed to the change.

In order to counteract this increase of flux into the page, we would need to produce some magnetic field out of the page. We saw in the previous chapter that electric current in a loop of wire does produce a magnetic field like this. Using the right-hand rule, we see that in order to create a field directed out of the page, the current would have to go counterclockwise around the loop. Therefore, the prediction of Lenz's law agrees with the current direction we figured out from the basic magnetic force in this case.

If you've gotten this far, you can now predict the direction of induced current for all kinds of electromagnetic induction. Besides the sliding rod trick, we might also use a permanent magnet to induce some current in a coil of wire. This time you'll need to remember something about the shape of the magnetic field lines around a bar magnet, a typical magnetic dipole. Also, don't forget that the field is stronger where field lines are closer together.

Imagine a circular loop of wire that is flat and stationary. If we bring a bar magnet close to the loop in the direction shown in Figure 21.3, then as it moves, the magnetic flux through the loop is downward and getting larger. In order to oppose this, a current will be induced in the loop that creates a magnetic field pointing upward. If the magnet stops moving, then the current stops. What's more, if we then move the magnet up and away from the loop, the downward flux through the loop will decrease. During that motion, there will be current induced in the loop in the opposite direction.

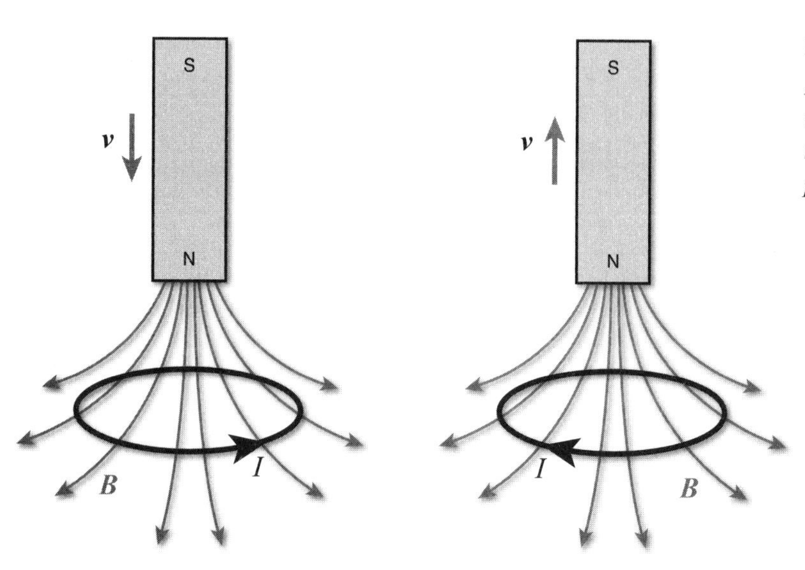

Figure 21.3

At left, the north pole of a permanent magnet is being brought closer to a loop of wire, and at right it is being pulled away.

The Generator

This motion-induced emf is the key to generating electricity on a large scale. Chemical batteries are bulky and very limited in the amount of energy they can produce, requiring extra efforts to recharge or replace them. Our modern technological society relies on larger amounts of electrical energy which must be available at all times and many places.

The bar magnet and wire loop we just talked about serve as a first example of an electric generator. All we need to do is arrange things so that the permanent magnet moves up and down repeatedly relative to the wire loop. This will naturally produce current in the loop that alternates in direction, just like the alternating current we described at the end of Chapter 19. If that current is delivered to some resistive circuit, then it is clearly producing electric power. Some agent will have to do mechanical work to move the magnet back and forth, so the overall result is that mechanical energy has been converted into electrical energy. Total energy is still conserved.

If you have a reasonably strong magnet and some way to detect electric current (a small light bulb will do), you could even try this at home. You will increase the amount of electrical energy produced if you wrap several coils of wire around the same loop, like the solenoid arrangement described earlier. Connect the coil to the light bulb, and move the magnet quickly back and forth as shown in Figure 21.3, and see if you can light the bulb. You may soon realize that moving a magnet back and forth like this is actually a lot of work and might not be the most practical way to generate electricity.

Better setups use rotational motion to continually change the magnetic flux in some coils to generate electricity. In addition, they are often arranged so that the magnet stays put and the coil moves instead. A very simple arrangement is shown in Figure 21.4.

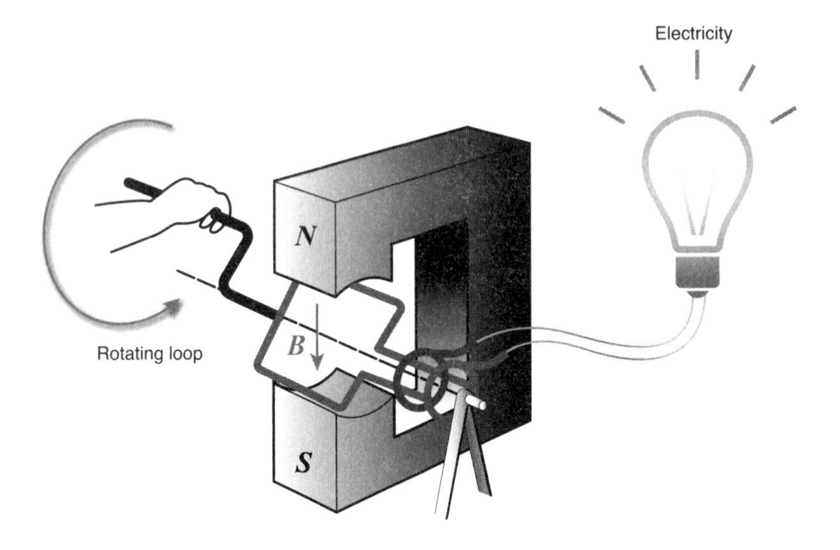

Figure 21.4

A single loop of wire is rotated in a strong magnetic field. The magnetic field is stationary and constant, but the periodic changes in flux through the loop cause current to flow.

Again, real generators will have many more loops of wire than the single one shown here, arranged in various ways to get the most energy out of each rotation. But the basic idea is the same. Every time the loop rotates, the amount of magnetic flux through it changes from zero up to a maximum (when the plane of the loop is perpendicular to the field lines) then back down to zero. Whenever the flux is changing, Faraday's law tells us that an emf is induced, and Lenz's law tells us which way current will flow. The flux will alternately increase and decrease, so the direction of the resulting current will also reverse every time around. A constant rate of rotation will yield alternating current with a constant frequency.

CONNECTIONS

It is interesting to note that the electric generator is conceptually just the opposite of an electric motor. The motor uses the force on moving charges to convert electrical energy into mechanical energy in order to do useful work. The generator uses motional induction to convert mechanical work into electrical energy. In fact, if properly designed, the same device can serve either purpose. If no electric power is supplied to a motor, applying a torque to turn the shaft can produce electric current.

Electric generators come in all sizes, and use various forms of energy to drive them. Muscle power can be used, either hand cranking or pedaling with feet and legs. Blowing wind can be used to produce the rotational motion. On a larger scale, falling water from behind a dam can turn a waterwheel or a turbine to get the generator spinning. Pressurized steam can also be used, adding a thermodynamic step where heat is turned to mechanical energy before the mechanical energy is turned into electricity. But in every case, electromagnetic induction is the key to producing electricity.

Mutual Inductance

If we use an electromagnet instead of a permanent magnet, there is yet another way to induce current in a coil of wire. Let's imagine two short solenoids, located right next to each other, so that their axes are lined up. Initially, there is no current flowing in either solenoid.

We connect a simple resistor to the second solenoid, and some kind of power supply to the first one. The power supply allows us to put current into the first solenoid in a controlled way. Imagine turning on the power supply, and slowly increasing the current in the first solenoid from its initial value of zero. We know that a dipole magnetic field will be created, the magnitude of which is proportional to the applied current. Some of this field extends into the second solenoid, so that its coils all see a magnetic flux that is increasing. What we observe is current flowing in the second solenoid, whenever the current is changing in the first solenoid. This is another case of electromagnetic induction, but this time, no motion was required.

The emf and resulting current were still induced by the changing magnetic flux, according to Faraday's law, and their directions will still be determined by Lenz's law. Since the two coils are similar to each other, we could just as well have applied current to the second one to induce current in the first. A situation like this is often described as mutual induction, and no motion of the coils is required.

It is critical to remember that a static configuration of charges, conductors, and magnetic fields cannot induce any electromotive force or current flow. Something must be changing in time—be it mechanical motion, changing current, or something else—for there to be any electromagnetic induction.

How much current gets induced in the second coil depends on a lot of details, including the resistance in the circuit to which it is connected. We can give a fairly simple expression if we ignore this resistance and just concentrate on the electromotive force induced in the second coil by the changing current in the first. For a fixed setup (nothing moving), the magnitude of the induced emf is simply proportional to the rate at which the current is changing:

$$\mathcal{E}_2 = M\left|\frac{\Delta I_1}{\Delta t}\right| \qquad \textbf{Equation 21.4}$$

The constant of proportionality M is called the *mutual inductance*, which is a property of the detailed physical setup (how large the coils are, how close together they are, etc.). In this expression, I_1 is the current in the first coil, the one we were controlling, and $\mathcal{E}_2$ is the resulting emf in the second coil. The direction of this emf will once again be determined by Lenz's law.

Mutual inductance is a measure of how much electromotive force is produced in one circuit due to a time-varying current in another. The standard unit is V s/A, also called the henry (H).

Mutual inductance is the basis for a very important electrical device called the transformer. Most practical transformers attempt to maximize the mutual inductance effect by having a pair of colocated, multiturn coils (i.e., really wound around the very same area, often with some iron in the center). This ensures that the two coils always share the exact same magnetic flux.

Let's imagine such an ideal transformer with two separate coils sharing the exact same flux area. One coil (which we will call the "primary") has a total number of turns equal to N_1. We'll call the other coil the "secondary," and say it has N_2 turns of wire. The usefulness of a transformer comes when some oscillating AC current is sent into the primary coil. The AC current is constantly changing, which means the magnetic flux in the primary is also changing at some rate. This of course induces an emf in the primary over any time interval Δt according to Faraday's law, as given in Equation 21.3:

$$\mathcal{E}_1 = N_1 \left| \frac{\Delta \Phi_B}{\Delta t} \right|$$

During this same time interval, the flux in the secondary will be changing at exactly the same rate, so there will be an induced emf in the secondary coil, equal to:

$$\mathcal{E}_2 = N_2 \left| \frac{\Delta \Phi_B}{\Delta t} \right|$$

Equation 21.5

Just a little algebra is required to reveal that $\mathcal{E}_2 = \frac{N_2}{N_1}\mathcal{E}_1$. Thus, if the two coils have an equal number of turns ($N_1 = N_2$), then any AC voltage that is applied to the primary will also appear on the secondary coil. But if the number of coils is different, then this is a useful device for changing the voltage of any AC power source. Transformers with $N_1 > N_2$ have the effect of reducing the voltage that appears in the secondary circuit, while if $N_1 < N_2$, the voltage is increased.

Although voltage is related to potential energy, we should note that the transformer does not violate the principle of conservation of energy. Recall that the rate of electrical energy flow is equal to the product of voltage and current. A transformer that triples the voltage of an AC source will also deliver one third of the current that was fed to the primary coil. The product VI is the same for both coils, so that in any time period, the same amount of energy goes in as comes out.

Because of the joule heating we mentioned earlier, power transmission over long distances is more efficient if we use very high voltage (and therefore low current) in power lines. But these high voltages are not safe for residential use, so there is a widespread need for changing voltage levels in the real world. Thanks to transformers, this is much easier to do with AC power rather than DC power. That fact, combined with the oscillation that naturally results from rotary generators, is the reason that household current is AC rather than DC.

Energy in a Magnetic Field

Let's go back and consider a simple solenoid for a moment, a single wire wound many times around in a cylindrical shape. Even if the wire of the solenoid is a very good conductor, you will find that it is not so easy to increase the current in that coil.

Say you have a nice 20 V battery available. If you attach it to the solenoid, it will take a little time for the current to build up to its final value. The reason for this is something called "self-inductance." As you increase the current, you are trying to increase the magnetic field inside the coil of wire. The changing magnetic flux in that region leads to an induced emf in the coil that opposes the change you are trying to make. If you ever get into electronics with AC circuits, you will have to worry about this self-inductance effect.

For right now, we are more interested in the fact that electrical energy had to go into that solenoid in the process of ramping up the current. What happened to that energy? The answer is that, like electric fields, magnetic fields contain energy. The best way to deal with the energy contained in a magnetic field is again to specify the energy density, which is the amount of energy per unit volume. In this case, magnetic energy density turns out to be:

$$u_B = \frac{1}{2}\frac{B^2}{\mu_o}$$

Equation 21.6

You should notice a close resemblance between this expression and the one we derived for electric energy stored in a capacitor. In both cases, the energy density is proportional to the square of the fields' magnitude.

Finally, we would like to reflect a little more deeply on the meaning of Faraday's law of induction. In trying to get a handle on this difficult topic, we concentrated on induced emfs and currents due to changing magnetic fields. But in a current loop, it seems very difficult to pinpoint where exactly the emf is located. The reality is that it is not localized at all. The emf idea is a convenient one for doing calculations, but at a more fundamental level, the induced emf itself is really a product of an electric field.

A current-carrying wire is still an electrically neutral object, so any electric field created by its negative charges is cancelled out by the positive charges that are also present. But as we have seen, the flow of current creates a magnetic field. The real explanation for electromagnetic induction is that whenever a magnetic field changes, it creates an electric field all by itself. It is a very fundamental feature of nature that changing magnetic fields always produce electric fields. And the reverse is also true: changing electric fields always produce magnetic fields. In the next chapter, we will see how important this symmetry really is.

The Least You Need to Know

- An electromotive force is created by any change in magnetic flux.
- If motion is used to change magnetic flux, then mechanical energy can be converted to electrical energy in a device called a generator.
- Mutual inductance is used in transformers to step down or step up AC voltage.
- Magnetic fields carry energy, and the energy density in a magnetic field is proportional to the square of the strength of the field.

PART

5

Light and Optics

A special relationship between electric and magnetic fields allows a disturbance in these fields to propagate at a very high speed without the need for any matter to be present. We call this phenomenon the electromagnetic wave, and that is where we will start this final part of our book.

Electromagnetic waves are fundamentally different from the mechanical waves we described in Part 2, but some of the same ideas and tools developed there can still be applied here. It turns out that visible light is composed of electromagnetic waves within a certain range of wavelengths. We will begin by looking at the properties of visible light, keeping in mind that these same properties apply to other forms of electromagnetic radiation as well.

Then we will introduce you to some basic optics. We will analyze what happens whenever light shines on an object that is large compared to its wavelength. This will include rules for reflection and refraction of light, and how these can be applied to understand how lenses and mirrors form and manipulate images.

CHAPTER

22

Light

In this chapter, we begin our formal study of light. We'll start by explaining how light waves, and electromagnetic waves more generally, are produced by the acceleration of electric charge. We'll tell you how fast light travels and how much energy it carries when it does. We will also introduce you to the so-called electromagnetic spectrum, a broad and continuous range of wavelengths spanning from the very long waves produced in radio antennas to those produced deep within the microscopic confines of atomic nuclei.

We'll also begin to explore what happens when light encounters objects in its path. While we'll save "big" things like lenses and mirrors for the next chapter, here we'll see what happens when light interacts with objects that are roughly the same size as its wavelength. This is when the wave nature of light is most apparent, which leads to important parallels with mechanical waves, the other type of wave we've encountered in this book.

In This Chapter

- The physical basis of light
- The (quite impressive) speed of light
- Radio waves vs. light waves vs. gamma rays
- Diffraction of light through and around small obstacles

Electromagnetic Waves

By now you know that a stationary electric charge will produce a static electric field in its vicinity. You have also seen that any constant flow of charges (i.e., a steady electric current) will produce a static magnetic field in its vicinity. But what would happen if you were to take an electric charge and let it accelerate back and forth in some periodic way?

The result of this charge oscillation would be a combined electric and magnetic field that travels through space in a direction perpendicular to the charge's movement.

What's more, at any point in space, the magnitude of those fields would rise and fall, and the frequency of the fields' oscillations would be the same as that of the oscillating charges. If you harken back to Chapter 11, you should recognize this phenomenon—a disturbance with oscillating magnitude that travels through space—as a wave phenomenon. And since this wave is composed of electric and magnetic fields, we call it an *electromagnetic wave.*

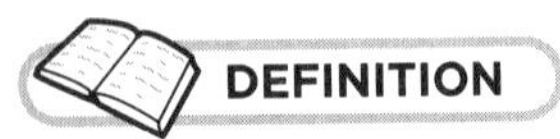

An **electromagnetic wave** is a synchronized oscillation of electric and magnetic fields that propagates through space, which can be produced by accelerating electric charges.

Scientists have been studying electromagnetic waves in the laboratory since the nineteenth century. In 1887, Heinrich Hertz (after whom the frequency unit is named) managed to measure the speed of electromagnetic wave propagation. His fateful discovery, that such waves move at the same speed light was known to travel, demonstrated with little doubt that light is in fact an electromagnetic wave.

In Chapter 11, we defined mechanical waves as "disturbances in some material medium that travel over time." If you take away the material medium, the mechanical wave will vanish. That's why there is no sound in space, for example. But if you think about it, there must be light in space. How else would sunlight make its way to Earth? One fundamental difference between electromagnetic waves and mechanical waves is that the former can travel even when there is no material medium present. This is because the underlying "disturbance" is in the electromagnetic field, which, as you know, can exist perfectly fine in a vacuum.

Let's return to our oscillating charges to get a better picture of electromagnetic waves. Consider a long, vertical wire in which you allow some charge to slosh up and down at some particular frequency. As the charge moves back and forth in the vertical direction, which we will call z, it creates an electric field that is constantly changing. Near the z-axis, the field alternately points up and down, pointing toward wherever the excess negative charge is located. The disturbance in the electric field travels away from the wire in all directions perpendicular to the wire. Let's call one of those directions of travel x.

Let's forget about the charge now, so that all we see is an electric field whose magnitude oscillates back and forth in the z-direction with a shape that travels in the x-direction. One consequence of the interaction between electric and magnetic fields is that the oscillating electric field will generate an oscillating magnetic field that travels along with it. In addition, the direction of magnetic field oscillation will be perpendicular to the electric field's oscillation. In terms of the coordinate system we established above, the magnetic field will oscillate back and forth along the y-direction, as indicated in Figure 22.1. Although no matter is moving, it is still true that electromagnetic waves are transverse waves.

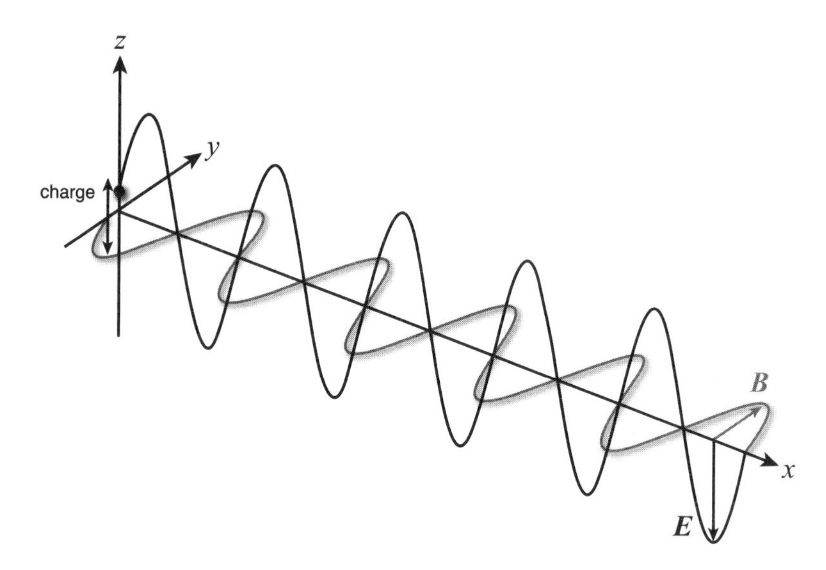

Figure 22.1

An electric charge oscillating along the z*-axis produces an electric field that oscillates in the* z *direction. This generates an oscillating magnetic field in the* y *direction. This conjoined electromagnetic field disturbance travels away from the charge along the positive* x *direction.*

As you can see from Figure 22.1, whenever the electric field is at a maximum (or minimum), the magnetic field is also at a maximum (or minimum). The electric and magnetic fields actually act to sustain one another, which explains why electromagnetic waves can travel so far from the accelerating charges that created them in the first place. And, the amplitude of the wave is inversely proportional to the distance from where it is produced. This means that it can weaken but will never vanish entirely—even after traveling great distances like from the sun to Earth, or from one galaxy to another.

A careful theoretical study of this phenomenon shows that the speed of light, which we will call c, is defined in terms of the two fundamental constants we encountered when studying electric and magnetic fields: the permittivity of free space (ε_0) and the permeability constant (μ_0). The exact relationship and the resulting value is as follows:

$$c = \frac{1}{\sqrt{\varepsilon_o \mu_o}} = 2.998 \times 10^8 \text{ m/s}$$

Equation 22.1

This may surprise you on two accounts. First, the speed of light is huge! Simple unit conversion will show that the speed of light is equivalent to about 1,000,000,000 km/h or 670,000,000 miles per hour, an enormous rate by everyday standards. The other interesting fact is that the speed of light does not depend in any way on how the wave is generated. It doesn't care about the frequency of oscillation or the nature of the charges that created it. The speed of light is a constant value. (At least, that is, in a vacuum; whenever light travels through a transparent medium like water or glass, its speed is somewhat less, as we will see in the next chapter.) What's more, it turns out that the amplitudes of the electric and magnetic fields in the wave are always in the same ratio, and the ratio is c. In other words, $c = E/B$.

Energy Carried by Electromagnetic Waves

When studying mechanical waves, we learned that these could carry energy, so it should be no surprise that the same holds true for electromagnetic waves. But how much energy? If you've been reading closely, you may realize that we've already given you the answer.

In Chapter 18, we discussed how much energy is stored in an electric field, concluding that the electrical energy contained per unit volume is $u_E = \frac{1}{2}\varepsilon_o E^2$. Likewise, in Chapter 21, we showed you that the amount of energy stored per unit volume in a magnetic field is $u_B = \frac{1}{2}\frac{B^2}{\mu_o}$. In neither case did we say that this relationship holds for static fields only, or for any particular situation. These are general relations for electric and magnetic fields, and the same holds for the time-changing variety that is at the heart of electromagnetic waves.

The only thing we have to be sure of is that we take into account both the magnetic and the electric energy density. After doing so, we see that the total energy per unit volume is given by the following:

$$u = \frac{1}{2}\varepsilon_o E^2 + \frac{1}{2\mu_o}B^2$$

Equation 22.2

One elegant consequence of the symmetry linking electric and magnetic fields is that, at any particular instant in any particular location, the energy density stored in the electric field is precisely the same as the energy density stored in the magnetic field. As a result, the two terms in the right side of Equation 22.2 are in fact equal to one another. We can substitute either one for the other to show that the instantaneous energy density of an electromagnetic wave is:

$$u = \varepsilon_o E^2 = \frac{1}{\mu_o}B^2$$

Equation 22.3

The Electromagnetic Spectrum

We've determined that light is an electromagnetic wave, and the frequency of such waves is set by the oscillation frequency of some electric charge. The example we used, however, charge oscillating up and down along a wire, is actually a better description for how radio waves are generated than light waves. The reason we could get away with this is that radio waves and light waves are both electromagnetic waves that differ only in frequency. In fact, there is a very broad range of frequencies that are possible, and this range is aptly named the *electromagnetic spectrum.*

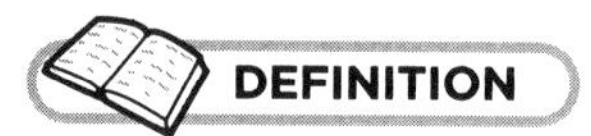

The **electromagnetic spectrum** is a broad and continuous range of electromagnetic waves permissible in nature.

Before we get into all that, though, it's worth remembering that all waves are characterized not just by frequency and speed, but by wavelength as well. Recall that the speed of any wave is given by the product of its frequency and its wavelength ($v = \lambda f$). Since all electromagnetic waves travel at the constant speed c, it follows that their wavelength is inversely proportional to their frequency and given by $\lambda = c/f$. In other words, the lower the frequency, the longer the wavelength (and vice versa).

TRY IT YOURSELF

46. The orange light produced by the common street lamp has a wavelength of 5.89×10^{-7} m. What is its wavelength in cm? What is its frequency?

We provide a glimpse of the electromagnetic spectrum in Figure 22.2. Here, you can see that radio waves sit at the low frequency/long wavelength end of the spectrum, while visible light is somewhere near the middle. While you can generate radio waves (with wavelengths spanning up to 1,000 meters) by moving charges in a wire, visible light waves are produced by the electrons within an atom. Since atoms are so much smaller than radio antennas, visible light must have a much shorter wavelength than radio waves. Indeed, it has a wavelength on the order of 1×10^{-6} meters or less.

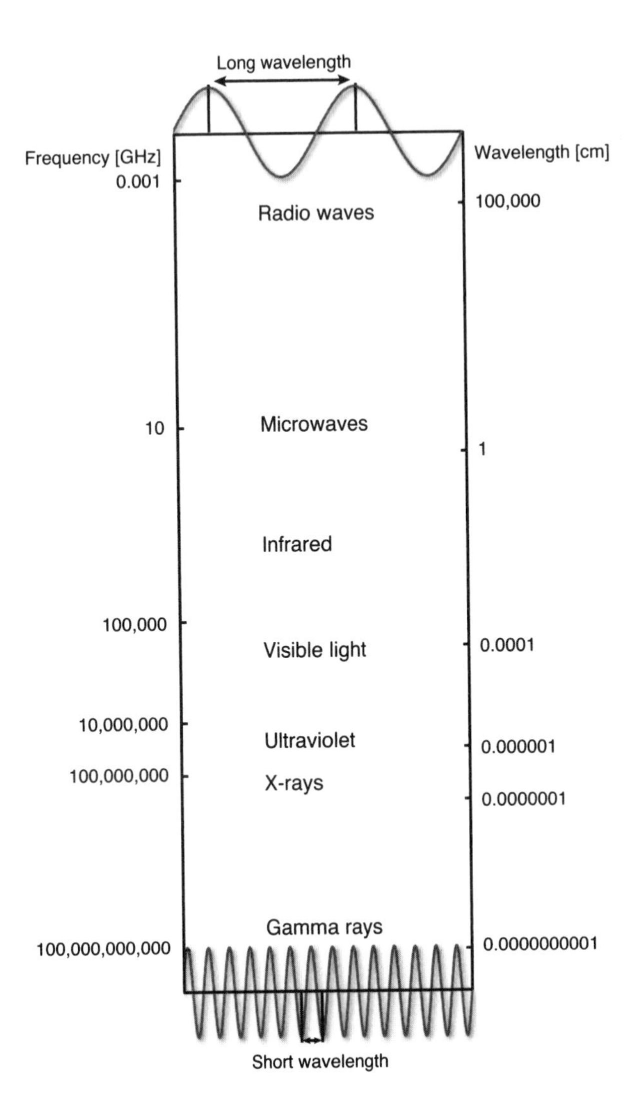

Figure 22.2

This chart shows the region of the electromagnetic spectrum spanning from radio waves to gamma rays. All possible frequencies in between are allowed. Note that one GHz is equal to one billion Hz.

If you were to put colors on this chart, the "visible light" portion would feature a small straight rainbow with red at the top and violet at the bottom. All the intermediate colors would span the region in between. These are the frequencies that can be detected by the human eye. Above and below the visible range, the eye is a poor detector and other devices are required to "see" the electromagnetic waves. Also, when red, blue, and green light are combined, the resulting light appears white. When speaking of light, these are the three primary colors.

Waves with wavelengths slightly longer than red light are considered to be infrared. Infrared waves are produced by the thermal oscillations of charged particles within matter—the same oscillations we discussed in the context of temperature. As a result, infrared waves are often associated more with heat than with light. Waves somewhat shorter than violet light sit on the ultraviolet portion of the spectrum, while x-rays are shorter still. These wave types are still produced by the electrons inside atoms; however, gamma rays, which have shorter wavelengths still, actually originate from processes that take place within atomic nuclei.

Although it may look like there are gaps in this chart between the various categories of waves, this is not the case. Electromagnetic waves can exist at any frequency (or wavelength) along the spectrum. Likewise, there are no sharp boundaries between one type and the next. In addition, it's worth noting that most real-world light sources, from the sun to the incandescent light bulb, actually emit light over a range of frequencies. Each light source is therefore said to have its own minispectrum sitting somewhere on the broader electromagnetic spectrum.

Polarization

Let's return to our vertical wire with oscillating charge. Recall that the waves that emanated had an electric field oscillating in a vertical plane. If we were to rotate the wire so that it was horizontal, the wave's electric field would move back and forth in a horizontal plane. Physicists deploy a fancy term when describing these two cases. If the electric field oscillates vertically, the wave is said to be vertically polarized. If the electric field moves back and forth in a horizontal plane, it is said to be horizontally polarized. *Polarization* is a property of an electromagnetic wave that describes how its electric field oscillates.

Polarization is a property of light that describes the direction in which its electric field is oscillating. For example, light traveling straight at you could have an electric field oscillating up and down (vertical polarization), side to side (horizontal), or somewhere in between.

Due to their simple geometry, electromagnetic waves produced by charges moving along a straight wire are naturally polarized in one direction. More typically, light sources produce light that is unpolarized. This doesn't necessarily mean that the electric field from any particular charge can change direction willy-nilly. Rather, it means that the waves produced by different charges in the source will be polarized in random directions relative to the direction of wave travel. In most sources (the sun, say, or an incandescent light bulb), the source charges oscillate independently of one another in this fashion.

The nice properties of vectors allow us to convert unpolarized light into polarized light. Remember that a vector can always be broken down into subcomponent vectors that are perpendicular to one another. For any particular wave that is neither vertically nor horizontally polarized, its electric field vector can be decomposed into one component in the horizontal direction and one in the vertical direction.

This doesn't help yet, since all we've done is to apply a mathematical transformation. However, if you could now introduce a material designed to absorb vertically polarized light while transmitting horizontally polarized light, only the horizontal components of the original wave would make it though. The result is polarized light, and its amplitude will be diminished after transmission unless it just happened to have been horizontally polarized from the start. Such materials have been given the rather unimaginative name of polarizers.

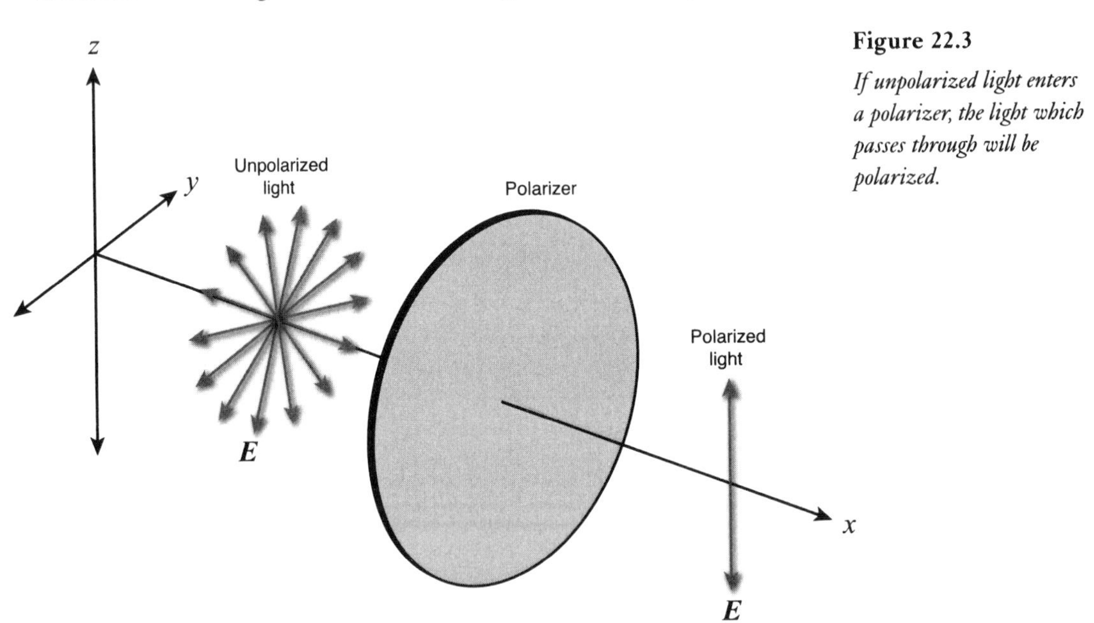

Figure 22.3

If unpolarized light enters a polarizer, the light which passes through will be polarized.

Of course, a similar story holds if you have a vertical polarizer—only vertically polarized light will get through. You can even rotate a polarizer along the axis of wave propagation to select out light at 45° or 60° or any angle you so desire. If the incident light is unpolarized, the intensity of the polarized light that emerges will be cut by half.

Polarized sunglasses, with which you may be familiar, take advantage of the fact that sunlight is horizontally polarized when reflected off of a smooth surface (e.g., the surface of a wet road or lake). The sunglasses themselves are vertical polarizers, so they block the bright sunlight (glare) that reflects directly from the surface.

The polarization we have introduced here (which has to do with the direction of an electromagnetic wave's electric field) is fundamentally different from the polarization we introduced in Chapter 17 (which had to do with a physical separation of charge within an object). They are both related to charge but otherwise have little in common. Fortunately you can almost always tell by context which kind of polarization is intended.

One way of making a polarizer is to embed a bunch of long-chained molecules onto some material, and then to really stretch that material in one direction. The result will be a material with stripelike features in one direction. When light passes through such a material, waves whose electric field is pointing perpendicular to the molecular "stripes" will pass through with very little interaction, while the electric fields that are pointed along the molecules will interact strongly and be absorbed.

Interference and Diffraction

In the next chapter, we will explore what happens when light is transmitted through or bounced off of objects that are much larger than its wavelength. Before we get to that, though, it is instructive to consider what happens whenever light encounters objects that are about the same size as its wavelength (which, you'll recall, is about 1×10^{-6} meters for visible light).

In Chapter 11, we introduced the concept of wave interference in the context of mechanical waves. For example, when two waves combine, the displacement of the resulting wave is simply the sum of displacements of the component waves. Also, two waves with equal but opposite displacements interfere destructively such that the disturbance in the medium vanishes entirely. Since light is composed of waves, we would expect that light sources could also interfere with one another.

For this to work, though, we need to throw in a few caveats. The first stems from the fact that most light sources produce waves over a range of different frequencies. Adding up waves with different frequencies has the effect of washing out interference effects due to all the averaging involved. So, for the purposes of this discussion we will consider only light sources that produce light at one single frequency, also called *monochromatic* light sources.

Second, it is important that the waves produced by the two sources are "lined up" in the sense that the peak from one source emerges at the same time as the peak from the other. When this relationship between the phases of the two sources holds, the sources are said to be *coherent.*

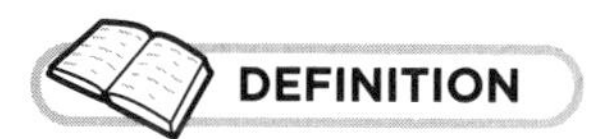

Monochromatic light sources produce light at a single frequency.

Coherent light sources produce waves for which the peaks (troughs) of one source's waves are produced at exactly the same time as the peaks (troughs) of the other.

Suppose now that you had two monochromatic, coherent point sources of light placed side by side. Due to the interference effects between the two waves produced, you would see a striped pattern emerge, much like that shown in Figure 22.4. Bright stripes (or "fringes") appear in places of constructive interference, while dark stripes are caused by destructive interference between the two sources waves.

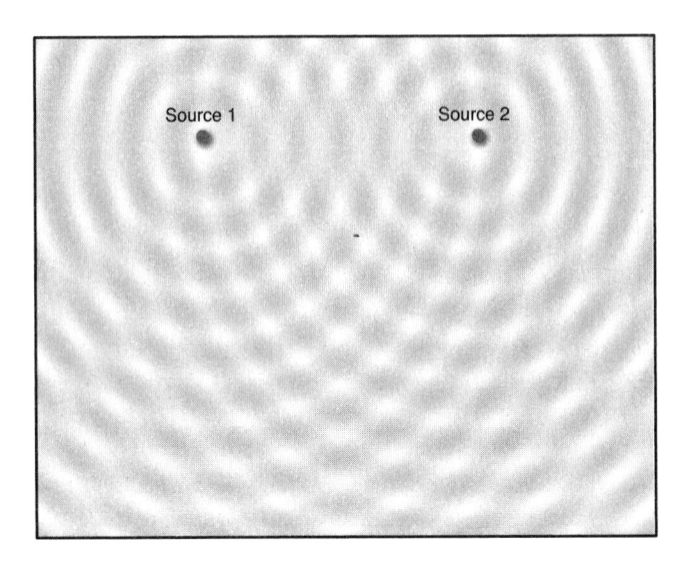

Figure 22.4

A diffraction pattern forms when the waves from two coherent, monochromatic light sources interfere with one another. Similar patterns can also be formed from the interference of one electromagnetic wave with itself.

Physicists refer to such striped patterns as *diffraction patterns.* Diffraction patterns can be observed any time two waves interfere with one another—be they mechanical waves, electromagnetic waves, or even sound waves. In fact, we showed the effect of two sound waves interacting with one another in Figure 12.2. It is no coincidence that that image and the one here share many similarities.

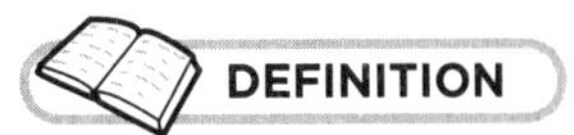

DEFINITION

A **diffraction pattern** is the image that appears when two or more waves interfere; bright stripes emerge in regions where the waves constructively interfere while dark stripes form where the waves destructively interfere.

We have considered the case of two coherent light sources side by side. Due to the coherence requirement, this is actually difficult to pull off in the lab. Instead, clever physicists have discovered that a simple way to simulate two coherent point sources is simply to shine a single electromagnetic wave through a pair of small, nearby slits (where by "small" we mean about the size of the light's wavelength). In this case, a diffraction pattern results from one electromagnetic wave interfering with itself.

CONNECTIONS

We've discussed here situations in which light waves exhibit interference effects. In fact, such phenomena are not limited solely to light, but have been witnessed in experiments with material objects as well. The discovery of so-called "matter waves," dating back to the 1920s, was another lynchpin in the foundation of quantum physics. This tenet of quantum physics tells us that material objects can behave like either waves or particles, depending on how we look at them, though never like both at the same time. The same holds for light, and in some cases light actually behaves more like a particle (called a photon) than a wave. This "wave-particle duality," as it is now known, is just one of the paradoxes that makes quantum physics so interesting.

By analyzing diffraction patterns, and in particular the spacing between the characteristic stripes that emerge, you can work out the wavelength of the light source without difficulty. Incidentally, this is how Heinrich Hertz measured the wavelength of the electromagnetic waves in his laboratory, from which he found that they were traveling at the speed of light.

The Least You Need to Know

- Light is an electromagnetic wave, and such waves can be produced by accelerating electric charges.
- The speed of light in any given medium is a constant value; its value in a vacuum is about 3×10^8 m/s.
- The energy of an electromagnetic wave is shared equally between its electric and magnetic fields.
- If the electric field vectors of all electromagnetic waves produced by a light source point in the same direction, the light is said to be polarized.
- A diffraction pattern is formed any time two coherent, monochromatic light waves interfere with one another.

CHAPTER

23

Optics

In This Chapter

- How light travels through various materials
- Rules for reflection
- Refraction of light in transparent media
- How lenses and mirrors work

Now that we've learned what light actually is, we're going to talk about some of the things that light does. In this chapter, we'll learn how light interacts with macroscopic objects, how it bounces off or gets absorbed, even how its path can be bent from its normal straight line. We'll get a glimpse of the physics that allows us to construct cameras, microscopes, telescopes, projectors, and all sorts of optical instruments. This same technology is even used by nature in the construction of animal eyes.

We will only be able to give a brief survey of optics, because again we are talking about a very broad field of study in and of itself. Optics plays an important role in many areas of our lives, some more obvious than others. Optical fibers are now used extensively for communication, and lasers produce a special kind of light that has proven to be extremely useful in a variety of ways. While this chapter focuses on visible light, keep in mind that this is only a small part of a wider spectrum, and that much of the physics discussed here also applies to other electromagnetic waves as well.

Basic Principles

Just by looking around, you know that different materials affect light in different ways. The first obvious distinction we should make is between materials that are opaque and those that are transparent. It is simple to state that transparent materials allow light to pass through them, while opaque materials do not. In reality, though, there are degrees of transparency. Even the clearest glass absorbs a little bit of the light as it passes through.

Recall that light is really electromagnetic waves, and that these waves always carry energy. The brightness or intensity of light can be quantified by referring to the amount of energy in the wave per cross-sectional area in the direction of travel. Then the relative transparency or opaqueness of a material is the fraction of that energy that is transmitted or absorbed as light passes through.

Another complication is that the transparency of a substance varies depending on the wavelength of the light. The same object may be transparent to some wavelengths, and opaque to others. Ordinary glass, for example, is transparent to the visible wavelengths, but does a pretty lousy job of transmitting ultraviolet and infrared radiation. Adding certain materials or coatings to the glass can make it even more selective, resulting in filters that only transmit certain colors. Light that we perceive as white often really has equal amounts of all colors (or wavelengths) in the visible range of the spectrum, as discussed in the previous chapter.

In the rest of this chapter, for the most part, we will ignore effects that depend on wavelength and concentrate on more general behaviors of visible light. For example, you are probably familiar with the phenomenon of shadows. If you have a single small source of light in a room, any opaque object will cast a sharp shadow on a wall or floor, and that shadow will have the shape of the outline of the object. This tells us that on normal human length scales, light travels in straight lines. This somewhat trivial observation turns out to be an important tool for understanding all of the principles of basic optics.

Let's be a little more careful here. Light only travels in straight lines as long as the medium it is traveling in is transparent and uniform. As we have seen, light doesn't actually need any medium in which to move. In fact, it travels best in a perfect vacuum, the absence of all matter. It also does pretty well in clean dry air, which is almost perfectly transparent as well. Physics tells us that light travels in straight lines whenever its wavelength is much, much smaller than the sizes of objects it is interacting with. (In contrast, we saw in the previous chapter that light can bend like a wave whenever it encounters very small obstacles.)

Reflection of Light

In this chapter, we will assume that light travels in a perfectly straight line unless the medium changes, and we will use actual lines (called rays) on diagrams to represent various paths for light. Light rays are just imaginary lines that are in the direction of travel of the waves which actually make up light. All of these different lines on a diagram can be confusing, so we will use arrows to indicate the direction of travel, and use different words to describe them. Physicists use the term *incident light* to refer to the initial ray of light, the one which first strikes an object or surface.

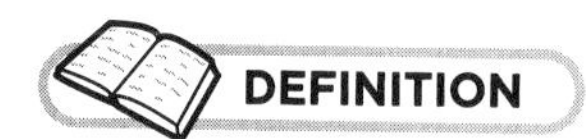

Incident light refers to an incoming ray of light that encounters a surface or object in its path.

When light is incident on the surface of an object, there are two things that can happen. The light can either bounce off of the surface, or enter through the surface into the object. In the latter case, when the light enters the object, it either continues to travel (if the object is transparent) or it gets absorbed. It is best to keep in mind the nature of light as a wave which carries energy. When we say that light is absorbed in an opaque object, what we really mean is that the energy in the light gets changed into some other form or forms, and is no longer in the form of a traveling electromagnetic wave.

When the energy of the incident light can't be absorbed or transmitted by the material object, it has no choice but to bounce off, continuing to travel in some other direction. This is what we call reflection. It is possible for both reflection and transmission to happen at the same surface. That just means that part of the energy in the light enters the material, and the rest of it bounces off. Next time you pass by the side of a car in daylight, you should be able to see this effect clearly. You will be able to see into the car, as well as see a reflection of yourself in the window.

Two different kinds of reflection are possible depending on how rough the surface in question is. If it is very rough, then even if light comes at the surface from a single direction, the reflected light is scattered off in many directions. This is what happens with most surfaces, and is called *diffuse reflection.*

Diffuse reflection occurs from surfaces that are relatively rough at a scale not much larger than the wavelength of the light, and results in the reflected light getting scattered in all directions.

Specular reflection occurs from smoothly polished surfaces or those that are naturally very smooth (like a still liquid). In this case, there is a definite relationship between the directions of the incident light and the reflected light.

If, on the other hand, the surface is polished and very smooth, then light is always reflected in a definite direction. This kind of reflection is called *specular reflection*, and it is what mirrors do. Because there is a definite direction for all of the reflected rays of light, specular reflection is able to form images.

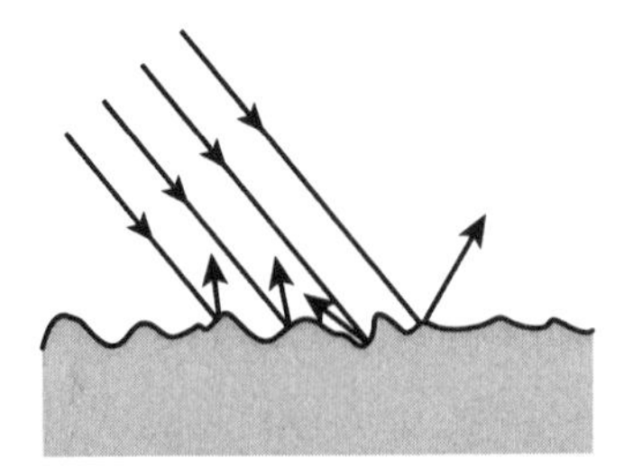

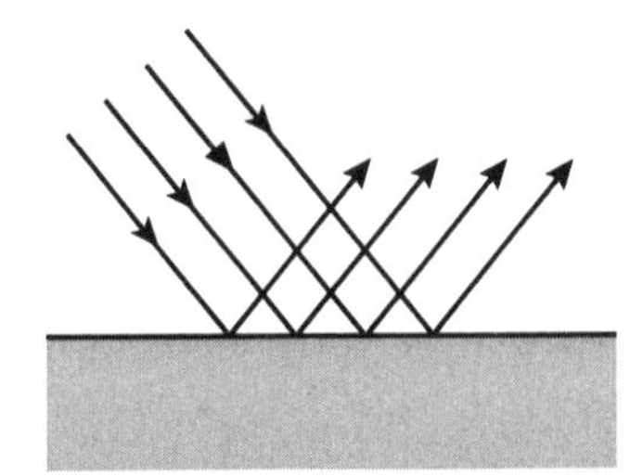

Figure 23.1

Light from one single direction (represented by parallel rays) strikes a flat surface. Left: a rough surface results in diffuse reflection. Right: a smooth surface results in specular reflection, like a mirror.

For simplicity, let's consider a perfectly flat, polished, reflective surface, like the right side of Figure 23.1. If you experiment with narrow beams of light, you will quickly discover the simple rule that governs how light is reflected. We again define an imaginary line called the "normal" to the surface, which is perpendicular to the plane of the surface. We then observe that any light ray that hits the smooth surface gets reflected such that the incident ray, the normal line, and the reflected ray all lie in a single plane. Within that plane, the angle that the reflected ray makes with the normal is the same as the angle the incident ray made with the normal. Once we agree to measure angles with respect to the normal direction, we can simply write $\theta_r = \theta_i$.

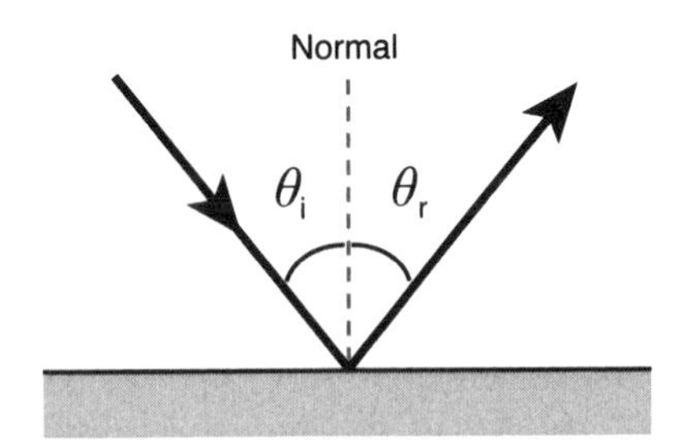

Figure 23.2

Light is incident on a smooth planar surface, from a single direction. The reflected light also leaves the surface in a single direction, determined by the angle of incidence.

Figure 23.3 shows how reflection from a plane surface leads to the formation of an image. The observer and some object are both on the same side of the mirror. But because of the law of reflection, it will appear to the observer that the object is located behind the mirror. If all the light rays are perfectly reflected, then the eye can't tell that there is not a real object behind the mirror, at the same distance behind that the real object is in front of the mirror. This is because all of the light reaching the observer is the same as it would be if there were a real object at that location. An image of this type is called a virtual image in optics.

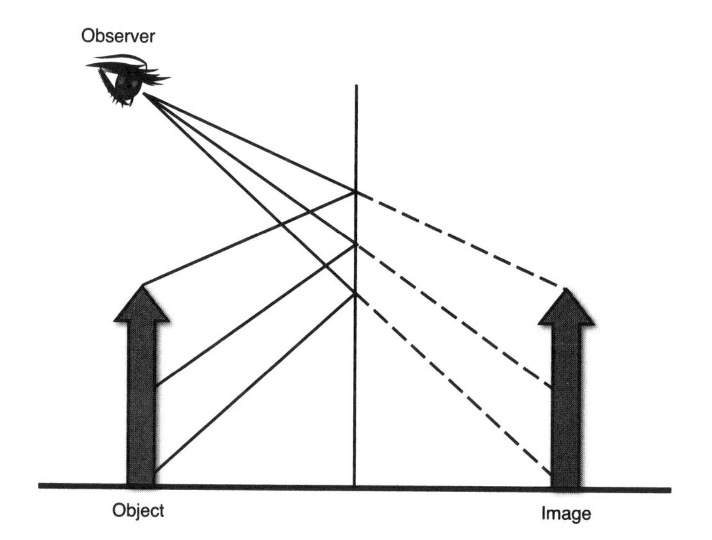

Figure 23.3

Due to the law of reflection, a plane mirror can form a virtual image of an object in front of it.

The simple law of reflection also applies to surfaces that are curved. At any point on a curved surface, you can still define a normal line that is perpendicular to the surface. Any ray of light that hits at that point is still reflected at an angle that is equal to the angle of incidence.

Let's think about circular mirrors, and imagine that no matter what, they will remain symmetric about an axis through the center that is perpendicular to the mirror. That means if we rotate the mirror about that axis, nothing changes. Then there are two main ways that a mirror can be curved: convex and concave. If the center of the mirror bulges out toward the observer, the mirror is called convex. If the center recedes and the edges are closer to the observer, then it is concave.

If the curvature is not too great, then we can qualitatively imagine what happens with curved mirrors. Recall the virtual image formed by a plane mirror. It was the same size as the object, and appeared to be the same distance behind the mirror as the object really was in front. For a convex mirror, the image becomes smaller than the object. Since the size of any mirror is limited, only a certain amount of area can be viewed in it. A convex mirror increases this "field of view," at the expense of making everything viewed appear smaller. You may have seen mirrors like this attached to trucks with trailers, or in stores where a wide-angle view is desired for security purposes.

A concave mirror has the opposite effect, producing a virtual image which is larger. Mirrors like this are useful when you need a magnified image of your own face, for example. They also form the basis for most astronomical telescopes, where they not only magnify images of distant objects, but also collect the light from a wide area and concentrate it, so that very faint objects can become clearly visible. The curvature of a concave mirror is specified by the location of its focal point, as illustrated in Figure 23.4. If light comes at the mirror in rays that are parallel to its axis, it will converge to a single point, called the focal point.

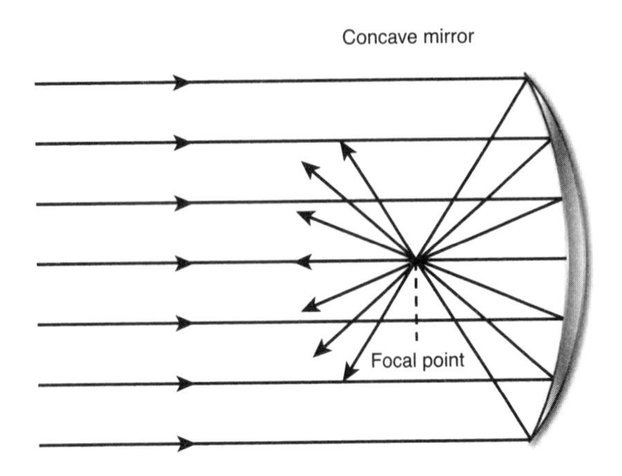

Figure 23.4

A concave mirror focuses light from a distant source to a small point called the focal point.

TRICKS AND HACKS

If you ever get into ray-tracing of optics, it will be useful to realize that rays indicating the actual path of light are reversible. For example, take a look at Figure 23.4 again. If a small source of light were placed at the focal point of the mirror, light rays would emanate from it in all directions. The ones that hit the mirror would be reflected along parallel paths to the left, just the reverse of the incoming light shown. This property is commonly used in flashlights, car headlights, and other cases where we want to direct light in a relatively narrow beam.

Refraction

We now turn our attention from reflection to the transmission of light through a transparent medium. To get some understanding of how this works, we need to consider what is happening at the atomic level. For simplicity, let's consider a solid material like glass or clear plastic.

Typical transparent materials are not good electrical conductors. Unlike in metals, electrons in transparent materials are bound in place. They can't travel around within the solid, but they can vibrate a little around their home positions. Since light is a traveling oscillation in electric and magnetic fields, when light enters the transparent substance it can cause the electrons to vibrate along with it. So, when a beam of light hits the surface of a transparent material, its energy is transferred to vibrations of the first layer of electrons. These vibrating electrons, being charges in motion, soon re-emit that energy in the form of light, which causes the next layer of electrons to vibrate, emit, etc.

How free the electrons are to vibrate, and how quickly they can re-emit the energy, depends on details of the atomic structure of the particular material. In most materials, the absorption and re-emission is delayed, there are shifts in frequency, and the light doesn't get very far before its energy is disbursed into different forms. But if everything is just right, most of the energy from the original light keeps getting re-emitted in the same direction with the same wavelength and frequency, until it comes out the other side of the material. Only those materials with the right structure at the atomic level can do this and therefore appear transparent.

The interesting wrinkle is that even in a transparent material, there is a very slight delay in going from layer to layer. The effect of this is that light travels more slowly in any transparent medium than it does in vacuum. Thus in any material, the speed of light is effectively less than c.

The actual speed varies depending on the material, and it's usually slower in more dense materials. For example, light travels in pure water at a speed of about $0.75c$, in typical glass its speed is about $0.66c$, and in diamond it is only about $0.41c$. As a direct consequence of this, the path of light may be bent when passing from one medium to another, if the effective speed of light is different between the two media. This phenomenon is called *refraction.*

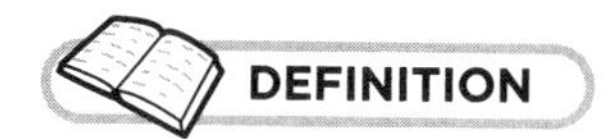

DEFINITION

Refraction is the change in direction of light when passing between two transparent media with different physical properties.

The **index of refraction** (n) is a characteristic of any transparent material, including liquids, gases, and solids. It is a dimensionless quantity equal to c/v, where c is the speed of light in a vacuum and v is the effective speed of light in the material.

To develop a quantitative description of refraction, it is easier to work with another parameter that is not the effective speed of light, but which is related to it. The *index of refraction* is the ratio of the speed of light in a vacuum to its speed in the medium. For the vacuum, then, $n = 1$, and for any transparent material, $n > 1$. The speed of light in air is very close to c, so we normally approximate the index of refraction for air as $n = 1$ also. We don't have to worry about any units for n because it is a ratio of two speeds and therefore a dimensionless quantity.

Again, many decades of careful observations have revealed a simple rule, called Snell's law, which tells us how much a beam of light will be bent when entering a medium with a different index of refraction. Let's label the first medium, in which the light is originally traveling, as a, and say that it crosses a boundary into a second medium labeled b. We again reference our angles to the normal line at the point where the light ray hits the surface. As illustrated in Figure 23.5, Snell's law states that:

$$n_a \sin(\theta_a) = n_b \sin(\theta_b)$$ **Equation 23.1**

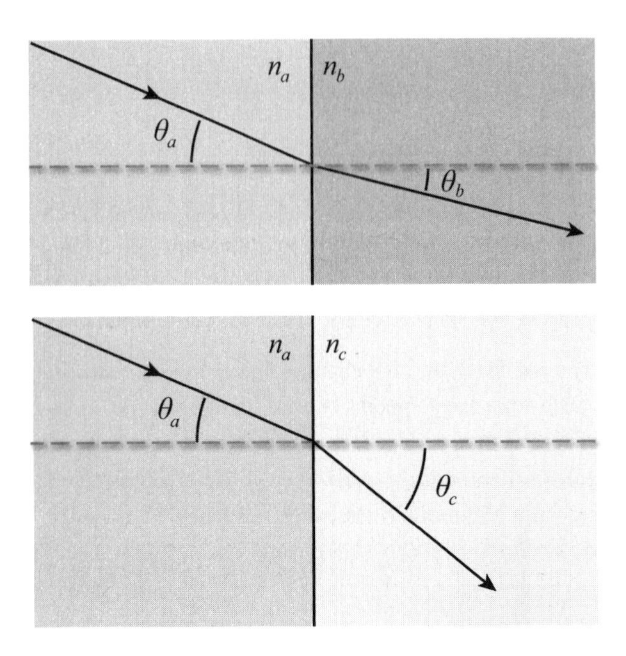

Figure 23.5

Light entering from the left is bent as it enters a new medium with a different index of refraction. In the first case, $n_b > n_a$*, and in the second case,* $n_c < n_a$*.*

This expression tells us that if you go from a lower *n* to a higher *n*, the light will get bent toward the normal direction, but if you go from higher *n* to lower *n*, it will be bent away from the normal. It also says that if light heads straight at the interface (with $\theta_a = 0$) that there will be no bending, no matter what the indices of refraction are.

You can see examples of refraction in many places, especially where water is involved. Put a straight straw into a glass of water, and it will appear to be bent. If you have ever looked carefully at an aquarium full of fish, you have probably noticed some distortions due to refraction there as well.

TRY IT YOURSELF

47. The effective speed of light in clear diamond is 1.23×10^8 m/s. What is the index of refraction in diamond?

48. A block of diamond with smooth sides is located in air. A beam of light from inside the diamond hits one of the sides of the block and emerges into the air. The light inside strikes the surface at an angle of 22° from the normal. At what angle does the light emerge?

Another observation about light may help us understand the nature of refraction. Take any two points in space. It turns out that if light travels from one point to the other, it will always take the quickest route. If the medium is uniform between the two points, then that path is obviously a straight line joining them. But if the two points are in regions with different indices of refraction, it means that there is a boundary between them, and the speed of light is slower on one side of the boundary than the other. As a result, the quickest path might not be a single straight line any more. In fact, Snell's law follows from the light taking a path that is longer in the medium where its speed is greater, and a shorter leg in the slower medium, in order to minimize the total travel time.

In reality, the amount of refraction depends on the wavelength of the light getting refracted. This is because the index of refraction in a given substance is different for different wavelengths. In a perfect vacuum, all wavelengths of light (and any other electromagnetic radiation) travel at exactly the same speed. But in transparent materials, the different wavelengths have different speeds, leading to this variation. This effect can be exploited to separate white light into its component colors. A piece of material shaped to do this is called a prism.

If the boundary between regions with different indices of refraction is smooth, then both specular reflection and refraction can occur at the same time. So far we have illustrated only one of these effects at a time, so as not to confuse the issue, but in general they both happen, with part of the energy being reflected, and the rest being refracted and transmitted. All of the energy must be accounted for somehow, so knowing what percent gets reflected can immediately tell you what percent gets refracted. This will be illustrated in Figure 23.6 in the next section.

There is one interesting case, however, where we know for sure what fraction of the light energy will be reflected. Imagine that light is traveling in a medium with a large index of refraction, say $n_a = 2$. Now it approaches the edge of this material, and outside is air, where $n_b \approx 1$. Because you are going from a larger n to a smaller one, Snell's law tells us that if there is refraction, the outgoing ray will be at a larger angle to the normal than the incident ray, as illustrated in the lower half of Figure 23.5.

Imagine gradually increasing the angle of the incident light. At some point, the outgoing refracted light will reach an angle of 90°. In our example, that will occur when $\sin(\theta_a) = \frac{1}{2}$ or $\theta_a = 30°$. A light ray that is 90° from the normal is skimming right along the boundary between the two media. If the incident light comes in at any angle greater than 30° it will be impossible for any refraction to happen! In that case, all of the light will definitely be reflected, and no light will pass through the boundary. This phenomenon is known as "total internal reflection."

Lenses

Even though window glass has an index of refraction $n \approx 1.5$, we don't normally worry about refraction when looking out of a window. Why not? For one thing, we usually look straight out, so that the angle of incidence is small. But even when we look through the window at an angle, the light we see is not changed very much by the window, if the glass is thin and its two sides are parallel to each other. This is illustrated in Figure 23.6, where we also show a second order reflection.

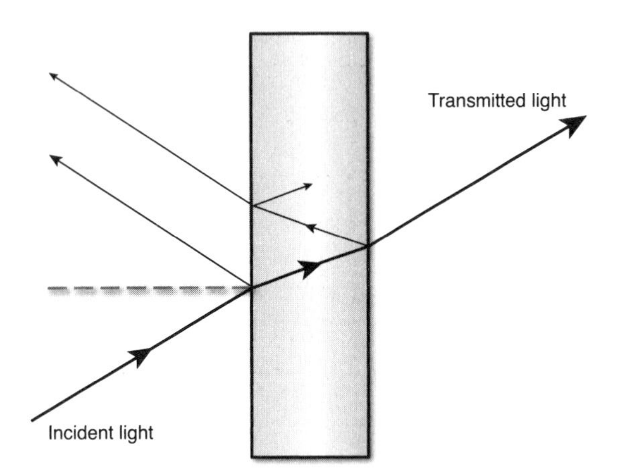

Figure 23.6

Refraction occurs both on entering and leaving a pane of glass. If the sides of the pane are parallel, the direction of the transmitted light is unchanged and the path displaced only slightly. Also, some fraction of the light will be reflected every time it reaches a boundary.

On the other hand, if the sides are not parallel, and if the glass surface is curved, then the story is completely different. This brings us into the wide world of lenses. "Lens" is a general term for a curved piece of transparent material that is used to alter the paths of light rays in certain predictable ways. The standard lens has rotational symmetry about a central axis (like the curved mirrors we discussed previously) and is made of a transparent material with an index of refraction greater than air. Lenses come in a wide variety of shapes and sizes, and are the workhorse component in the vast majority of optical instruments.

All lenses rely on refraction to bend light rays for whatever purpose they may have. Applying the terms we used previously, concave and convex lenses behave in fundamentally different ways. We assume that light enters the material of the lens from a region with lower index of refraction (like air). Because of the curvature, light entering the lens at different locations will be bent by different amounts. A convex lens will tend to bring parallel light rays together, and lenses with this property are also called "converging" lenses. A concave lens will tend to spread parallel rays apart, and so is called a "diverging" lens.

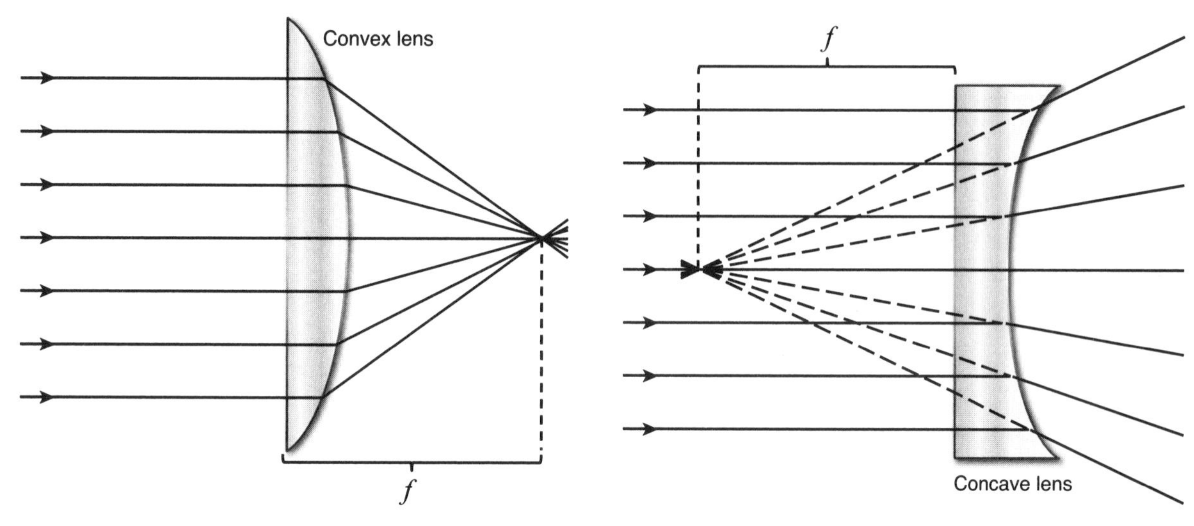

Figure 23.7

In both of these cases, parallel light rays enter a lens from the left. In the case of the convex lens, the rays converge to a point at a distance f *beyond the lens. In the case of the concave lens, the rays appear to diverge from a point a distance* f *from the lens on the other side of it.*

Both convex and concave lenses are characterized by a parameter called the focal length f. This is similar to the case of curved mirrors, and indeed, lenses are able to form images in much the same way that mirrors do. For a convex lens, the focal length is the distance from the lens where parallel rays are brought to a point, as shown on the left of Figure 23.7. A weakly converging lens will have a relatively long focal length, while a strong lens will have a short focal length. For the concave lens, there is a virtual point from which the rays appear to originate. The distance between the lens and this point is also a focal length, but is usually given as a negative number to distinguish the fact that it is somehow opposite to the focal length of a converging lens.

CONNECTIONS

Galileo Galilei didn't invent the telescope, but he was among the first to point one into the night sky. After doing so, he carefully documented imperfections in the surface of the moon, discovered that several moons orbit the planet Jupiter, and observed that the planet Venus exhibits phases—akin to those we observe in the moon. Galileo's very basic observations put a serious strain on the prevailing notion that Earth was the center of the universe and that it was surrounded by heavenly perfection. Thus, basic optics not only revolutionized the field of astronomy, it was also instrumental in overturning our cosmological worldview.

Using focal lengths and straight lines for light rays, it is possible to figure out how light passes through and images are formed from all sorts of combinations of lenses. That process in general is called "ray-tracing," and it's a method that can be used to design all sorts of optical instruments. Lenses can be combined to produce magnified images of distant objects (as in binoculars and telescopes) or nearby small objects (as in magnifiers and microscopes). Lenses in eyeglasses are able to adjust the effective focal length of the lens in your eye, to restore sharp focus of the images after age causes the eyeball to change its shape.

Through the Lens and Far Beyond

In this book, we have tried to cover all of the topics essential for a basic understanding of the physical world, and we hope you'll agree it's amazing just how much of the natural world can be covered within the span of 23 chapters. Physics is always everywhere in our daily lives. Optics helps explain what we see, sound determines what we hear, and the electric force underlies our very sense of touch.

Physics holds a special place amongst the sciences, because it is at the root of all of them. You can certainly have chemistry without biology, but you cannot have either without physics. And even if we haven't convinced you to become a physicist, we hope that we've awakened an appreciation for the hidden order that physics brings to the world around us. And, if we're really lucky, we've also managed to awaken a new sense of wonder that so few principles can bring such a broad understanding of the physical world.

You will certainly agree that all good things must come to an end. We hope you'll feel this book belongs in that category. Over the last 300 odd pages, we have introduced you to the most basic fundamentals of physics—from mechanics to energy to sound to temperature to electricity to magnetism to light. We've covered a lot of topics, but the truth of the matter is that, to keep this book of manageable thickness, we had to omit a great many more. We hope that in the process we have whetted your appetite to pick up where we left off, and explore even further the wonderful world of physics.

The Least You Need to Know

- Light travels in straight lines, to a good approximation, until it encounters an object or until the medium in which it is traveling changes.
- When a polished surface reflects light, the angle of incidence equals the angle of reflection.
- Light bends when entering transparent materials because it changes its effective speed, an effect called refraction.
- Lenses use refraction to form and manipulate images for various purposes in a wide variety of optical instruments.
- Physics is, quite simply, very cool.

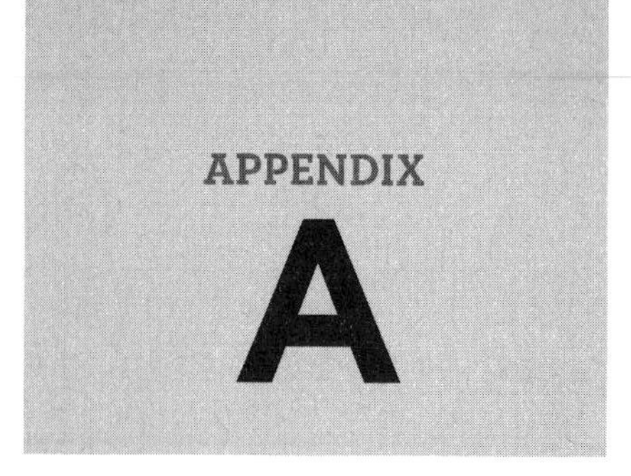

Glossary

absolute zero The lowest temperature that any physical system can theoretically attain. Its value is approximately -273.15°C.

acceleration The time rate of change of an object's velocity. Its symbol is $\boldsymbol{a}$ and its standard units are m/s^2.

amplitude The maximum displacement from equilibrium when an object undergoes oscillatory motion.

angular acceleration The rate at which the angular velocity of a rotating body changes, with a sign that indicates whether the angular velocity is getting more positive or more negative. Its symbol is α and its standard units are rad/s^2.

angular momentum For a rotating body, the product of its moment of inertia and its angular velocity. Its symbol is L and its standard units are kg m^2/s.

angular position The angle between a line fixed to a rotating body and a fixed reference line in space, in a plane perpendicular to the axis of rotation. Its symbol is θ and its standard units are radians.

angular velocity The rate at which the angular position of a rotating body changes with time. It is the rate of rotation about a specific axis, with the algebraic sign representing the direction of the rotation. Its symbol is ω and its standard units are rad/s.

Avogadro's number A convenient, dimensionless quantity used when considering the large number of atoms present in everyday objects. Its value is $N_A = 6.02 \times 10^{23}$, and it is defined as the number of atoms in a 12-gram portion of carbon-12.

boiling point The temperature at which a substance transforms between the liquid and gaseous state when heated or cooled.

buoyant force A force exerted by a fluid on an object any time that object is partially or completely submerged, assuming there is gravity present. It is directed vertically upward, and it has a magnitude that is exactly equal to the weight of the fluid displaced by the object.

capacitance A fixed characteristic of any particular capacitor, given by the ratio of the amount of positive or negative charge on either plate of the capacitor, and the resulting potential difference between the two plates. Its symbol is C and its standard units are C/V, also called the farad (F).

capacitor An electronic device that is capable of storing electrical energy. It is typically constructed from a pair of conducting surfaces placed near to one another, across which a voltage may be applied.

Cartesian coordinate system Any coordinate system based on simple rectangular coordinates whose axes are mutually perpendicular straight lines. This term can be applied to both two-dimensional and three-dimensional systems.

Celsius The temperature scale used in the SI system of units. It is set by the fact that water freezes at 0°C and that water boils at 100°C.

center of mass The weighted average location of all of the mass making up a system.

centripetal acceleration The type of acceleration that occurs when a velocity changes direction.

charge The property that allows objects to feel the electric force. Uncharged objects are called "neutral" and are not affected by the electric force. Its symbol is Q or q and its standard units are coulombs (C).

condensation point See *boiling point.*

conduction (thermal) The transfer of heat through a substance from a region of higher temperature to a region of lower temperature. It relies on direct contact, and the energy is transferred mainly through collisions of the material's constituent particles.

conductor (electrical) A material in which electrons are able to move very freely, thus carrying an electric current.

constructive interference The combination of two or more waves in the same medium, for which the superposition of waves results in a wave with a larger magnitude than either of the component waves.

contact force A force exerted on one object by another object at their point of contact.

convection The transfer of heat through a fluid (liquid or gas), which is aided by currents established by temperature-induced density gradients in the fluid.

coordinate system A concept in geometry that allows us to assign a unique number or set of numbers to any position in space. It consists of an origin point and a set of coordinate axes.

current (electric) The rate at which electric charge passes through a particular point or area. Its symbol is I and its standard units are C/s, also called the ampere (A).

density A property of anything that has mass, given by its mass divided by its volume. Its symbol is ρ and its standard units are kg/m^3.

destructive interference The combination of two or more waves in the same medium, for which the superposition of waves results in a wave with a smaller magnitude than either of the component waves.

dielectric constant A dimensionless scalar that quantifies the ease with which an electrically insulating material can be polarized. It is equal to 1 for a pure vacuum, and is larger for other substances depending on how easily they are polarized.

diffuse reflection The reflection of light from rough surfaces, which results in the reflected light getting scattered in many different directions.

dimension The particular type of physical property associated with a numerical quantity (e.g., length, mass, energy, etc.).

elastic collision A collision in which the sum of kinetic energies of all bodies remains unchanged.

elastic potential energy See *spring potential energy.*

electric current See *current.*

electric field A vector function of position in space that accompanies any distribution of electric charges, which is equal to the electric force that a test charge would feel if introduced at any particular location, divided by that charge. Its symbol is $\boldsymbol{E}$ and its standard units are N/C.

electric potential See *potential.*

electromagnetic induction A process by which an electromotive force is induced by a magnetic field that is changing with respect to some charge.

energy The capacity to do work. A conserved physical quantity which takes many different forms, measured in units of joules (J).

energy density The amount of energy per unit volume in some region of space. Its symbol is u and its standard units are J/m^3.

entropy A quantitative measure of the amount of disorder in a system. Changes in entropy are given by the ratio of heat flow to temperature. Its symbol is S and its standard units are J/K.

equilibrium (static) Any state where the total or net force on an object, and the net torque on that object, are both equal to zero.

equilibrium (thermal) The final, balanced state achieved whenever two objects are brought into thermal contact with one another. Their temperatures will approach one another until they reach exactly the same temperature.

Fahrenheit A temperature scale set by the fact that water freezes at 32°F and boils at 212°F.

first law of thermodynamics A law stating that the internal energy of a system can be increased either by adding heat to the system or by doing mechanical work on the system. It is the most general form of the principle of conservation of energy.

force The general term for any physical cause that has the potential to change the motion of an object with mass. A vector quantity, its symbol is $\boldsymbol{F}$ and its standard units are newtons (N).

free-body diagram A sketch showing all of the forces that act on a single mass.

free fall The state wherein a massive object is moving solely under the influence of gravity near Earth's surface.

frequency For an oscillating object, the number of complete cycles of motion which occur per unit of time. It is given by the mathematical inverse of the period, $f = 1/T$. Its symbol is f and its standard units are cycles per second, or hertz (Hz).

freezing point See *melting point.*

friction A type of interaction, between an object and a surface with which it is in contact, which produces a force acting opposite the direction of motion and which converts mechanical energy into heat.

gauge pressure The difference between the absolute pressure at a certain location and atmospheric pressure.

gravitational potential energy The potential energy of an object due to its location in a gravitational field.

heat The amount of thermal energy transferred when two objects at different temperatures are placed in contact. Its symbol is Q and its standard units are joules (J).

heat capacity The ratio of heat added to an object to the resulting rise in temperature. Its symbol is C and its standard units are J/K.

heat engine A device designed to convert thermal energy into mechanical energy.

heat pump A device designed to use mechanical energy to transfer heat from a colder object or area to a warmer object or area.

impulse The change in linear momentum caused by a force. It is a vector quantity with the same dimensions as momentum.

incident light Term used in optics to denote the incoming light that encounters a surface or object in its path.

index of refraction A dimensionless quantity of any material medium equal to $n = c/v$, where c is the speed of light in a vacuum and v is the effective speed of light in the material.

inductance (mutual) A measure of how much electromotive force is produced in one circuit due to a time-varying current in another. Its symbol is M and its standard units are V s/A, also called the henry (H).

inelastic collision A collision in which the total kinetic energy after the collision is different (usually less) than it was before the collision.

infrasound The range of sound with frequency too low for normal human hearing.

insulator (electrical) A material in which current does not flow due to the fact that its internal electrons are immobile.

insulator (thermal) A material which impedes the flow of heat, or thermal energy, and is therefore a poor conductor of heat.

intensity (sound) The power per unit area carried by a sound wave. Its symbol is I and its standard units are W/m^2.

internal energy of a system The total kinetic and potential energy embodied by the particle constituents of the matter within the system. It excludes all kinetic and potential energy of the macroscopic system (e.g., the energy of linear or rotational motion or gravitational potential energy).

Kelvin A temperature scale that provides information about absolute temperature, which says that water freezes at 273.15 K and boils at 373.15 K.

kinetic energy The energy possessed by any mass that is in motion. Its symbol is K and its standard units are joules (J).

kinetic friction A force exerted by a surface in contact with an object, parallel to the surface, which occurs whenever the object is moving relative to the surface.

laws of thermodynamics See *zeroth*, *first*, *second*, or *third law of thermodynamics*.

lever arm The closest distance between the line along which a force acts on an object and the object's axis of rotation.

longitudinal wave A wave in which the parts of the medium undergo small back-and-forth movements in a direction parallel to the direction of wave travel.

magnetic dipole A source of magnetic field formed by any magnetized element with one north pole (N) and one south pole (S).

magnetic flux The total strength of magnetic field passing through any given area, multiplied by that area. Only the component of the magnetic field that is perpendicular to the area contributes to the flux. Its symbol is Φ and its standard units are T m^2, also called webers (Wb).

magnetic moment A vector parameter that indicates the orientation and strength of a magnetic dipole.

mass A measure of how much matter is contained in the object and also how difficult it is to change the motion of the object.

mechanical wave A disturbance in some material medium that travels over time.

medium The background distribution of matter that gets disturbed when a mechanical wave travels.

melting point The temperature at which a substance transforms between the solid and liquid state when heated or cooled.

model A well-defined mathematical system that can be used to predict how a physical system will change under different conditions.

mole The amount of any substance formed by 6.02×10^{23} atoms (or molecules) of that substance. It is sometimes abbreviated "mol."

moment of inertia A property of a rigid body that measures its resistance to rotation about a specific axis. Its symbol is I and its standard units are kg m^2.

momentum (angular) See *angular momentum.*

momentum (linear) A vector quantity possessed by any massive object in motion, given by the product of its mass and its velocity. Its symbol is $\boldsymbol{p}$ and its standard units are kg m/s.

node A fixed location in a standing wave pattern where destructive interference results in no motion.

normal The direction perpendicular to a surface at a point on that surface.

normal force A contact force exerted on an object by a flat surface, in a direction perpendicular to the surface.

origin The point on a coordinate system from which all other locations are measured.

oscillating motion Any motion that repeats itself at regular time intervals, typically in a back-and-forth fashion.

period The time it takes an oscillating object to complete one full cycle of its repeated motion. Its symbol is T and its standard units are seconds.

phase angle For an object undergoing simple harmonic motion, the argument of the trigonometric function used to describe the position of the oscillating mass as a function of time. For motion at frequency f, it is an imaginary angle equal to $2\pi ft$ and is measured in radians.

phase change A transformation of a substance from one physical state to another, normally as a result of heating or cooling.

position A vector quantity that gives the location of an object relative to a set of coordinate axes. In general, its symbol is $\boldsymbol{r}$ and its standard units are meters.

position coordinate A number (with units) that indicates how far from the origin something is located along one coordinate axis.

potential (electric) A scalar function of position in space that accompanies any distribution of electric charges, which represents the potential energy that would be felt by a test charge introduced at any particular location, divided by that charge. Its symbol is V and its standard units are J/C, also called the volt (V).

potential difference The arithmetic difference between the electric potential at two points. Its symbol is ΔV, and its standard units are volts.

potential energy A position-dependent energy associated with any conservative force. When a potential energy is defined for every point in space, the force is in the direction that corresponds with the most rapid decrease in potential energy. Its symbol is U and its standard units are joules (J).

power The time rate of energy transfer or conversion. Its symbol is P and its standard units are J/s, also called the watt (W).

pressure A force divided by the area of the surface over which it is applied. Its symbol is P and its standard units are N/m^2, also called the pascal (Pa).

projectile motion The specific kind of motion that occurs whenever a massive object is moving solely under the influence of gravity near Earth's surface.

radian A natural unit for measuring angles, based on the length along a circular arc and abbreviated "rad." A full circle corresponds to 2π radians.

range (horizontal) The horizontal distance traveled by a projectile during the time it takes to rise and then return to the same height from which it was launched.

refraction The change in direction of light when passing between two transparent media with different physical properties.

resistance (electric) A property of an object that describes precisely how difficult it is for charges to move from one end to the other. Its symbol is R and its standard units are ohms (Ω).

scalar The class of physical quantities that have only a magnitude and do not have any direction.

second law of thermodynamics A law stating that the entropy of an isolated system never decreases with time.

SI The abbreviation for Système International, French for the international system of units and standards.

solenoid A device formed by winding many loops of a single wire into a cylindrical coil. When an electric current runs through the wire, a large dipole magnetic field is created.

specific heat The ratio of heat added to an object to the resulting rise in temperature, per unit mass. Its symbol is c and its standard units are J/(kg K).

specular reflection The reflection of light from smooth surfaces, in which there is a definite relationship between the directions of the incident light and the reflected light.

spring (or elastic) potential energy The potential energy stored in a spring, or other elastic object, due to the amount that it is stretched or compressed relative to its natural length.

static friction A force exerted by a surface in contact with an object, parallel to the surface, which occurs whenever the object is stationary at the point of contact.

superposition A mathematical principle that tells us the net displacement of multiple waves is determined by the simple sum of their displacements from equilibrium.

temperature A measure of how hot or cold an object is, relative to some standard value. It is closely related to the amount of internal energy within the object.

tension force A pull exerted by a rope or cable or similar thing attached to an object.

thermal contraction The decrease in size of a solid or liquid due to cooling.

thermal energy The form of energy transferred from one body to another as a result of a difference in temperature.

thermal expansion The increase in size of a solid or liquid due to heating.

thermal radiation The transfer of heat via electromagnetic waves, which stems from the thermal motion of charged particles within the hotter body.

third law of thermodynamics A law stating that the lowest achievable entropy for any physical system occurs with a perfect crystal at absolute zero.

torque A physical quantity that corresponds to a tendency to cause rotation, given by a force multiplied by the lever arm. Its symbol is τ and its standard units are N m.

transverse wave A wave in which the back-and-forth motion of elements of the medium is perpendicular to the direction of the wave's travel.

trigonometry The branch of mathematics that studies the relationships between lengths and angles in triangles.

ultrasound The range of sound with a frequency too high for normal human hearing.

units The reference standards in which quantities are measured.

vector The class of physical quantities for which direction is essential information.

velocity The vector quantity that gives the time rate of change of an object's position. Its symbol is $\boldsymbol{v}$ and its standard units are m/s.

velocity (instantaneous) The limiting value of the average velocity as the averaging time interval is made vanishingly small.

voltage See *potential difference.*

wavelength The minimum distance over which a wave pattern repeats itself. It is also the distance between two successive peaks (or troughs) of the wave.

wave speed The speed at which a wave pattern travels.

weight A quantity representing the force of gravity acting on a mass. On Earth, the magnitude of the weight force is the mass multiplied by the acceleration due to gravity: mg.

work The scalar product of force and displacement. Its symbol is W and its standard units are joules (J).

zeroth law of thermodynamics A law stating that if Object A is in thermal equilibrium with Object B, and Object B is in thermal equilibrium with Object C, then it follows that Object A is in thermal equilibrium with Object C (even if A and C are not in contact with each other).

APPENDIX

B

Conversion Factors and Fundamental Constants

We have avoided making you do a lot of conversions in this book, so that we could concentrate more on the actual physics. If you ever want to do calculations in the real world, however, there will probably be many occasions when you have to convert a number from one system of units to another. So here in this appendix we have gathered a lot of the factors needed to do such conversions.

In the list below, we group the factors by the dimension of the quantity being converted, starting with the basic dimensions and moving to the compound ones. After all the conversion factors, we provide a one-stop consolidation of all the physical constants that we've introduced, for handy reference.

First, however, we want to revisit the mass vs. force issue. Some lists like this, or online calculators, will offer a conversion factor between pounds and kilograms. This merely perpetuates the confusion most people have between two totally different physical quantities. Kilograms measure mass, and in the U.S. customary system, pounds are used to measure weight. But you now know that weight is a force, not a mass. The mass unit in the U.S. system is called the slug, which is rarely used. We just assume that the acceleration due to gravity is always 9.8 m/s^2 (about 32.2 ft/s^2), in which case weight is simply proportional to mass. In any case, it is perfectly correct to say that 1.0 kilogram weighs about 2.205 pounds at the surface of Earth.

Conversion Factors

Length

Metric System

1 millimeter = $^1/_{1,000}$ meter

1 centimeter = $^1/_{100}$ meter

1 kilometer = 1,000 meters

1 light year = 9.46×10^{15} meters

American or British System

1 inch = $^1/_{36}$ yard = $^1/_{12}$ foot

1 foot = $^1/_{3}$ yard = 12 inches

1 yard = 3 feet = 36 inches

1 mile = 1,760 yards = 5,280 feet

Conversion Factors

1 inch = 2.54 centimeters

1 meter = 39.37 inches

1 foot = 0.305 meters

1 yard = 0.914 meters

1 meter = 1.094 yards

1 kilometer = 0.62 mile

1 mile = 1.609 kilometers

Area

Metric System

1 square centimeter = $^1/_{10,000}$ square meter

1 hectare = 10,000 square meters = 2.47 acres

1 square kilometer = 1,000,000 square meters

American or British System

1 square inch = $^1/_{144}$ square foot

1 square foot = $^1/_{9}$ square yard

1 acre = 4,840 square yards = 43,560 square feet

1 square mile = 3,097,600 square yards = 640 acres

Conversion Factors

1 square inch = 6.45 square centimeters

1 acre = 0.405 hectares

1 hectare = 2.47 acres

1 square kilometer = 0.386 square miles

1 square mile = 2.59 square kilometers

Volume and Capacity (Liquid and Dry)

Metric System

1 cubic centimeter = $^1/_{1,000,000}$ cubic meter = 1×10^{-6} m^3

1 milliliter = $^1/_{1,000}$ liter = 1 cubic centimeter

1 centiliter = $^1/_{100}$ liter

1 cubic meter = 1,000 liters

American or British System

1 cubic inch = $^1/_{1,728}$ cubic foot

1 cubic foot = $^1/_{27}$ cubic yard

1 U.S. fluid ounce = $^1/_{128}$ U.S. gallon = $^1/_{16}$ U.S. pint

1 pint = $^1/_{8}$ gallon = $^1/_{2}$ quart

1 quart = $^1/_{4}$ gallon

1 U.S. gallon = 231 cubic inches

1 dry pint = $^1/_{64}$ bushel = $^1/_{2}$ dry quart

1 dry quart = $^1/_{32}$ bushel = $^1/_{8}$ peck

Conversion Factors

1 cubic inch = 16.4 cubic centimeters

1 cubic yard = 0.765 cubic meters

1 fluid ounce = 29.6 milliliters

1 U.S. quart = 0.946 liters

1 liter = 1.06 U.S. quarts

1 U.S. gallon = 3.785 liters

1 imperial gallon = 1.2 U.S. gallons = 4.5 liters

1 dry quart = 0.95 liters

Speed

American or British System

1 mile/minute = 60 miles/hour = 88 feet/second

1 mile/hour (MPH) = 1.47 feet/second

Conversion Factors

1 foot/second = 0.305 meters/second

1 kilometer/hour = 0.6214 miles/hour

1 mile/hour = 0.447 m/s = 1.609 kilometers/hour

Mass

Metric System

1 milligram = $\frac{1}{1,000,000}$ kilogram = $\frac{1}{1,000}$ gram

1 gram = $\frac{1}{1,000}$ kilogram

1 metric ton = 1,000 kilograms

Conversion Factor

1 slug = 14.59 kg

Force

Metric System

1 newton = 1 kg m/s

American or British System

1 ounce = $\frac{1}{16}$ pound

1 pound (lb.) (basic unit of force)

1 ton = 2,000 pounds (also called "short ton")

Conversion Factors

1 newton = 0.2248 pounds

1 pound = 4.448 newtons

Pressure

1 pascal = 1 N/m^2 = 1.45 × 10^{-4} lb/in^2 = 0.0209 lb/ft^2

1 bar = 10^5 pascals

1 pound/square inch = 1 psi = 6,895 pascals

1 torr = 1 mm Hg = 133.3 pascals

1 inch Hg = 3,386 pascals

1 atm = 1.013 × 10^5 pascals = 1.013 bar = 14.7 lb/in^2

Energy

1 joule = 0.239 calories = 10^7 ergs

1 calorie = 4.186 joules

1 foot-pound = 1.356 joules

1 Btu = 1,055 joules = 252 calories = 778 foot-pounds

1 kilowatt-hour (kWh) = 3.6 × 10^6 joules

1 food Calorie = 1,000 calories = 1 kcal

Power

1 watt = 1 joule/second

1 horsepower = 746 watts = 550 foot-pounds/second

1 Btu/hour = 0.293 watts

Angle

180° = π radians

1 revolution = 360° = 2π radians

1 radian = 57.3°

Temperature

T_F is a temperature in Fahrenheit degrees

T_C is a temperature in Celsius degrees

T is an absolute temperature in Kelvin

$T_C = \frac{5}{9}(T_F - 32)$ $T_F = \frac{9}{5}T_C + 32$ $T = T_C + 273.15$

Fundamental Constants

Name	Symbol	Value
acceleration due to gravity	g	9.8 m/s^2
atmospheric pressure	P_{atm}	1.013×10^5 Pa
Avogadro's number	N_A	6.02×10^{23}
Boltzmann's constant	k	1.38×10^{-23} J/K
Coulomb's constant	K	8.99×10^9 N m^2/C^2
electron charge	e	1.602×10^{-19} C
gravitational constant	G	6.67×10^{-11} N m^2/kg^2
permeability constant	μ_0	1.26×10^{-6} T m/A
permittivity of free space	ε_0	8.85×10^{-12} C^2/(N m^2)
universal gas constant	R	8.31 J/K/mol

APPENDIX

C

Powers of Ten and More

In this appendix, we review some basic mathematical conventions that are used in science and the rest of this book.

Scientific Notation

Scientific notation is a compact way to write out very large or very small numbers. In fact, any number can be written in scientific notation in the form: $A \times 10^B$. The value of B tells you how many digits you would have to move the decimal point in A. If B is positive you move it B places to the right, while if it is negative you move it to the left B places. This works best when the number A has only one number before the decimal point, though this is not a strict requirement.

For example, the wavelength of orange light is approximately 0.0000006 meters = 6.00×10^{-7} meters = 600×10^{-9} meters = 600 nanometers = 600 nm. Conversely, its frequency is about 500,000,000,000,000 hertz = 5.00×10^{14} hertz = 500×10^{12} hertz = 500 terahertz = 500 THz.

Here you'll notice that we inserted some different prefixes depending on the values of B that were used. For certain values of B, there is a nice standard way of doing this, with which you should get familiar. For example: 1 kilogram = 1×10^3 grams. Here we put this all together in one place, starting with the large numbers:

Decimal vs. Scientific Notation

Decimal Notation	Scientific Notation	Prefix	Symbol
1,000,000,000,000	1×10^{12}	tera-	T
1,000,000,000	1×10^{9}	giga-	G
1,000,000	1×10^{6}	mega-	M
1,000	1×10^{3}	kilo-	k
0.01	1×10^{-2}	centi-	c
0.001	1×10^{-3}	milli-	m
0.000001	1×10^{-6}	micro-	μ
0.000000001	1×10^{-9}	nano-	n
0.000000000001	1×10^{-12}	pico-	p
0.000000000000001	1×10^{-15}	femto-	f

You probably know that we use the word "thousand" for 10^3. You should also know that a "million" is 10^6, a "billion" is 10^9, and a "trillion" is 10^{12}.

Precision and Uncertainty

Even at the introductory level, it is important to understand that scientific data are never perfectly precise. Physics deals with the real world, and everything we know about the real world is ultimately based on measurements. The accuracy of any measurement is limited in its precision. There is always some small amount by which our measurement could miss the mark without us knowing it, an amount which is characteristic of the measurement device we used. We call this quantity the uncertainty.

A measured quantity whose value is equal to A will properly be expressed as $A \pm a$, where a represents the uncertainty in A. Alternatively, the relative uncertainty may be given, usually as a percent. The relative uncertainty is $100\% \times (a/A)$.

In this book, we've saved space by ignoring uncertainty, but you should not conclude that all the parameters we used are perfectly exact. An alternate way to implicitly express uncertainty is with a concept known as "significant figures." When you express a number using significant figures, you only include as many digits as you are reasonably sure of. For example, if you write a number like 2.07863, which has six significant figures, you are implying that you know that last digit is probably a 3. In any case, you are sure it's not greater than 4 or less than 2. That would make the largest possible relative uncertainty of this number equal to 0.00002/2.07863, or about 0.001 percent. That is extraordinarily precise, also corresponding to one part in 100,000. The general, run-of-the-mill data you will actually work with at the introductory level is much less precise, and three (or at most four) significant figures will normally be sufficient.

Zeroes that appear in a number are sometimes significant, sometimes not, which can be a source of confusion for students. It all works very naturally if you combine significant figures with scientific notation. Zeroes whose only role is to place a decimal point are not considered significant. For example, none of the zeroes in the number 0.0012 are significant. By writing this as 1.2×10^{-3}, it is easier to see that there are only two significant figures given.

So, a difference of one in the least significant digit that you write (i.e., the rightmost one) should roughly correspond to the uncertainty in the number. If you are just doing problems or calculations using numerical input data, realize that your answer can't be more precise than the data used as input. It is generally good practice to give your answers with the same number of significant figures as the data used in the calculations, even though your calculator can give a lot more. Round up or down as required to get a reasonable number of figures. As a rule of thumb, three significant figures correspond to a relative uncertainty of about 1 percent.

Abbreviations Used for Units

Throughout the chapters, we have introduced you to a lot of units, and for each we provided an abbreviation. For the SI units, here is a summary table for easy reference or review.

Frequently Used Units and Abbreviations

Unit	Abbreviation	Dimension
meter	m	length
centimeter	cm	length
kilometer	km	length
gram	g	mass
kilogram	kg	mass
second	s	time
hour	h	time
liter	l	volume
newton	N	force
pascal	Pa	pressure
atmosphere	atm	pressure
joule	J	energy
calorie	cal	energy

Unit	Abbreviation	Dimension
watt	W	power
kilowatt-hour	kWh	energy
hertz	Hz	frequency
radian	rad	angle
Celsius degree	°C	temperature
kelvin	K	temperature (absolute)
mole	mol	amount of a substance
coulomb	C	electric charge
volt	V	electric potential
tesla	T	magnetic field
ohm	Ω	resistance
farad	F	capacitance
henry	H	inductance
weber	Wb	magnetic flux

Exponents and the Logarithm

Exponents and logarithms are handy devices used for compacting numbers that span a very broad range. Logarithmic scales are common on plots and graphs, and this is closely related to exponents and powers of ten, also known as orders of magnitude.

There are two different "bases" commonly used for logarithms, base 10 and the natural base, which uses the irrational number *e*. This *e* should not be confused with the elemental charge of the electron. It is a dimensionless number equal to approximately 2.718282.

The logarithm is simply the inverse of raising something to a power. So for base 10, if $a = 10^x$, then $\log(a) = x$. For base *e*, if $b = e^y$, then $\ln(b) = y$. The symbol ln is referred to as the "natural log." By definition, $\ln(e^x) = x$.

The following rules are expressed for the natural log, but they work just as well for logarithms in base 10:

$$\ln(a \times b) = \ln(a) + \ln(b)$$

$$\ln\left(\frac{a}{b}\right) = \ln(a) - \ln(b)$$

$$\ln(a^n) = n\ln(a)$$

There is no simpler expression for $\ln(a + b)$.

Additional Resources

If you've read over this whole book from page one, you know that you've covered an awful lot of territory and hopefully learned some pretty neat physics. One thing's for sure, though—there is plenty more physics out there to discover. In this guide we tried to cover the most important parts of introductory physics, but we had to carefully select which topics to include and which to leave out, and often we could only go so far in the space available. If your appetite is not yet satisfied, we'd like to leave you with a number of recommendations for further learning.

In this appendix we list the best physics references we know of, many of which were used by ourselves when putting together this book. We've included both old-fashioned references in book form as well as some very useful internet sites. With the latter we keep our promise by providing a few good sites that illustrate the wonders of waves in motion. Enjoy!

Books

Bloomfield, Louis A. *How Things Work: The Physics of Everyday Life.* 5th ed. New York: John Wiley & Sons, 2013.

Cutnell, John D. and Kenneth W. Johnson. *Physics.* 9th ed. New York: John Wiley & Sons, 2012.

Daintith, John. *A Dictionary of Physics.* 6th ed. Oxford: Oxford University Press, 2010.

Gonick, Larry and Art Huffman. *The Cartoon Guide to Physics.* New York: HarperPerennial, 1991.

Hewitt, Paul G. *Conceptual Physics.* 12th ed. Addison-Wesley, 2014.

Hobson, Art. *Physics: Concepts and Connections.* 2nd ed. Upper Saddle River: Prentice Hall, 1999.

Jones, Roger S. *Physics for the Rest of Us: Ten Basic Ideas of 20th Century Physics.* New York: B&N, 2009.

Motz, Lloyd and Jefferson Hane Weaver. *The Story of Physics.* Berlin: Springer, 1989.

Serway, Raymond A. and John W Jewett. *Physics for Scientists and Engineers.* 9th ed. Stamford: Cengage Learning, 2013.

Serway, Raymond A. and Chris Vuille. *College Physics.* 10th ed. Stamford: Cengage Learning, 2014.

Shamos, Morris H. *Great Experiments in Physics.* New York: Dover, 1956.

Websites

Focus on Physics: Waves
animatedscience.co.uk/blog/wp-content/uploads/focus_waves/introduction.html

Animations for Physics and Astronomy, Catalog for Wave Animations
phys23p.sl.psu.edu/phys_anim/waves/indexer_wavesB.html

Hyperphysics
hyperphysics.phy-astr.gsu.edu/hbase/hph.html

Computer Animations of Physical Processes
physics-animations.com

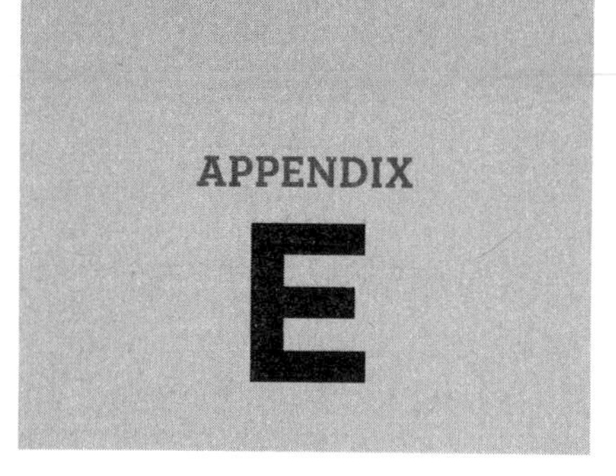

Solutions to the Exercises

These pages contain solutions to the "Try It Yourself" exercises scattered throughout *Idiot's Guides: Physics.* Those exercises are numbered in one sequence to make the solutions easy to find. In what follows, we first state the answer for every exercise, then explain how the answer is found. There are no exercises in Chapter 1 of the book.

Chapter 2

1. $\Delta t = 4{,}000$ seconds, or 1 hour and 7 minutes

   The distance to travel is 20,000 m and your speed is given as 5 m/s. Use Equation 2.3 to find the time required to go that far: $\Delta t = \Delta x/v = (20{,}000 \text{ m})/(5 \text{ m/s}) = 4{,}000$ seconds, which is about an hour and 7 minutes. If you don't remember Equation 2.3, it follows from the basic definition of velocity $v = \Delta x/\Delta t$ with just a little algebra.

2. $a = 11.07 \text{ m/s}^2$; $v = 94 \text{ m/s}$

   For the first part of this one, Equation 2.7 will do nicely, since distance and time are known, and $v_1 = 0$. The distance to travel is $\Delta x = 400$ m, and the time allowed is $\Delta t = 8.5$ s. Substituting into Equation 2.7 gives $400 = 36.125(a)$, so divide 400 by 36.125 to get the acceleration $a = 11.07 \text{ m/s}^2$. For the second part, just multiply this acceleration by the elapsed time: $v = 11.07 \times 8.5 = 94$ m/s. Note how this follows from the definition of average acceleration given in Equation 2.5. Since the question implied a constant acceleration, there is no difference between this constant value and the average acceleration during that time.

Chapter 3

3. $v_x = -15.43$ m/s

   The magnitude of the velocity vector here is 24 m/s (speed). The x component of any vector is the magnitude multiplied by $\cos(\theta)$, and here θ is given as 130°. Therefore, the answer we are looking for is just 24 cos(130°) = -15.43 m/s, as in the first part of Equation 3.2. Thus, the x component of the motorcycle's location is getting more negative at a rate of 15.43 m/s.

4. $R = 20.1$ m

   For this problem, you just need to use the definition of centripetal acceleration, $a_c = v^2/R$. First, you might as well solve this equation symbolically for what you're looking for, the radius of your path: $R = v^2/a_c$. Your speed v and acceleration a_c were given, so $R = (13)^2/8.4 = 20.1$ m.

Chapter 5

5. $F_N = 42.4$ N

   With no friction, the only two forces acting are the weight force and the normal force, as indicated in Figure 5.2. As with any problem like this, you should draw a free-body diagram that shows the forces as vector arrows. In a direction perpendicular to the plane, you know that there is no acceleration. Therefore, normal to the plane, Newton's second law would look like this: $F_N - mg\cos(\theta) = 0$. Thus, the normal force magnitude $F_N = mg\cos(\theta)$. Put in the 5 kg mass for m, the 30° for θ, and the usual 9.8 m/s² for g, and the result is $F_N = 42.4$ N.

6. $\theta = 50°$

   Since the rock is moving in a horizontal circle, there is no vertical acceleration, so the net force in the vertical direction is zero. Therefore, the tension force must have a vertical component that is upward but equal to the weight of the rock: $T\sin(\theta) = 4.9$ N, where θ is the angle you are looking for. You already know that the horizontal component of the tension is $T\cos(\theta) = 4.11$ N. Dividing the vertical component by the horizontal tells you that $\tan(\theta) = 4.9/4.11 = 1.19$. Since $\tan(\theta) = 1.19$, it follows that $\theta = 50°$. To get this last answer, you need a calculator that can do the inverse tangent function, also sometimes called the "arctangent". If your calculator is set to give angles in degrees, the result of $\tan^{-1}(1.19)$ is 50°.

7. $k = 457$ N/m

   Again, even though this is a simple problem, a free-body diagram will be useful. In this case there are only two forces, and both are completely vertical. The force from the spring F_s is directed upward, and the weight force F_g, as always, is down. Use $|F_s| = kx$ for the spring force. Since there is no acceleration, Newton's second law says $F_s + F_g = 0$, or $kx = mg$. In this case, x is the amount that the spring is stretched, given as 0.15 m, and the spring force is balancing the weight force of $7(g) = 68.6$ N. Thus, $k = mg/x = 68.6/0.15 = 457$ N/m.

Chapter 6

8. $r = 3.83 \times 10^8$ m

   This one is all about $\Sigma \boldsymbol{F} = m\boldsymbol{a}$. There is only one force, the force of gravity, so you don't need to worry about taking a sum on the left. Also, the acceleration (centripetal) is in the same direction as the attractive force, so you can just use this equation with the magnitudes of the vectors. The force of gravity between Earth and the moon is given by Equation 6.4. Set the force of gravitational attraction equal to the mass of the moon times its centripetal acceleration, so that Newton's second law becomes: $\frac{Gm_{\text{earth}}m_{\text{moon}}}{r^2} = m_{\text{moon}}\frac{v^2}{r}$. The speed v is given by $v = \frac{2\pi r}{T}$ where T is the time it takes to complete an orbit. The radius of the orbit r is the center-to-center distance you are looking for. So your expression becomes

$$\frac{Gm_{\text{earth}}}{r^2} = \frac{4\pi^2 r^2}{T^2 r} \quad \Rightarrow \quad r^3 = \frac{Gm_{\text{earth}}T^2}{4\pi^2}$$

   Now substitute in the values for G, Earth's mass, and the orbital period T. In order to get the answer to come out in the correct units, you should first convert the period T from days to seconds: $T = 27.3$ days $\times$ 24 hours/day $\times$ 3,600 seconds/hour $= 2.36 \times 10^6$ s. Your result (don't forget to square this time T) should be $r^3 = 5.624 \times 10^{25}$. Take the cube root of this, and you get $r = 3.83 \times 10^8$ m for the final answer.

9. $F = 2 \times 10^{20}$ N

   You can use either $F_g = m_{\text{moon}}\frac{v^2}{r}$ or $F_g = \frac{Gm_{\text{earth}}m_{\text{moon}}}{r^2}$. Either way, the answer is about 2×10^{20} N.

Chapter 7

10. There are 3.6×10^6 J in 1 kWh

    Recall that 1 watt = 1 joule/second, and a kilowatt is 1,000 watts. Thus a kilowatt-hour is $1000\,\mathrm{J/s} \times 3600\,\mathrm{s} = 3.6 \times 10^6\,\mathrm{J}$.

11. $v_{min} = 12.3$ m/s

    This is one of those problems that is relatively easy to do if you use conservation of energy. You can arbitrarily set the gravitational potential energy to be zero at 1.8 m above the ground, the place where the ball leaves your hand. Then that potential energy is *mgh* at the roof, if *h* is the 7.7 m difference between the starting 1.8 m height and the 9.5 m height of the roof. Equate the kinetic energy at the moment of launch with the desired increase in gravitational potential energy: $\frac{1}{2}mv_{min}^2 = mgh$. The mass of the ball cancels, and $v_{min} = \sqrt{2gh} = 12.3$ m/s. If you throw it with this speed, it will just get to the roof with zero speed remaining.

Chapter 8

12. 29.4 J of kinetic energy was lost

    With the combined mass moving at 2.8 m/s, the kinetic energy after the collision is $\frac{1}{2}(5\,\mathrm{kg})(2.8\,\mathrm{m/s})^2 = 19.6\,\mathrm{J}$. The initial kinetic energy was all in bead 1; with its mass and 7 m/s speed you saw it was 49 J. Subtracting 19.6 J from the initial 49 J, you see that 29.4 J of kinetic energy was lost.

Chapter 9

13. $\omega_{avg} = 2.09$ rad/s

    To find the average angular velocity, you just need to know the total angular change, and how much time it took. The only trick is to get these quantities in the customary units. Every full revolution is 2π radians, so 20 revolutions = 125.7 rad. And, of course, 1 minute = 60 s. Therefore, $\omega_{avg} = 125.7/60 = 2.09$ rad/s.

14. $\Delta t = 73.3$ s

Recall the definition of angular acceleration: $\alpha = \frac{\Delta\omega}{\Delta t}$. You are given α, so change this to $\Delta t = \frac{\Delta\omega}{\alpha}$. The angular speed decreases to zero, so the change is $\Delta\omega = -22$ rad/s. Therefore, $\Delta t = -22/(-0.3) = 73.3$ s is the time it takes to stop.

15. 807 radians, or 128 full rotations

You know that the average angular speed was halfway between the starting speed (22 rad/s) and the ending speed (0), which means $\omega_{avg} = 11$ rad/s. Multiply this by the time interval you just calculated in Exercise 14, and you will see that the total rotation was 11(73.3) = 807 radians. You could also describe this as about $807 \div (2\pi) = 128$ full rotations.

16. $\tau = 6.56$ N m

Since the direction of the force is at a right angle to the wrench, the lever arm is (16 cm)sin(90°) or 0.16 m. The magnitude of the torque is then just $\tau = Fd = 6.56$ N m.

17. $I = 0.083$ kg m^2

Here you use the analog of Newton's second law for rotation: torque $\tau = I\alpha$. First solve this for what you are looking for: $I = \tau/\alpha$. If you can find the torque and the angular acceleration, then you can solve the problem. The problem tells you by how much the angular velocity changes in 3 seconds, so the angular acceleration $\alpha = \Delta\omega/\Delta t = 13/3 = 4.33$ rad/s^2. The string comes off of the wheel automatically at right angles to the radius, so the lever arm is the radius: $d = 0.12$ m. The torque is $\tau = Fd = 3(0.12) = 0.36$ N m. Therefore $I = 0.36/4.33 = 0.083$ kg m^2.

18. $K = 7.02$ J; average power = 2.34 W

The wheel started with no kinetic energy, so the added kinetic energy is the same as the final kinetic energy = $\frac{1}{2}I\omega^2$. Use the I you found in Exercise 17, and the $\omega = 13$ rad/s. Substituting these values gives a kinetic energy of 7.02 J. Since it took 3 seconds to add this energy, the average power was 7.02/3 = 2.34 W.

Chapter 10

19. $T = 0.77$ s

Period is the inverse of frequency. As you saw from Equation 10.1, the frequency in this case is $f = \frac{1}{2\pi}\sqrt{\frac{k}{m}}$. So just use the given mass and the spring constant $k = 800$ N/m to find that $f = 1.3$ Hz. Taking the inverse of this gives the period equal to 0.77 s.

20. Maximum speed is 1.71 m/s

You just showed that the oscillator reaches its maximum speed when it goes through $x = 0$, and $v_{max} = A\sqrt{k/m}$. Therefore you need the amplitude $A = 0.21$ m in addition to k and m, which have the same values as in Exercise 19. Substituting yields 1.71 m/s for the maximum speed.

21. $g = 9.65$ m/s^2

You'll use Equation 10.4 for the period, $T = 2\pi\sqrt{\frac{L}{g}}$, but solve it for g: $g = 4\pi^2\frac{L}{T^2}$. If 10 oscillations take 14.3 s, then clearly $T = 1.43$ s. Inserting this, along with $L = 0.5$ m, gives $g = 9.65$ m/s^2.

Chapter 11

22. $\Delta t = 1.22$ s

Equation 11.1 tells you how the wave speed depends on the mass/length of the rope and the tension force on it. Using the data given, the mass per unit length $\mu = 12/100 = 0.12$ kg/m. Then the speed of the pulse $v = \sqrt{\frac{F_T}{\mu}} = 81.65$ m/s. At that speed, it only takes $100/81.65 = 1.22$ s for a wave pulse to travel the 100 m length of the rope.

23. $\lambda = 16.3$ m

First solve $v = \lambda f$ for the wavelength: $\lambda = v/f$. With $f = 5$ Hz and the speed from the previous problem, $v = 81.65$ m/s, you can determine that $\lambda = 16.3$ m. Here the amplitude of the wave is irrelevant.

24. $\lambda = 2.33$ m

A drawing will help you see what is going on. The pattern that has two nodes is the third one from the top in Figure 11.6. You see that in this case one full wavelength takes up $\frac{2}{3}$ of the length of the rope, so $\lambda = (\frac{2}{3})3.5 = 2.33$ m.

Chapter 12

25. $\lambda = 0.314$ m or 31.4 cm

You have just seen that for any traveling wave, the wave speed $v = \lambda f$. You can solve this for the unknown in this case, which is the wavelength. Using the speed of sound in m/s, and the given frequency, $\lambda = \frac{v}{f} = \frac{345}{1100} = 0.314 \text{ m}$ or 31.4 cm.

26. Intensity is 2×10^{-6} W/m^2

We will use the expression for sound intensity $I = P/(4\pi r^2)$. You have a choice of two methods here. Since you are given the intensity at a distance $r = 14$ m, you can use this to calculate the power P at the source (the baby). Thus $P = 4\pi r^2 I = 0.0443$ W. Now use this P with the new $r = 42$ m in the first equation, to find that the new intensity is 2×10^{-6} W/m^2. You can take a shortcut by writing $I = P/(4\pi r^2)$ twice, once for 14 m and again for 42 m. Dividing one of these by the other will give you the ratio of intensities in terms of the ratio of the distances. For this example, you'll find that the new intensity is $\frac{1}{9}$ of the intensity at 14 m, because you are now three times further away from the source.

Chapter 13

27. $P = 2.48 \times 10^5$ Pa

We use $P = P_{atm} + \rho g d$ for the absolute pressure. The second term $\rho g d = 1000 \times 9.8 \times 15 = 1.47 \times 10^5$ Pa. The absolute pressure is this much greater than the atmospheric pressure: $1.013 \times 10^5 + 1.47 \times 10^5 = 2.48 \times 10^5$ Pa.

28. Gauge pressure is 1.47×10^5 Pa

Since gauge pressure is just the difference between the absolute pressure and atmospheric pressure, it is simply $\rho g d$: 1.47×10^5 Pa.

29. 32.6% of the volume is above the surface

    As stated in the text, the volume fraction *below* the surface is the same as the density ratio between the wood and the fluid: 600/890 = 67.4%. Therefore, the fraction above the level of the oil is 100% – 67.4% = 32.6%.

30. $P = 2.46 \times 10^5$ Pa

    If you call the lower level where the water enters $y = 0$, then at the third-floor kitchen faucet, you would say $y = 14$ m. As the water exits the faucet, it is at atmospheric pressure. The desired speed of the water as it exits is given, so you have all you need to calculate the Bernoulli quantity at the third-floor faucet: $P_{atm} + \frac{1}{2}\rho v^2 + \rho g y$, or $1.013 \times 10^5 + 8{,}000 + 1.372 \times 10^5 = 2.465 \times 10^5$ Pa. Next, you have to use the continuity equation to find the speed of the water at the inlet, which is 0.7334 m/s, so that $\frac{1}{2}\rho v^2 = 269$. Subtract this from the Bernoulli quantity to get the pressure at the bottom $= 2.46 \times 10^5$ Pa.

Chapter 14

31. $T = -128.6°F$, or 184 K

    To convert from Celsius to Fahrenheit, use Equation 14.1 with the given $T_C = -89.2°C$: $T_F = 1.8(-89.2) + 32 = -128.6°F$. For absolute temperature on the Kelvin scale, all you need to do is add 273.15 to the Celsius temperature, so -89.2°C = 184 K.

32. $\Delta V = 1.456$ mm^3, so the new volume is 81.46 mm^3

    This is an example of thermal volume expansion, where you have seen that the *change* in volume $\Delta V = \beta V_0 \Delta T$. The temperature change $\Delta T = 100°C$, so with the given value of β, $\Delta V = 1.456$ mm^3. Add this to the original volume to get 81.46 mm^3 for the volume at 100°C.

Chapter 15

33. $Q = 2.54 \times 10^5$ J removed

    You have a mass of material, so you need the formula that employs *specific* heat, Equation 15.2: $Q = mc(\Delta T)$. The specific heat $c = 4.186$ J/g/°C for water or milk. So for a 16°C change, $Q = 2.54 \times 10^5$ joules of heat will have to be removed.

34. $Q = 248{,}600$ J

The boiling point of water is 100°C, so we will break the additional heat up into three parts. Q_1 will raise the liquid water from 70°C to 100°C, where it is still liquid. Q_2 will be the heat required to change the phase of the water from liquid to gas, the Q_3 will be added to raise the temperature of the gaseous water vapor from 100°C to the final temperature of 150°C. The calculations of Q_1 and Q_3 will both use Equation 15.2 as in the previous exercise, but with different specific heats. To calculate Q_2, you just multiply the mass by the given latent heat of vaporization. All three calculations use the same mass of water, 100 g. $Q_1 = 100(4.186)30 = 12{,}558$ J, $Q_2 = 100(2260) = 226{,}000$ J, $Q_3 = 100(2.01)50 = 10{,}050$ J. The total heat, therefore, is $Q_1 + Q_2 + Q_3 = 248{,}600$ J.

35. $Q = 0.032$ J (in 1 second)

The temperature difference between the ends is still 500 - 20 = 480°C and the physical dimensions are the same, so you just use Equation 15.3 again with the different k value. This equation tells you that the rate of heat flow is 0.033 J/s/m/°C × 0.0004 m^2 × 480°C ÷ 0.2 m = 0.0317 joules/s, so in 1 second, only 0.032 joules of heat would be transferred.

Chapter 16

36. $P = 2$ atm $= 2.0265 \times 10^5$ Pa

Since the problem specifies the amount of helium as a number of moles, you want to use the form of the ideal gas law that also has the number of moles in it, Equation 16.2: $PV = nRT$. Let's call state 1 the original state, with 40.9 mol of helium at atmospheric pressure, and state 2 the new state, where you have added another 40.9 mol of the same gas. Write the ideal gas relationship for both states (just the symbols, no numbers) and divide the state 2 equation by the state 1 equation. Since V, R, and T don't change, you find that $P_2/P_1 = n_2/n_1$. The second state has twice as many moles of helium as the first did, so the absolute pressure in the container is now twice atmospheric pressure, or $P_2 = 2.0265 \times 10^5$ Pa.

37. Efficiency $e = 0.67$; $Q_c = 50$ J

Equation 16.3 tells you that the efficiency is the mechanical energy divided by the heat input. In this case, $e = 100/150 = 0.67 = 67\%$. Since 150 J of heat came into the engine, and it only did 100 J of work, the other 50 J had to flow into the cold reservoir.

Chapter 17

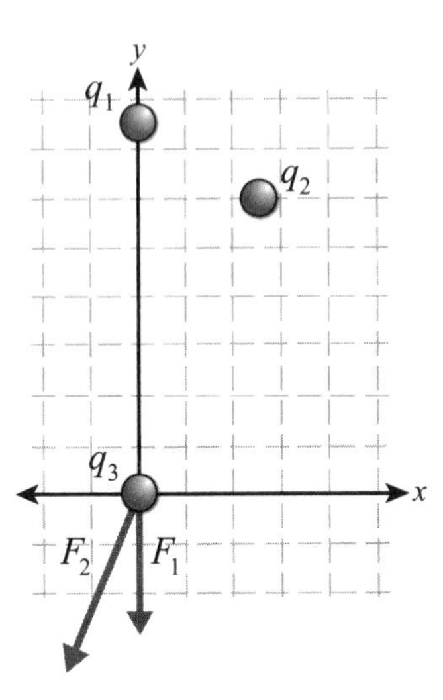

38. Net force = 20.4 N at an angle of -102.2°

Let's call $\boldsymbol{F}_1$ the electric force on q_3 due to q_1; likewise, $\boldsymbol{F}_2$ is the force on q_3 due to q_2. The vectors corresponding to these forces are shown as the down arrows in the figure above. Since all of the given charges are positive, they repel each other. The vector $\boldsymbol{F}_1$ is straight down in the negative y direction. $\boldsymbol{F}_2$ makes an angle of 67.4° with the negative x-axis ($\tan^{-1}\frac{12}{5}$). First use Coulomb's law to get the magnitudes of each force. Using Equation 17.1 with the 15 cm separation between q_1 and q_3, you find that the magnitude $F_1 = 9.59$ N. For F_2, you'll need the distance between q_2 and q_3, which the Pythagorean theorem tells you is 13 cm. Therefore, $F_2 = 11.17$ N in magnitude. Now add these two vectors in the usual way to get the total force.

Let $\boldsymbol{F}$ be the total force.

Then $F_x = F_{1x} + F_{2x} = 0 - 11.17\cos(67.4°) = -4.29$ N

$F_y = F_{1y} + F_{2y} = -9.59 - 11.17\sin(67.4°) = -19.9$ N

These components would be enough to completely specify the total electric force $\boldsymbol{F}$, but you could also find its magnitude and direction in the usual way (see Equation 3.3). Doing so, you would find that the magnitude of the net electric force $F = 20.4$ N, and that its direction is -102.2° from the positive x-axis.

39. 6.79×10^6 N/C in the same direction as the force

    The electric field is in the same direction as a force felt by a positive test charge. Since the force you calculated in Exercise 38 was for a positive charge q_3, all you have to do to find the electric field is divide that force by q_3. Thus, the electric field has a magnitude $E = F/q_3$, using $F = 20.4$ N from the previous solution. The result is 6.79×10^6 N/C, in the same direction as the net force.

Chapter 18

40. $V = -2.25 \times 10^9$ V

    For a single point charge, Equation 18.6 gives you the value of the potential at any distance *r*. The potential is $Kq/r = -2.25 \times 10^9$ V.

41. $\Delta V = 153$ V

    First, solve Equation 18.9 for the unknown potential difference: $\Delta V = \sqrt{\frac{2U_E}{C}}$. Then put in the given values for electric potential energy and capacitance, and the result is $\Delta V = 153$ V.

Chapter 19

42. $I = 20$ A

    Apply Ohm's law, Equation 19.1, but solve it for the unknown current: $I = \Delta V/R = 120/6 = 20$ A of current.

43. $E = 1.44 \times 10^5$ J

    Recall that power = energy/time, so to find the energy produced you just have to figure out the power of the heater and multiply that by the one minute of time. As you have seen in Equations 19.2 and 19.3, the power is either $P = I^2R$ or just $P = IV$. Either way, the power is 2,400 watts. Since each watt is 1 J/s, just multiply this power by 60 s to find that 144,000 J = 1.44×10^5 J of heat energy was produced.

Chapter 20

44. F_B = 12,000 N straight down vertically

For the magnitude of the force, use Equation 20.2. The wire makes an angle of 90° with the magnetic field, so you just need the product of the current, the field strength, and the length of the wire. Thus F_B = 12,000 N. For the direction, go back to the right-hand rule, curling the direction of the current (west) into the direction of the magnetic field (north). The result is that the force is straight down vertically.

Chapter 21

45. emf = 0.452 V

According to Equation 21.3, if you want to know the induced emf, you need to know how much the magnetic flux through the solenoid changes in the five-second time period. Since the magnetic field starts at zero, so does the flux. After five seconds, the flux is given by the final field (12 T) multiplied by the cross-sectional area of the solenoid, which would be πr^2. From the diameter you are given, you know that r = 2 cm. Putting this together, you find that the change in flux $\Delta\Phi_B$ = 0.01508 Wb, so the emf induced during the rise is 0.452 V.

Chapter 22

46. $\lambda = 5.89 \times 10^{-5}$ cm; $f = 5.09 \times 10^{14}$ Hz

For light (and any other electromagnetic radiation), the wavelength is related to the frequency through the speed: $c = \lambda f$. The conversion from meters to centimeters is accomplished by simply multiplying by 100, so $\lambda = 5.89 \times 10^{-5}$ cm. If you solve the equation above for frequency, then $f = c/\lambda$. To get the frequency in standard units, you need to have the same length units in both c and λ, so let's use $c = 3 \times 10^8$ m/s and the given wavelength in meters. Then the frequency $f = 5.09 \times 10^{14}$ Hz.

Chapter 23

47. $n = 2.44$

    By definition, the index of refraction is the ratio between the speed of light in a vacuum and its effective speed in the material, $n = c/v$. For the case of diamond, $n = (3 \times 10^8)/(1.23 \times 10^8) = 2.44$.

48. $\theta = 66°$

    For this you will use Snell's law for refraction, given in Equation 23.1. If you start inside the diamond, then $n_a = 2.44$, as you just calculated in Exercise 47, and $\theta_a = 22°$. You'll use the approximation that $n = 1$ for air, so this will be the value of n_b, and the angle you're looking for is θ_b. Snell's law tells you that $\sin(\theta_b) = 2.44\sin(22°) = 0.914$. Take the inverse sine of this number, $\theta_b = \sin^{-1}(0.914) = 66°$.

Index

A

B

C

D

E

F

G

H

I

J-K

L

M

N

O

P

Q

R

S

T

U